GAS MIGRATION

GAS MIGRATION

Events Preceding Earthquakes

Leonid F. Khilyuk

George V. Chilingar

John O. Robertson Jr.

Bernard Endres

Gulf Publishing Company

Houston, Texas

Gas Migration—Events Preceding Earthquakes

Gulf Publishing Company
Book Division
P.O. Box 2608 □ Houston, Texas 77252-2608

10 9 8 7 6 5 4 3 2 1

Library of Congress Cataloging-in-Publication Data
Gas migration : events preceding earthquakes / Leonid Khilyuk . . . [et al.].
p. cm.
Includes bibliographical references and index.
ISBN 0-88415-430-0 (alk. paper)
1. Volcanic gases. 2. Earthquakes I. Khilyuk, Leonid.
QE527.75.G37 2000
551.22—dc21 00-032107

Printed in the United States of America.

Printed on acid-free paper (∞).

Dedication

This book is dedicated to His Majesty Mohammed VI, King of Morocco, in recognition of His dedication to His people and great concern in prediction of earthquakes.

This book is also dedicated to Dr. Rudolf W. Gunnerman, scientist, inventor, humanitarian, and entrepreneur in the field of environmental protection and to the Nobel Prize winner, Dr. George Olah, Head of Hydrocarbon Research Institute at the University of Southern California, for his many outstanding contributions to the chemistry of hydrocarbons.

Contents

PART II

Events Preceding Earthquakes

Foreword

Gas Migration/Events Preceding Earthquakes discloses how the phenomenon of natural gas escaping to the earth's surface via fractures and faults from petroleum reservoirs becomes a potential health and safety risk in metropolitan areas. A series of disastrous explosive, migrating gas events occurring in Los Angeles are reviewed in the context of what was accomplished by investigating the cause and effect of the problem and the ensuing litigation in the courts. The documented cases clearly emphasize the extent of the destructive power generated by the ignition of entrapped natural gas and its long-term effect on an urban environment. It is important for forensic engineers, engineering geologists, architects, zoning commissions, and health departments to not only be aware of this potential geohazard, but to ensure that design guidelines and regulations are established to mitigate gas migration problems.

The authors' early engineering investigations suggested that the monitoring of non-hydrocarbon gas concentrations and emission rates in tectonically active areas could be the pacing parameters for predicting seismic events. At present, scientific research in the prediction of earthquakes is to characterize and evaluate accumulated stresses and corresponding strains in and near active faults. This approach has not met with much success; however, it has increased our comprehension about the mechanisms that lead to earthquakes. Unfortunately, in many cases the fault that becomes seismically active is not known either to exist or to have been active in recent memory. A unique method that showed promise in characterizing the relative intensity of seismic activity in fault systems was presented by Arab et al. (1994). I am suggesting that by coupling the geochemical and stress-strain monitoring with fractal analysis would enhance their definitive earthquake prediction methodology. If either is proven successful, then there will be a substantial reduction in the loss of life resulting from large seismic events. The authors' proposed gas migration detection methodology in my opinion is scientifically sound, based on years of geoscience and engineering experience, and innovative concepts.

Information contained in this book about gas emissions in seismically active regions must be of great interest and importance to city planners and managers, city councils, economic development boards, and state and national governments. It is up to these constituents to lead the way in seeing that this technology is developed and proven in the field. The authors have formulated, convincingly, an earthquake prediction approach based on the migration concept.

Herman H. Rieke, Ph.D.
Professor and Head
Petroleum Engineering Department
University of Louisiana at Lafayette
Lafayette, Louisiana, USA

*Arab, N., Kazi, A. and Ricke, H. 1994. Fractal geometry of faults in relation to the 12 October 1992 Cairo Earthquake. *Nat. Hazards,* 10:221–223.

Preface

Gases migrating to the earth's surface provide crucial information regarding the state and evolution of the structure and tectonics of our planet. By listening to and analyzing "whispering" gases, locations of intense tectonic activity can be identified and areas of potential natural disasters delineated, such as volcanic eruptions, tsunamis, and earthquakes. Like an experienced doctor who can determine the health conditions of a patient by listening to his breathing, a knowledgeable scientist can recognize a disorder in the earth's body by analyzing these migrating gases. Vernadskiy (1934), a prominent Russian scientist and academician metaphorically called this phenomenon the "gaseous breath of the Earth."

The migration of gas to the surface from oil and gas formations is a problem that greatly affects those surface areas where human activity exists. Underground gas storage facilities and oil fields have demonstrated a long history of environmental gas migration problems. Experience has shown that the migration of gas to the surface creates a serious potential risk of explosion, fires, noxious odors, and carcinogenic chemical emissions. These risks are examined for migrating gas in urban areas.

This book presents the mechanics of gas migration, paths of migration, and a review of several environmental risks that should be considered when operating an oil and/or gas facility in an urban area. Gas can migrate to the surface in a continuous or discontinuous stream through porous, water-filled media. The primary force for this gas migration is the difference between specific weights of gas and water.

Using remote measuring techniques, such as satellites equipped with laser and radar transducers, global and local gaseous images of the earth can be constructed (by constructing distributions of various gas concentrations over the earth's surface and monitoring their changes). Global images can be used for a general description of the earth's health and for forecasting large-scale, long-term changes in the earth's global climates. Changes in local images can signal future natural disasters, such as earthquakes. To signal future earthquakes these images must be related to those regions of intense tectonic activities that are usually associated with the edges (boundaries) of tectonic blocks and plates. These boundaries are characterized by intense folding, faulting, and fissuring that provide the pathways for migration of gas to the earth's surface.

The block and plate boundaries represent so-called "weakened" zones (Sadovskiy, 1979) and are the most probable locations for earthquake occurrence. Additionally, these highly fractured zones are the most conducive for migration of gases from deep layers of the earth to the surface. In seismically active areas, these zones and their properties (mainly permeability) are continuously changing in the process of preparation and occurrence of earthquakes. These changes are reflected in the varying rates of migration and relative composition of the gases migrating to the earth's surface.

Our worldwide investigation attributes pronounced changes in the rates of gas migration (methane, carbon dioxide, helium, radon, etc.) to the sequence of tectonic events preceding an earthquake (Chilingar et al., 1996). These studies are extremely useful in developing earthquake prediction techniques, especially when considering the enormous losses associated with large earthquakes. Historically, studies of gas migration rates, however, have not attracted sufficient attention from the scientific community nor received the proper priority in funding. One reason for this is that studies on gas migration have been scattered among numerous diverse scientific fields such as geology, geophysics, geochemistry, petroleum engineering, and environmental protection. Additionally, the efforts of researchers in seismic studies have primarily concentrated on the concept of accumulated stresses.

Extensive knowledge on natural gas migration from petroleum reservoirs has accumulated in petroleum science, environmental engineering, and geologic literature, providing researchers with the necessary theoretical concepts and methods for applied research on gas migration. In this book, studies on gas migration were developed to the level at which they provide ready theoretical concepts for describing the migration phenomenon. They also provide engineering methods for calculating the parameters of the gas migration process. Cumulative knowledge on gas migration gained as a result of these studies must be further developed to the level of interdisciplinary theory.

Although the need is extreme, this book is the only work available that has accumulated and analyzed interdisciplinary knowledge on gas migration and demonstrated its great potential for vital applications. Our goal was to analyze a broad spectrum of research works on gas migration and to develop a theoretical basis that covers the questions of origin of gases, avenues of migration, rates of migration, their relationship with tectonic activity, as well as to develop a conceptual basis for earthquake prediction. This book provides a theoretical and informational foundation for further research on gas migration, especially in relation to seismic forecasting.

Gas migration studies are also important for identifying, predicting and mitigating other environmental hazards, such as gas toxicity and explosions, ground fissuring, and faulting. Many researchers have established a strong correlation between peak gas concentrations and faulting. Through field measurements, Jones and Drozd (1983) demonstrated the migration of methane, helium, and hydrogen gases along the San Andreas Fault in Central California. Maximum gas concentrations were observed directly over deep faults, confirming the migration of gas.

Problems pertaining to gas migration and their relation to earthquake prediction are discussed in this book. The recent findings in crustal studies conclusively demonstrated that the structure of the earth's crust consists of rock blocks of various sizes. Based on this fact, we will reconsider the main theoretical concept of earthquake preparation and occurrence.

The static deterministic image of the continuous "bending and cracking" of the crustal medium layers (based on the conventional theory of accumulated stresses) must be replaced by a dynamic stochastic image of random "shifting and sticking" of the crustal blocks (based on the discoveries of discrete crustal structure and "flickering stresses"). This shifting rearranges a block assemblage participating in earthquake development and dramatically changes the conditions of gas migration to the earth's surface. The latter is reflected in the pronounced changes in the rates of gas migration to, and its relative composition at, the earth's surface.

Thus, the rates of gas migration (and relative composition) can be used as integral parameters reflecting the tectonic activity over an area of interest. They are stable with respect to a small dislocation of the measurement point, which makes them preferential informational precursors with respect to measured stresses, which are extremely sensitive to the shift of the monitoring point.

This book describes and analyzes the main approaches to identifying gas sources ("geochemical fingerprinting"), the properties of gases, the origins and pathways of gas migration to the earth's surface, and provides environmental specialists with the conceptual basis and methodology for evaluating migration paths and rates. It illustrates the possible applications for the use of the rates of migration and relative composition of gases in earthquake forecasting and for identifying and predicting associated environmental hazards.

This book is directed at scientists and engineers who need a good theoretical basis on gas migration and its associated application methods. It also provides a source of references on the subject. We hope this work will attract the attention of scientific planners who are responsible for the most efficient allocation of research funds, given the subject's serious nature and its consequences. The book is also recommended to curious readers interested in broadening their concepts of the earth's crustal structure, tectonic activity, and its manifestations pertaining to the causes and precursors of environmental disasters. Due to the enormous theoretical interest in gas migration and its important practical applications, it is easy to predict a surge in scientific research in the near future accompanied by many future books on this subject.

REFERENCES

Chilingar, G. V., Khilyuk, L. F., and Katz, S. A., 1996. Pronounced changes in upward natural gas migration as precursors of major seismic events. *J. Pet. Sci. Eng.,* 14:133–136.

Jones, L. M., 1995. *Putting Down Roots in Earthquake Country.* Dept. of Interior, U.S.G.S. and the Southern California Earthquake Center (SCEC), University of Southern California, University Park, CA, 30 pp.

Jones, V. T. and Drozd, R. J., 1983. Predictions of oil or gas potential by near-surface geochemistry. *Am. Assoc. Petrol. Geol. Bull.,* 67(6):932–952.

Sadovskiy, M. A., 1979. Natural fragmentation of rocks. *Dokl. Akad. Nauk SSR,* 247(4):829–831.

Vernadskiy, V. I., 1934. *Outline of Geochemistry.* Gosgeonefteizdat Publishers, Moscow (in Russian).

Vernadskiy, V. I., 1940. *Living Matter and the Biosphere.* Nauka Publishers, Moscow (in Russian).

PART I

Tectonics and Gas Migration

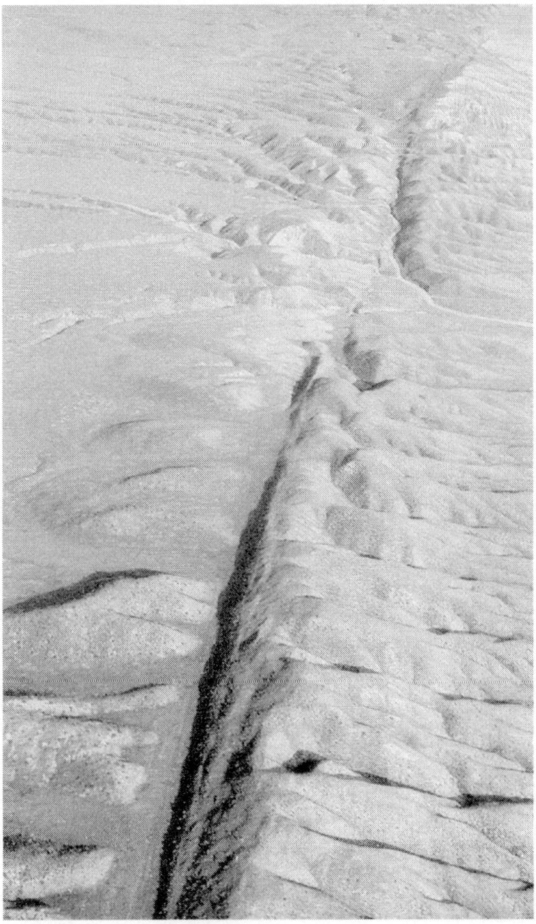

Surface trace of the San Andreas Fault, California (photo modified after R. E. Wallace, USGS; in Kious and Tilling, 1997).

CHAPTER 1

Tectonics and Earthquakes

INTRODUCTION

The earth's crust is composed of several different layers. The upper layer (higher standing level), known as the continental crust, has a mean land level of 840 m above water level (Kious and Tilling, 1997). The oceanic crust has a mean depth of 3,700 m below sea level. The continental layer is higher. It floats upon the ocean basin layer because it is composed of lighter materials (e.g., silicic or granitic rocks). Granitic rocks are rich in quartz and feldspar (orthoclase and plagioclase). The oceanic crust consists of mafic or basaltic rocks rich in pyroxene and olivine (together with orthoclase and plagioclase feldspars), which are heavier (denser) than the continental granitic crustal rocks. Both the continental and oceanic crusts are lighter than the mantle, which is composed of olivine and pyroxene with feldspar being virtually absent. The Mohorovicic discontinuity, or Moho, separates the oceanic and continental crusts from the underlying mantle. This is a relatively sharp compositional boundary between the lighter crust materials and denser mantle (Figure 1-1).

The granitic crust literally floats upon the denser basaltic ocean crust because its lower density makes it buoyant, thereby supporting the higher topography. Mountains rise above the continental plateaus because they have deeper low-density roots than other portions of the continental crust (Figure 1-1). As a result, thicker regions of the crust (i.e., mountains) float higher than thinner regions, much like icebergs in the ocean. Topography is a reflection of the hydrostatic equilibrium for the crustal materials. This concept comprises the theory known as isostasy.

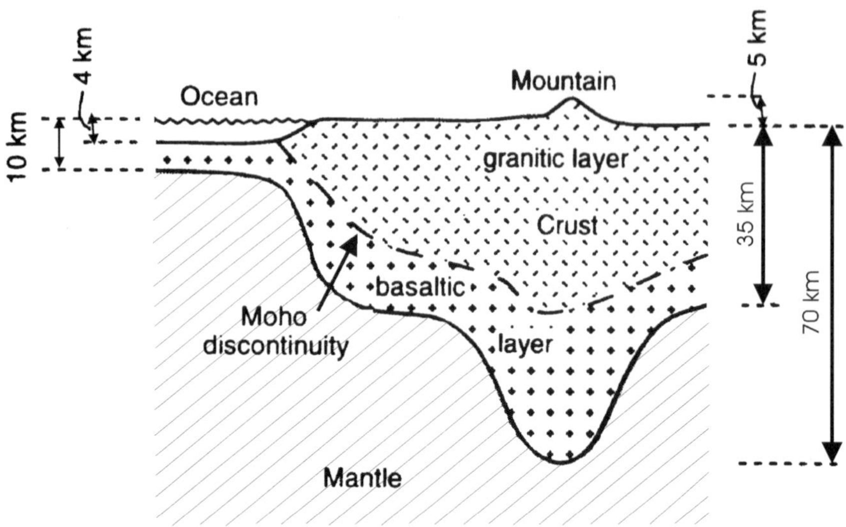

Figure 1-1. Schematic showing cross-section of earth's continental and oceantic crust, Moho discontinuity, and mantle. *(Modified after Yeats et al., 1997, fig. 1-1, p. 9.)*

GEOLOGIC SETTING FOR EARTHQUAKES

Olsen (1999) has suggested that 200 million years ago, the largest sustained volcanic eruption in the Earth's history split the ancient super continent of Pangaea. The eruption occurred in a very brief period of time. The extent of the basaltic lava flows covered over 2.7 million square miles (Figure 1-2). The eruption opened rifts through the middle of Pangaea that fractured the ancient landmass. It also set the fractured landmasses adrift and, by wedging them apart, gradually opened the gulf that created the Atlantic Ocean. This eruption, called the Central Magmatic Province (CAMP), appears to have been centered in Florida, U.S.A.

Intensive surface mapping conducted by the World-Wide Standard Seismographic Network (WWSSN) in the 1960s and 1970s, including the mapping of the seafloor, helped develop the understanding that the surface of the earth is composed of approximately a dozen continental and oceanic rigid crustal slabs (Figure 1-3). This conceptual framework, known as "plate tectonics," recognized that rocks and geological strata, making up the outer layers of the earth (crustal portion), are broken into a patchwork of ever-moving tectonic plates.

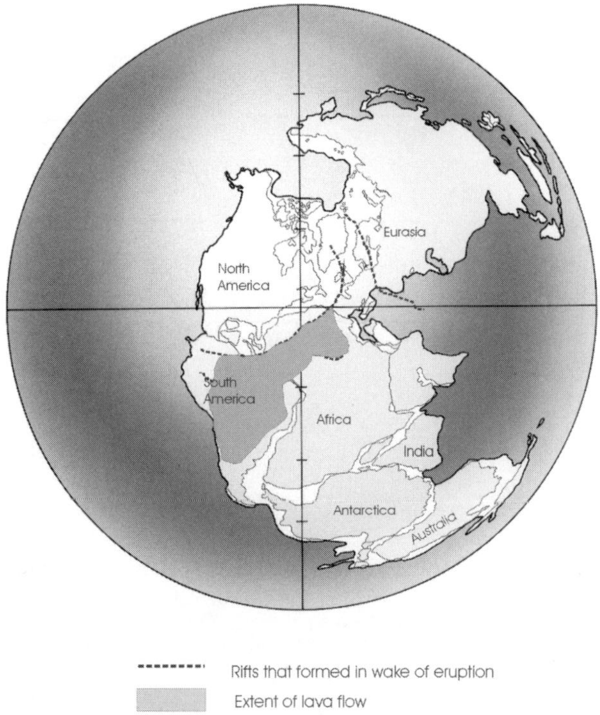

Figure 1-2. The continent of Pangaea showing the approximate location of the massive 2.7 million square mile lava flow and the rifts that were formed at the time of volcanic eruption. It has been suggested by P. E. Olsen (1999) that this eruption may have been fueled by a giant plume of molten rock rising through the earth's mantle.

All plates are not of the same size or shape. For example, the Pacific Plate comprises much of the Pacific Ocean and is more than 10,000 km wide, while the Scotia Plate only spans a few hundred kilometers. One factor that helps identify an individual plate from the other plates is its movement. Plate motion is slow, on the order of a few centimeters (or less) per year. This slow motion over millions of years can carry a plate from the equator to the poles, rip continental landmasses apart, or assemble crustal fragments into continents. As a plate moves, it will grind or rub against other plates and in many instances—especially along the Pacific Ocean coastal boundaries—override or underride other plates.

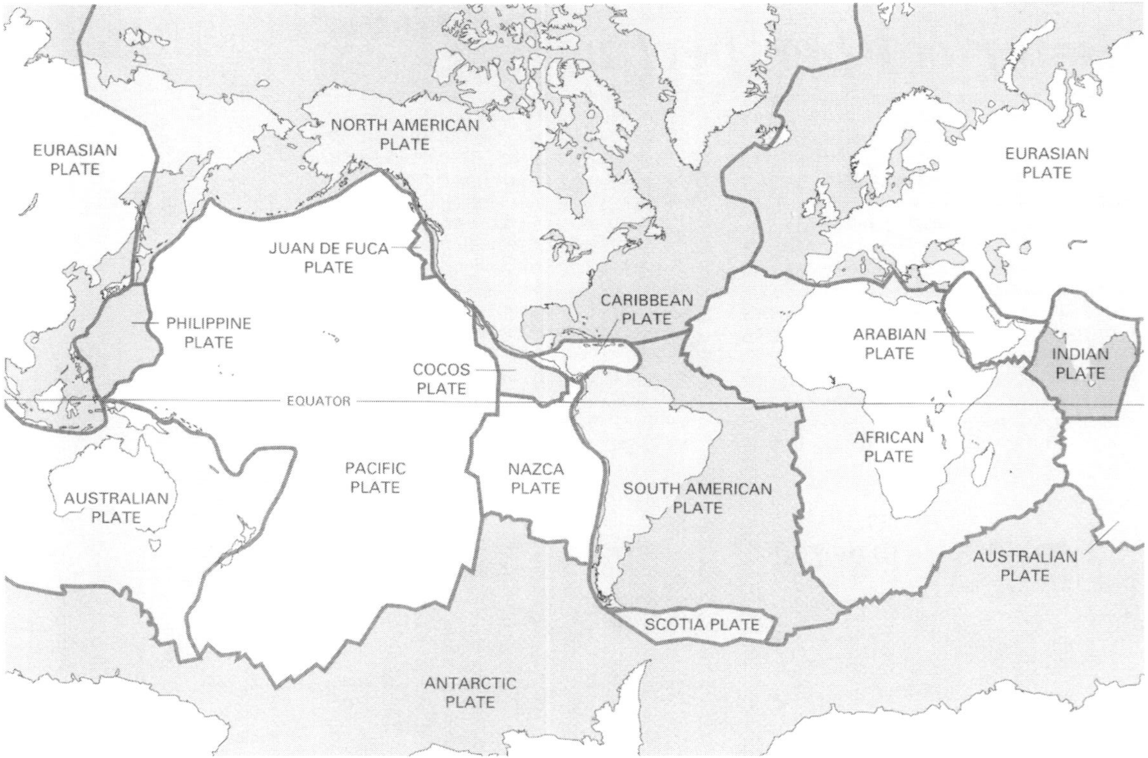

Figure 1-3. Approximately a dozen plates or slabs that are moving relative to one another as identified by seismic mapping in the 1960s ad 1970s by the World-Wide Standard Seismographic Network. *(Modified after Kious and Tilling, 1997, p. 2.)*

Dr. David G. Howell (U.S.G.S., Menlo Park, California) likens movement of continents (as the plates join and separate over and over throughout earth's history) to the motion of "lithospheric bumper cars." One should remember, however, that electric bumper cars at amusement parks can each move independently, rather than being parts of an integrated system. In addition, their average speeds are at least 500 million times faster than those of tectonic plates (Kious and Tilling, 1997, p. 56).

Earthquake Geology

Based on studies by the WWSSN, at approximately 120 seismographic stations in 60 countries, many of the continental slopes, including most of the slopes surrounding continents facing the Atlantic, Indian, and Arctic oceans, have very few earthquakes. Other continental slopes, including those bordering continents facing the Pacific Ocean, have many seismic events relating to active tectonics. This tectonic activity has been responsible for the formation of high mountains, deep oceanic trenches, and volcanic activity. These margins are also sometimes related to subduction zones that can be the source for very powerful earthquakes.

With a few significant exceptions, identifying the most likely locations of earthquake occurrence is synonymous with the identification of the active tectonic plate boundaries, oceanic as well as continental margins. About 90% of all the world's earthquakes occur along two belts of active tectonic plate boundaries (Figure 1-4). One extends along the plate boundaries of the Pacific Ocean (*Pacific Ring of Fire*) and the other stretches from the Pacific Ocean westward through Indonesia, China, Central Asia, the Near East, and the Mediterranean Sea. The band of seismic activity stretches out across the ocean basins, particularly in the western Pacific Ocean far from the continents. Deep-sea trenches and active volcanoes are often associated with these bands of seismic activity. Studies by the WWSSN show that most earthquakes occur at active mountain belts on continents, mid-ocean spreading ridges, and deep-sea trenches (see Kious and Tilling, 1997, p. 38.).

Plate Tectonics

The earth's surface consists of a series of tectonic plates, with each plate consisting of the crust and the more rigid part of the upper

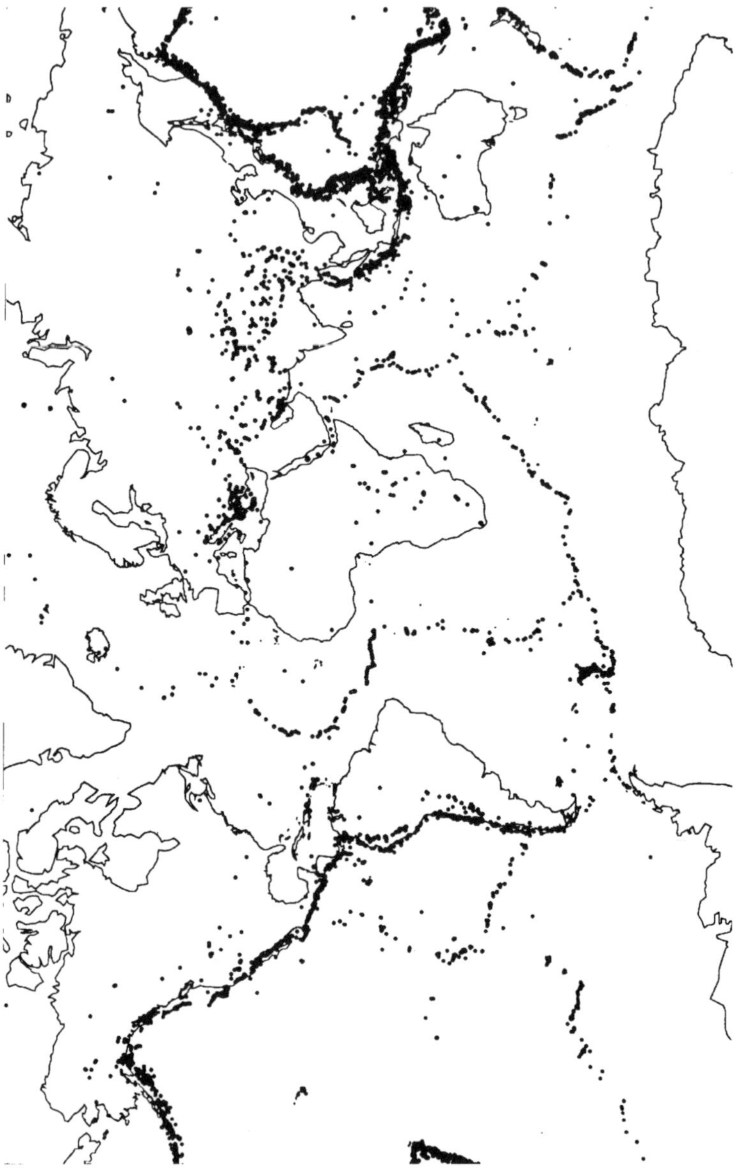

Figure 1-4. Epicenters for earthquakes shallower than 50 km and with magnitude equal to or greater than 5.5 for the period of 1963 to 1987, from the catalog of the National Earthquake Information Center. *(Modified after Gordon and Stein, 1992; in Yeats et al., 1997, fig. 1-2, p. 10.)*

mantle, termed "lithosphere" (Figure 1-5). A weaker or more plastic zone called the "asthenosphere" underlies the lithosphere. Two indicators for the lithosphere/asthenosphere boundary are temperature and composition (Yeats et al., 1997).

As indicated in Figure 1-4, there is a tendency of earthquakes to occur in two relatively narrow belts along active plate boundaries. The earth's surface consists of tectonic plates, with each plate consisting of a portion of the crust and a more rigid lithosphere. Most earthquakes occur within the lithosphere (Figure 1-5). The asthenosphere (underlying the lithosphere) is semi-solid and yields by way of plastic flow in compensation for the motion of the lithospheric plates. The lithosphere can be as thick as 100 km beneath older continental and oceanic crust. At the spreading ridges, however, it may be only a few kilometers thick.

Most earthquakes occur at the boundaries of the active lithospheric plates, which are in motion with respect to one another. Figure 1-6 schematically illustrates the motion of plates along transform, divergent, and convergent plate boundaries. "Oceanic spreading ridges" are formed by the upwelling of hot magma that cools and solidifies as

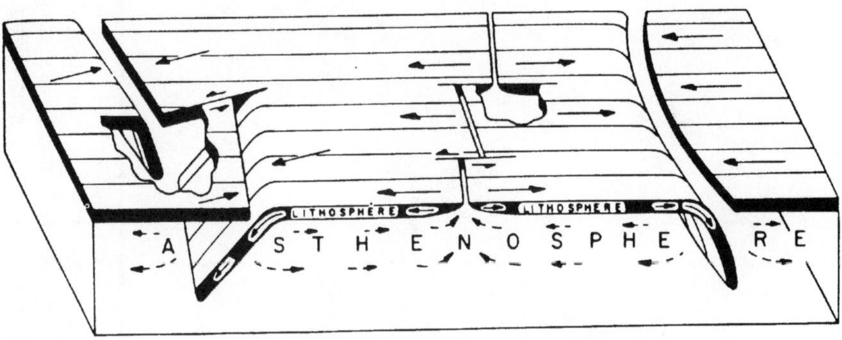

Figure 1-5. Arrows show relative motion of lithosphere and asthenosphere. New lithosphere is formed at mid-ocean ridges and thickens as it cools and moves away from the ridge. At points of collision, plates can often descend (subducted) beneath other plates. These collision zones are the sites of active volcanoes and trenches and can be the source of large earthquakes. Both mid-ocean ridges and trenches may be offset along zones where plates move parallel to each other along transform faults. *(Modified after Isacks et al., 1968; in Yeats et al., 1997.)*

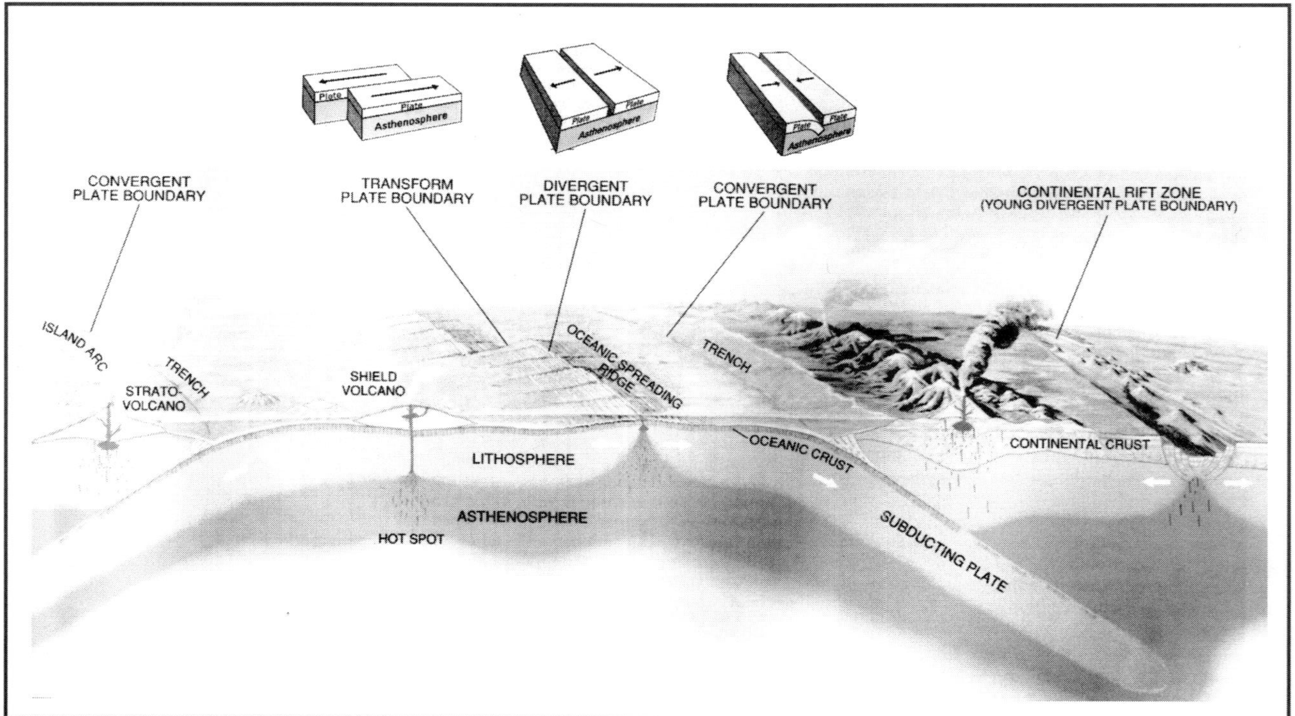

Figure 1-6. Schematic illustrating the main types of plate boundaries. *(Modified after cross-section by José F. Vigil from* This Dynamic Planet*—a wall map prepared jointly by the U.S. Geological Survey, the Smithsonian Institution, and the U.S. Naval Research; in Kious and Tilling, 1997, p. 30.)*

lithospheric plates move away from the divergent plate boundary (Figure 1-7). The best known of the divergent plate boundaries is the Mid-Atlantic Ridge. The rate of spreading for this ridge averages about 2.5 centimeters per year (Kious and Tilling, 1997). Newly-formed oceanic crust is carried down the flanks of the ridge (Figure 1-6). The history of ocean-floor spreading is recorded in the magnetization of basaltic crust into magnetic stripes that are symmetric about the ridge axis. Earthquakes on spreading ridges are limited to the divergent plate boundary, where new crust is being formed. These earthquakes tend to be relatively small in magnitude and occur at shallow depths near the surface where the cooler thin crust is brittle. A global map illustrating the development of mid-ocean ridges around the world is shown in Figure 1-7.

Figures 1-5 and 1-6 illustrate deep-sea trenches where the lithosphere is subducted (converges), forcing one slab beneath the other. A zone of earthquakes (Wadati-Benioff zone) marks where the lithosphere extends into the mantle to depths up to 700 kilometers and helps define the subduction zone. The shape of the subduction zone

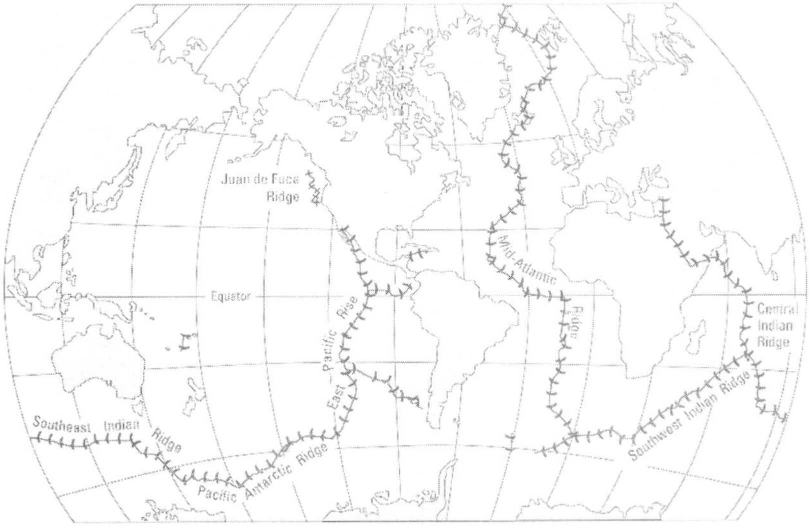

Figure 1-7. The mid-ocean ridge—a submarine mountain chain winding its way under the ocean between the continents. On average, this chain rises 4,500 meters above the sea floor. *(Modified after Kious and Tilling, 1997, p. 15.)*

may be determined seismically by mapping the dipping zone of earthquakes beneath the active volcanic areas. Subduction zones dip at angles less than 90°, resulting in a broad belt-like pattern for earthquakes around the Pacific Ocean. Yeats et al. (1997) noted the prominence of the earthquake zone around the Pacific (where a large percentage of the seismic energy released, worldwide, is generated) at subduction zones.

Active transform faults can also cause earthquakes along zones where lithospheric plates move parallel to one another. Two types of transform faults are shown in Figure 1-6. The oceanic spreading ridge is offset along transform faults such that lithospheric plates move past one another between the offsets. For relatively long transform faults, the lithosphere far from either offset spreading ridge has cooled enough to be relatively thick and rigid and is thereby able to generate large earthquakes. East-west trending zones of earthquakes along the Mid-Atlantic Ridge mark transform faults offsetting the ridge. The second type of transform fault shown in Figure 1-5 separates trenches, one subducting in one direction and one subducting in the opposite direction.

The global distribution of earthquakes, shown in Figure 1-4, defines a worldwide mosaic of crustal plates (Figure 1-3) up to 100 km thick, which are moving with respect to one another. It should be noted that the continental slopes do not necessarily correspond to plate boundaries. As mentioned earlier, the continents are crustal slabs riding on oceanic plates and much of the dynamic activity (earthquakes) involves the oceanic crust at spreading ridges and subduction zones. Spreading rates between the plates range from 12 to 160 mm/yr, whereas the convergence rates at trenches vary from 10 to 20 mm/yr (DeMets et al., 1990). Gordon and Stein (1992) have confirmed these rates by measurements averaged over a few years using space geodetic techniques.

Yeats et al. (1997) noted that the diffuse distribution of earthquakes across plate boundaries within continents shows that the assumption of perfectly rigid lithospheric plates is incorrect. Specifically, in central Asia, the western United States, and within some ocean basins, such as the Indian Ocean between India and Australia, plate motion appears to be distributed across a zone hundreds or even thousands of kilometers wide. Gordon and Stein (1992) have noted that although the earthquake belts described earlier in this chapter may only occupy as much as 15% of the earth's surface, these diffuse plate boundary zones contain most of the earth's great earthquakes.

Some plate boundaries contain relatively rigid micro-plates, the relative motion of which with respect to the major plates must be calculated to understand slip rates across major faults within the region (Yeats et al., 1997). For example, the San Andreas Fault is one of the few transform faults exposed on the surface of the land (Figures 1-8, 1-9, and 1-10). The displacement rates across the San Andreas Fault of California are lower than the rate one would expect understanding the relative motion between the Pacific and North American plates, producing a slip-rate discrepancy.

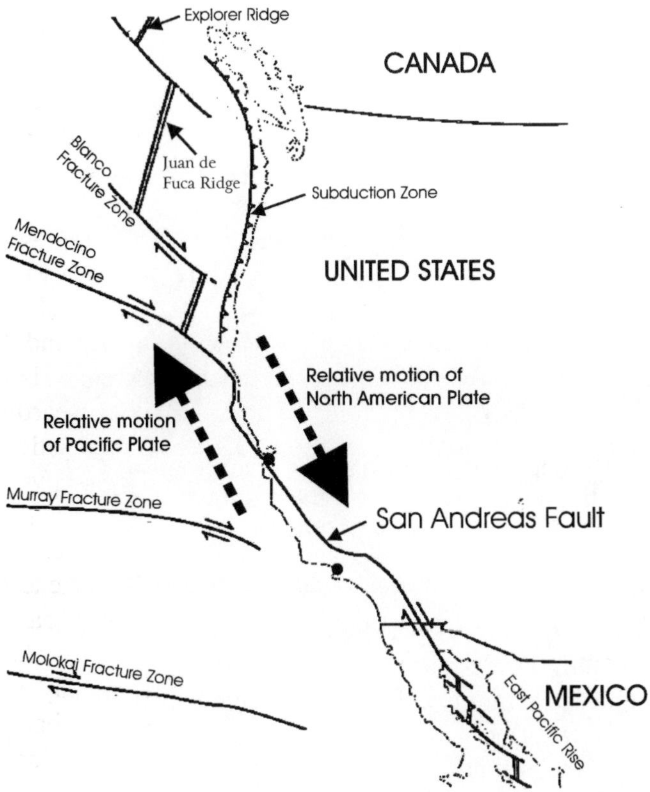

Figure 1-8. The San Andreas Fault is one of the few transform faults exposed on land. Several offshore fracture zones (transform faults) are also shown as part of the fault system between the Pacific and North American Plates. *(Modified after Kious and Tilling, 1997, p. 43.)*

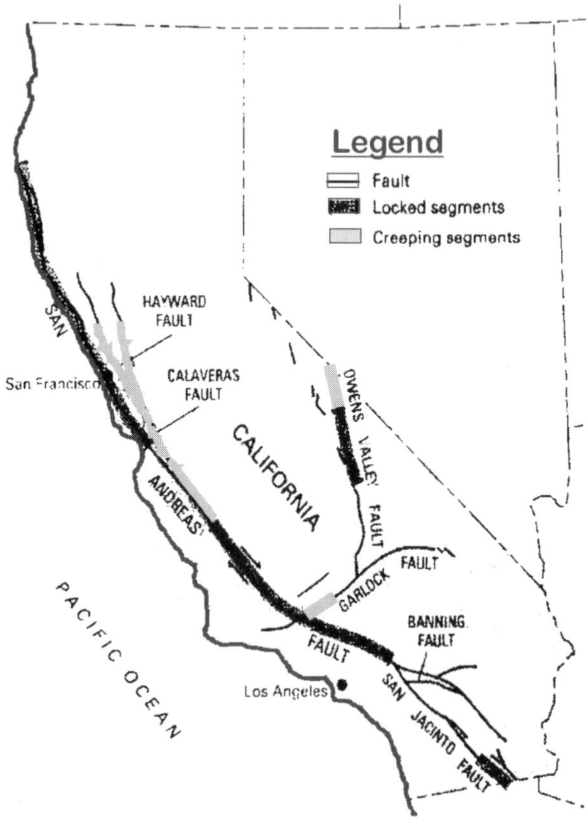

Figure 1-9. Map of the San Andreas Fault system, California, showing portions of the fault system that are locked and those that are unlocked. *(Modified after Kious and Tilling, 1997, p. 60.)*

The island of Hawaii has frequent earthquakes, even though it is entirely within the Pacific Plate, far from a plate boundary. Hawaii is the emergent portion of several huge basaltic shield volcanoes that rise from the ocean floor. Many of these earthquakes are related to the injection of magma into the volcanic edifice, whereas others are likely due to gravitational collapse of the flanks of the volcano. The volcanoes are related to a thermal plume or "hotspot" within the mantle, which warms and elevates the oceanic crust in addition to producing the volcano. This hot spot is similar to that suggested earlier, which developed the volcanic eruption shattering the continent of Pangaea.

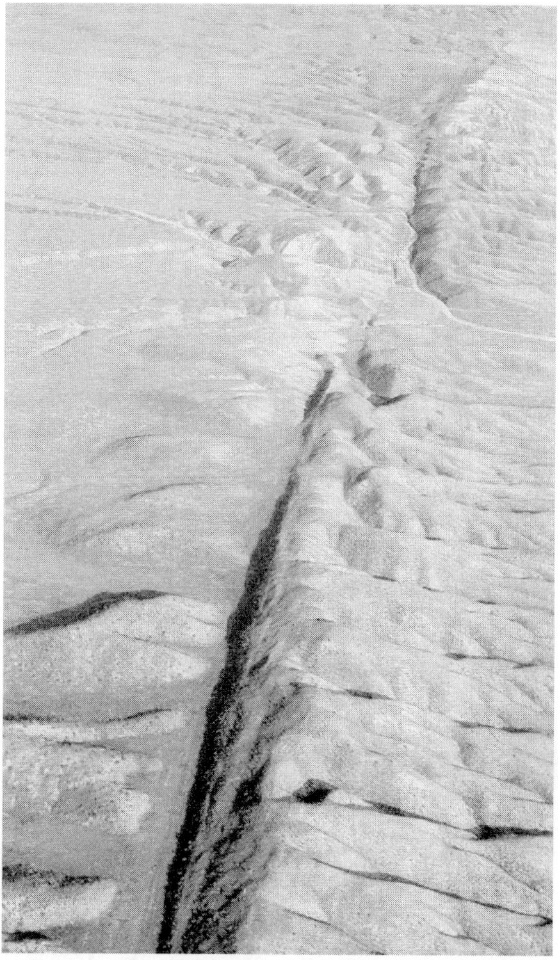

Figure 1-10. Aerial view of the San Andreas fault slicing through the Carrizo Plain in the Temblor Range east of the city of San Luis Obispo, California *(Modified after photograph by Robert E. Wallace, U.S.G.S.; in Kious and Tilling, 1997, p. 44.)*

The thermal plume appears more stationary with respect to the lower mantle than the Pacific Plate, which is moving west-northwest with respect to the center of the earth (Figure 1-11). The other islands of the Hawaiian chain west-northwest of Hawaii have progressively older volcanoes and represent the period when that part of the Pacific Plate

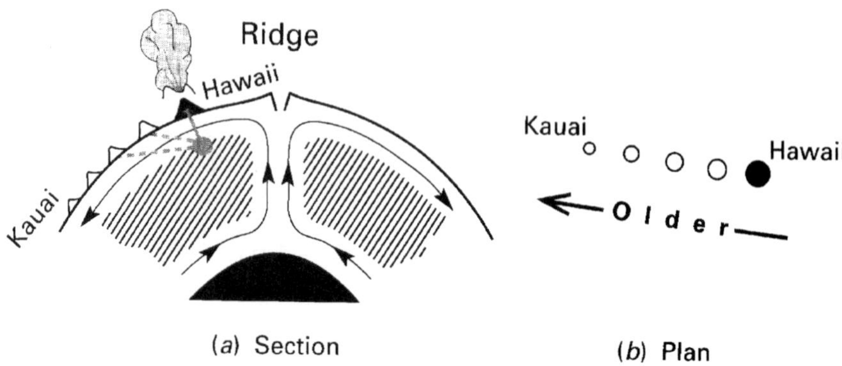

Figure 1-11. Schematic of the movement of the Pacific Plate over the fixed Hawaiian "hot spot" illustrating the formation of the Hawaiian Ridge-Emperor Seamount Chain. *(Modified after Kious and Tilling, 1997, p. 48.)*

passed over the stationary mantle plume. In this way, the volcanic islands record former positions of the plate with respect to the plume. Duncan (1991) suggested that a similar age pattern of hotspot tracks relate the Reunion and Kerguelen hotspots in the southern Indian Ocean to the basaltic lava fields in the Indian peninsula.

Yeats et al. (1997) noted that Yellowstone Park in the western United States provides an example of a mantle hotspot rising beneath a continent. Yellowstone Park is characterized by young volcanic rocks, hot groundwater manifested as geysers, reflecting anomalously high heat flow through the crust, and by elevated topography. If the Yellowstone hotspot is stationary with respect to the center of the earth, it should have left a track of older volcanic rocks west of it as North America moved westward away from the Mid-Atlantic Ridge. Volcanic rocks of the Snake River Plains of southern Idaho and the Columbia Plateau of Washington and Oregon may be ancient expressions of the Yellowstone hotspot.

SUMMARY

The origin of earthquakes is ultimately the result of relative movement of tectonic plates and blocks, which produces the stresses within the lithosphere that must be relieved. Over 90% of all major earthquakes occur within two well-delineated belts of seismic activity that stretch along the boundaries of tectonic plates.

CHAPTER 2

Gas Migration at Plate Boundaries

SEEPS

The edges or boundaries of moving lithospheric plates (fault zones) can serve as pathways or conduits for the seepage of gases to the surface formed by changes in the permeability of these zones resulting from tectonic movement. When there are hydrocarbon deposits present along these fault zones through the earth's crust, oil and gas can migrate up the faults toward the surface. Surface seeps are common along active lithospheric plates where there is a presence of hydrocarbons. Worldwide, there is a correlation between these seepages and earthquake activity, with the majority of the seeps being close to plate boundaries where such earthquake activity is the highest. Figure 1-4 illustrates the areas of worldwide seismic activity. Link (1952) observed that most surface seeps are found in younger, recently uplifted sediments. In the young and small basin areas of California where uplift, folding, and faulting have occurred, many structures that once held oil and gas have been ruptured, resulting in seeps at the earth's surface.

Geologists have long used hydrocarbon surface seeps as an indicator of subsurface petroleum deposits. In 1965, Chilingar pointed out that about 90% of all major oil fields in the world were discovered because of surface seeps. In the Middle East, the Zoroastrians built temples at the site of gas seeps. These gas seeps fed the fires in the temples. For example, "the eternal fire" of Baba Gurgur (an ancient Zoroastrian temple) is located over the big Kirkuk Field of Iraq. Masjed-e-Soleiman oil field in Iran, one of the largest, was also discovered at the site of an old Zoroastrian temple.

Thompson (1933) pointed out that practically all the great oil fields of the world were marked by oil and gas seeps near the crest of anticlines or the apices of domes. DeGolyer (1940) reported that oil and gas seeps have resulted in the discovery of more oil and gas fields than has any other single method of oil prospecting. Dickey and Hunt (1972) noted that visible seeps are most common around the margins of petroliferous basins where the oil-bearing sediments crop out. Surface seeps commonly occur at the outcrop of a permeable rock unit that constituted an oil reservoir until the overlying impermeable stratum was breached by folding, faulting, and/or erosion. Gulf Oil Corporation entered Kuwait and Italy on the basis of the presence of seeps (Pyre, 1977), which were directly related to the Burgan (Kuwait) and Ragusa (Sicily) oil fields. "Drilling on oil seeps and trends is as good a way of prospecting today as it was then and down the years more oil has probably been found by the use of these simple principles than from all the new techniques of geology and geophysics combined" (the centennial issue of the *Oil and Gas Journal,* January 28, 1979). The age of the faulting and subsequent gas migration to the earth's surface for a seep implies that the age of the pathway (fault; fractured zones) is that of the youngest geologic deposits (Bernard, 1980).

Surface soil surveys help identify the type of gases migrating to the surface. They also help identify the surface location of the faults. Jones and Drozd (1983) cautioned that these studies of surface soil must be used with great care in identifying the presence of subsurface hydrocarbon deposits. This is due to the fact that shape and location of a geochemical anomaly is controlled more by the local tectonic structure of the region than by the position and shape of the actual hydrocarbon deposit. In addition, the magnitude of seep anomalies strongly depends on the tectonic activity of an area. The chemical compositions of near-surface hydrocarbon soil gases are generally determined by the hydrocarbons in the underlying hydrocarbon-bearing formations. Soil-gas compositions typical of seeps over oil fields differ from those occurring over gas fields.

Present geochemical methods cannot predict whether an anomaly is associated with a commercial deposit, because "commercial" is an economic concept. Geochemical prospecting can be used to verify the presence of petroleum hydrocarbons and to predict whether oil or gas is likely to occur in a potential structure. Conventional geophysical and geological mapping and drilling, however, may be used to estimate the dimensions of a potential gas/oil reservoir.

The history of petroleum production from oil wells in the Middle East began in the 1890s when de Morgan, a French geologist, wrote about the numerous oil and gas seeps in western Iran. The numerous seeps in Iran (Persia) and Iraq (Mesopotamia) called attention to the possibilities of oil in the region and, as late as 1949, every field in Iran was associated with surface oil or gas seeps (Link, 1952). Throughout the Middle East, early exploration and drilling of oil were done near surface seeps. The Middle East surface seeps of migrating oil and gas are found along the Arabian and Eurasian Plate boundaries in the countries of Turkey, Iran, and Iraq (Figure 2-1). The activity (movement) of these two adjacent plates is demonstrated by the heavy concentration of earthquakes within this zone. Prior to the collision of the Arabian and Eurasian Plates, gas and oil had accumulated in formations under an impermeable caprock during the Cretaceous period. The collision of the Arabian and Eurasian Plates ruptured (fractured) this caprock in the area of plate contact, creating pathways

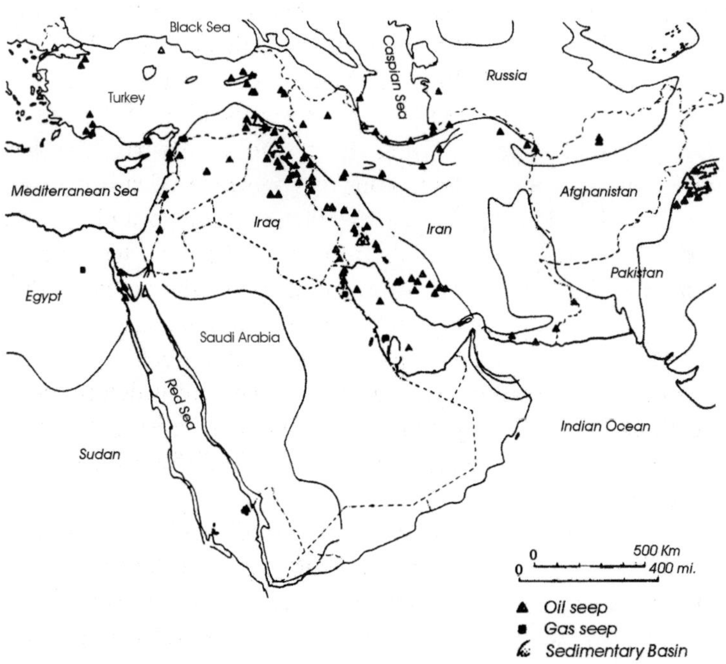

Figure 2-1. Distribution of oil and gas seeps in the Middle East. *(Modified after Link, 1952, fig. 28, p. 1523.)*

(faults) for gas and oil migration to the surface. In the southern and western parts of Saudi Arabia, with low earthquake activity, there are very few surface seeps, even though this area contains some of the largest oil fields of the Middle East.

South American seeps in Venezuela, Trinidad, Colombia, Ecuador, Peru, and Argentina attracted geologists and incited interest in the area. Surface seeps are found along the western coast of South America in Ecuador and Peru (Figure 2-2), illustrating the relationship of seeps to tectonic activity. These seeps follow the "earthquake belt," including the subduction zone of the Nazca Plate beneath South America. Seeps are also found on the Mesopotamian geosyncline, the Monagas Basin of eastern Venezuela, and the eastern foothills of the Andean mountain chain from Colombia to Cape Horn near the edge of lithospheric plates. The hydrocarbon-rich area of Lake Maracaibo, in the western portion of Venezuela, has over 200 oil and gas surface seepage areas. Many of these surface seeps occur along the flanks of the Venezuelan

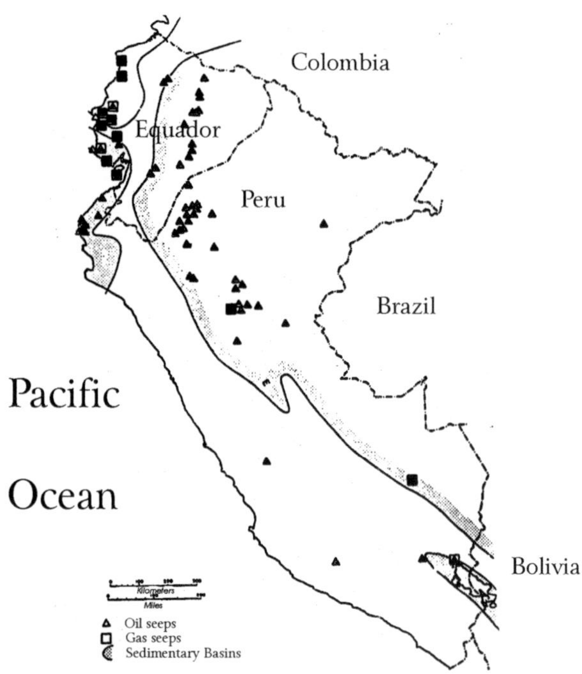

Figure 2-2. Distribution of oil and gas seeps in Colombia, Equador, and Peru. *(Modified after Link, 1952, fig. 33, p. 1528.)*

Andes and the Perija Mountains. In this area, seeps are particularly numerous along the margins of basins, where unconformities and oil-producing formations are exposed at the surface.

Oil and gas seeps in Indonesia and Burma were responsible for significant discoveries. The birthplace of Shell Oil Company was in the Netherlands East Indies (Indonesia) in the 1890s. The first production was established near oil seeps in the Langkat District of North Sumatra. In Burma, shallow hand-dug wells in surface seep areas have yielded gas and oil for centuries. These seeps are located on the edge of the Indo-Australian-Eurasian Plate boundary (Figure 2-3).

In 1543, Herman de Soto noted seeps in the Gulf of Mexico. Early Indian inhabitants called these petroleum substances "Chapopote." The many seeps located on the edge of Mexico along the Gulf of Mexico led to the development of the Mexican oil fields (Figure 2-4).

California, Alaska, the Philippines, Europe, and Japan are other areas where numerous surface seeps are directly related to the migration of fluids along faults at the plate boundaries.

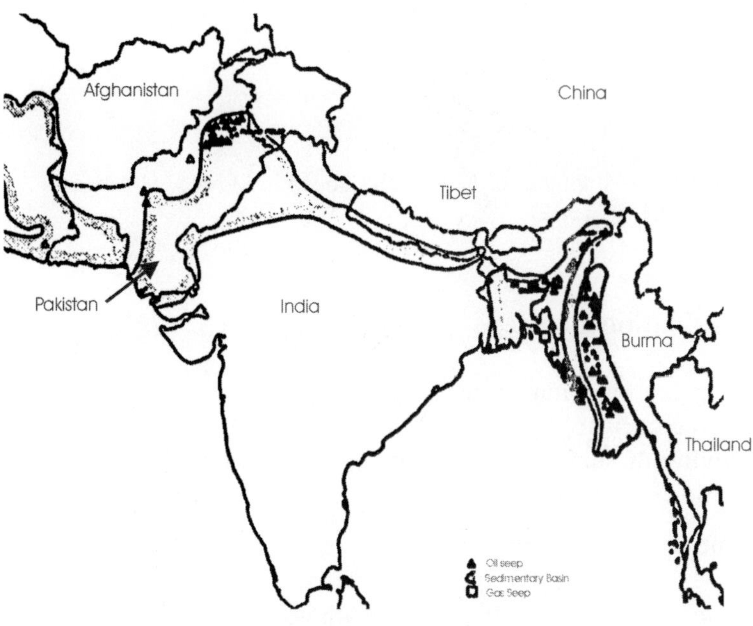

Figure 2-3. Distribution of oil and gas seeps in Burma, India, and Pakistan. *(Modified after Link, 1952, fig. 37, p. 1533.)*

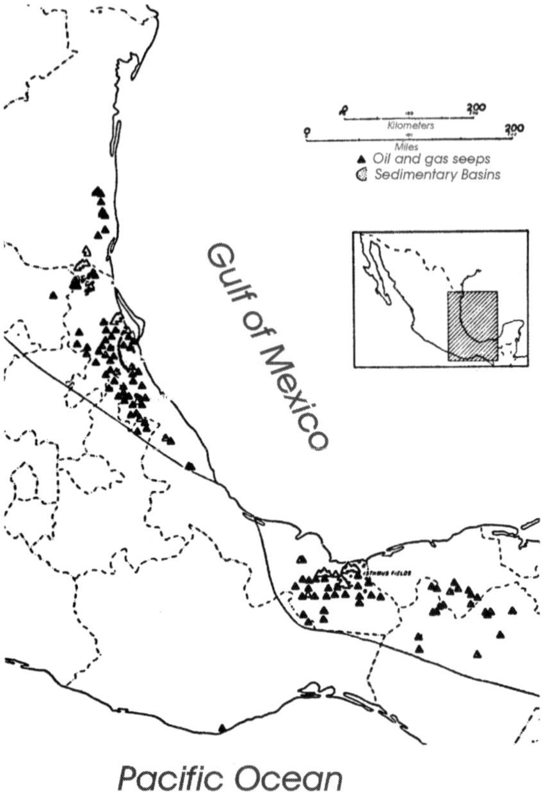

Figure 2-4. Distribution of oil and gas seeps in Mexico. *(Modified after Link, 1952, fig. 42, p. 1538.)*

Classification of Seeps

Seeps are most numerous on the margins of basins and in sediments that have been folded, faulted, and eroded. Link (1952) developed a cataloging of surface hydrocarbon seeps by placing them into one of the following categories:

Type 1—Surface seeps emerging from hydrocarbon-bearing homoclinal beds, the ends of which are exposed at the surface. In general, these seeps are not prolific, but are persistent, because the oil and gas moves slowly toward the surface along permeable formations. Manitoulin Island, Ontario,

Canada, is an excellent example of this type of seep (Figure 2-5a). Another example is the Los Angeles oil field that outcrops at the surface within the Los Angeles Basin.

Type 2—Surface seeps associated with beds and formations in which the oil was formed (direct linkage with source bed). This type of seep yields small volumes of oil. The fracturing and shattering of these formations by tectonic activity liberates the oil and gas in small quantities. An example of this type of seep is located in Sixaola Valley, Costa Rica. There is very little migration of oil and gas and, therefore, little surface indication of a seep (Figure 2-5b).

Type 3—Surface seeps from large subsurface hydrocarbon accumulations (oilfields, etc.) where the formations have been ruptured by faulting or folding due to the movement of crustal plates. This class of seep often represents migration of large volumes of oil and gas. A variety of seeps from around the world are presented in Figures 2-5c-f and 2-5a-d.

Type 4—Surface seeps at the outcrops of unconformities where the path of migration is along the unconformity. Oil and gas migrate upward along the other types of paths (Types 1 through 3 seeps) until the hydrocarbons reach an unconformity. The hydrocarbons then migrate along the unconformity, along the path of least resistance, to the surface. Because of the indirect path of migration, it is often difficult to recognize the subsurface location of the source as the outcrops frequently branch out like limbs on a tree, leaving deposits of tar sands and asphalts in the subsurface soil. The Athabaska tar sands likely have their origin due to this type of migration. Examples of these types of seeps are found in Peru, Venezuela, and Utah as shown in Figures 2-6e-f and 2-7a-b. The La Brea Tar Pits in the Fairfax area of the Los Angeles Basin is another example where the primary oil and gas migration occurs along the Sixth Street Fault prior to emerging into an unconsolidated zone near the surface.

Type 5—Surface seeps associated with intrusions (i.e., igneous intrusions, piercing salt domes, and mud volcanoes). These are

(text continued on page 27)

Schematics of Various Types of Seeps

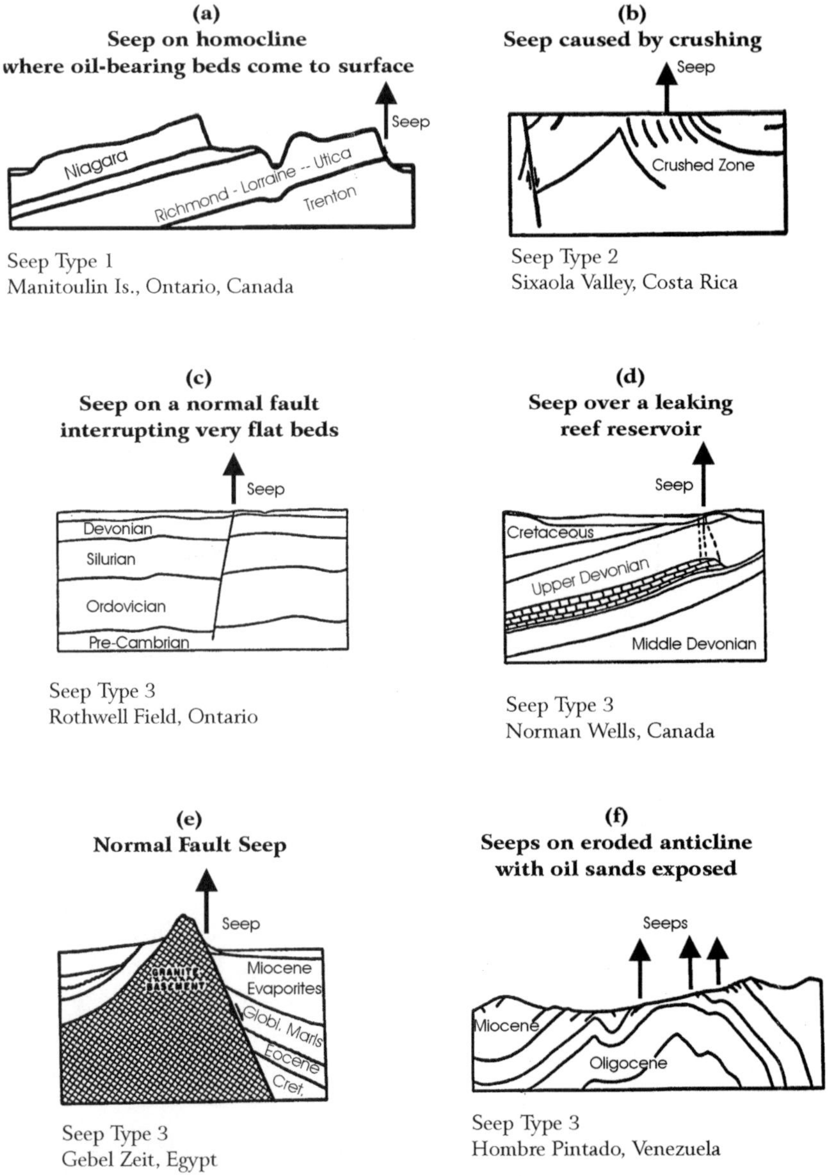

Figure 2-5. Various types of seeps. *(Modified after Link, 1952, figs. 1–6, pp. 1507–1508.)*

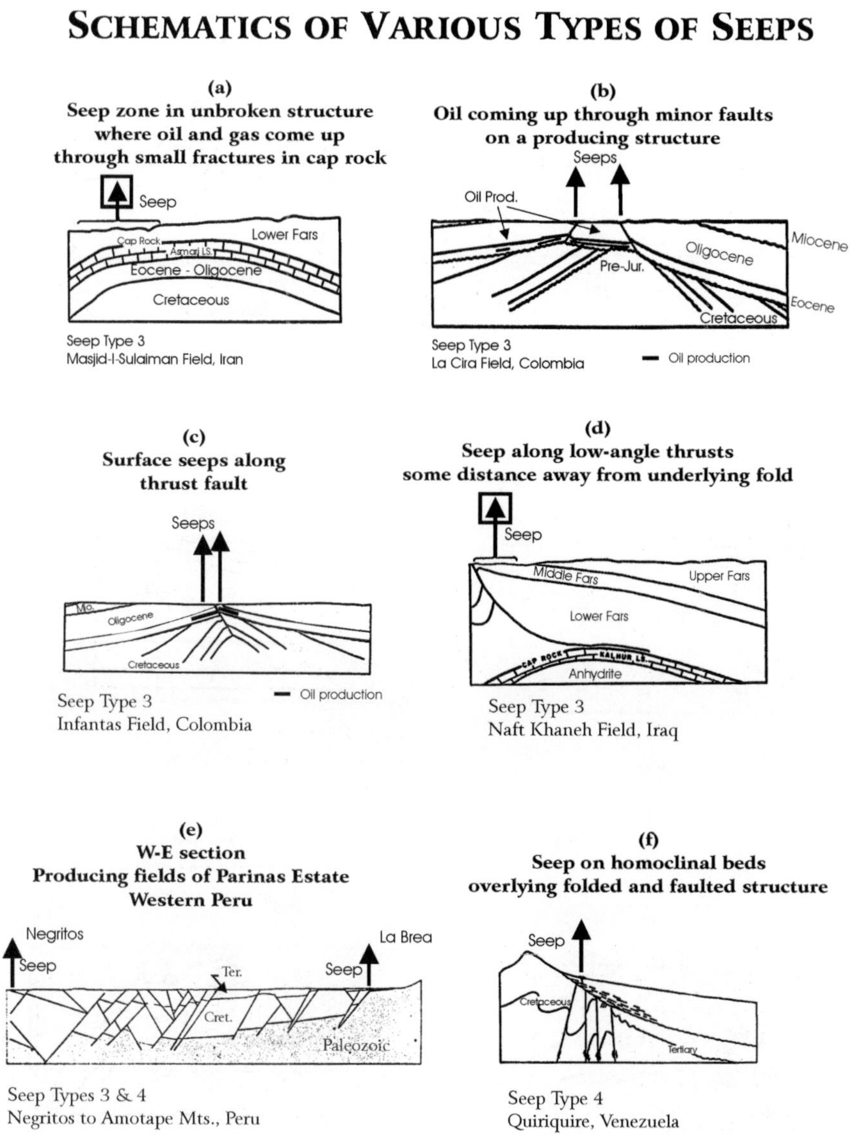

Figure 2-6. Various types of seeps. *(Modified after Link, 1952, figs. 7–12, pp. 1508–1511.)*

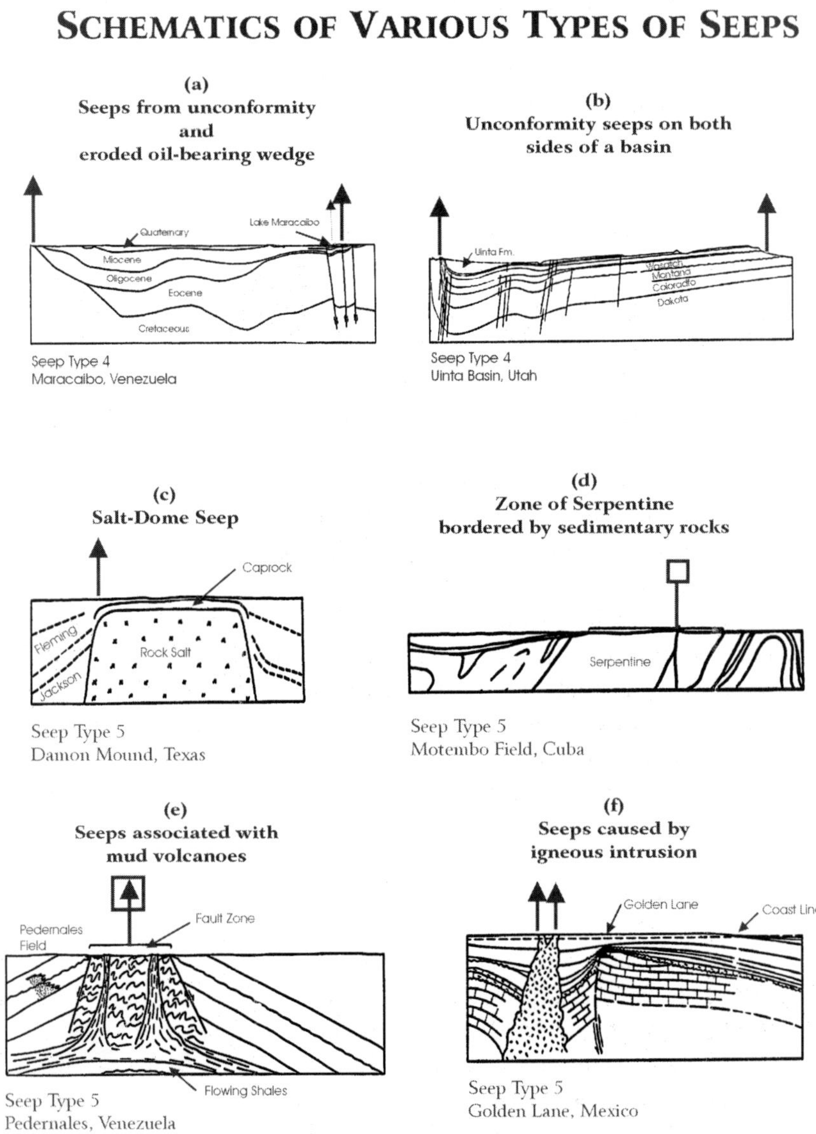

Figure 2-7. Various types of seeps. *(Modified after Link, 1952, figs. 13–18, pp. 1512–1514.)*

(text continued from page 23)

seeps where the path of migration is along fractures and faults generated by tectonic activity. This type of seep may or may not be associated with ruptured hydrocarbon-bearing formations (Figure 2-7c-f).

Summary

Oil and gas seeps to the Earth's surface are important in exploration because they can indicate the presence of hydrocarbon source rocks and deep faults. When oil seeps to the surface, it undergoes various stages of decomposition including vaporization of the light-end hydrocarbons, microbial degradation, polymerization, and oxidation. This decomposition tends to thicken the migrating oil, resulting in lower °API gravity oil as it reaches the surface. The liquid oil is gradually converted to asphalt, asphaltite, and eventually into a pyrobitumen. In the extreme, the crude oil is converted into a thick asphaltic substance deposited near the surface. Buoyancy is the primary driving mechanism, by which the oil and gas are moved to the surface through permeable zones, including faults and fractures.

Geochemical fingerprinting can be used to confirm the depth of origin of the gas that has migrated to the earth's surface. From an environmental viewpoint, surface seeps must be carefully evaluated—especially in an urban setting—because of the risks of explosions and fires, as well as the presence of hazardous, toxic, and carcinogenic chemicals in the oil and/or gas.

CALIFORNIA OFFSHORE SEEPS

Offshore oil and gas seeps have long been known in California. Archeologists have established the fact that prehistoric Indians used tar from seeps to waterproof baskets and coat canoes at least 7,000 years ago. Pleistocene fossils removed from the La Brea Tar Pits in Los Angeles, California, prove that this famous seep has been in existence for more than one million years. The California Division of Oil and Gas has prepared Table 2-1, listing California offshore seeps that have been visually documented. One of the earliest known records of an offshore

(text continued on page 34)

Table 2-1
Listing of Offshore Oil and Gas Seeps for California

	Location	Lat. & Long.	References	Remarks
1	Coronado	32° 28'20"N	Oil Age, 1923 117° 21'30"W	Questionable—may be a spill
2	Catalina, west end	33° 28'20"N 118° 36'50"W	Western Oil & Gas Assoc., 1931	One-half acre oil slick sighted from steamer, 1923 Frequently sighted by fishermen; now inactive
3	Catalina, west end	33° 29'10"N 118° 36'30"W	See 2	See #2
4	San Pedro Channel	33° 35'40"N 118° 20'30"W	U.S. Bureau of Mines, 1923 Calif. Academy of Science, 1917	Oil droplets & slabs of tar as large as 8" × 5" × 1"
5	Huntington Beach	33° 40'10"N 118° 01'50"W	Signal Oil and Gas Co.	Gas bubbles from outcrop, probably of marsh Gas sands
6	Palos Verdes	33° 48'10"N 118° 24'40"W		
7	Redondo Beach	33° 49'00"N 118° 26'10"W	U.S. Bureau of Mines, 1923	Oil and gas, about 3.5% sulfur Intermittent flow in 1927; no estimate made Oil slick about 50 acres in 1928
8	Redondo Beach	33° 50'10"N 118° 25'30"W	University of Southern California	Oil and gas On line of Palos Verdes fault

9	Manhattan Beach	33° 52'10"N 118° 27'40"W	Seismological Soc. America, 1920 Standard Oil Co. of California and W. T. Knowlton, 1928	Oil and gas, about 4% sulfur 9 & 10 together estimated to be discharging about 10 barrels of oil per day in 1928 200- to 400-acre oil slick On line of Palos Verdes fault
10	Manhattan Beach	33° 52'20"N 118° 29'20"W	See 9	See 9
11	San Miguel Island	34° 04'30"N 120° 25'20"W	Western Oil & Gas Assoc., 1931 Oil and Gas Journal, 1937	First noted by U.S. Coast & Geodetic Survey in 1875 Oil and gas
12	Santa Cruz Island	34° 03'20"N 119° 54'10"W	Papers written about the island in French in 1893 and translated by Dr. Redwine, Union Oil Co.	Mostly tar
13	Santa Cruz Island	34° 00'40"N 119° 53'40"W	Emery, 1960	Mostly tar
14	Santa Barbara Channel	34° 02'20"N 119° 16'10"W	Emery, 1960	Discovered by continuous presence of oil, gas bubbles, and tar on the water
15	Santa Barbara Channel	34° 08'00"N 119° 41'24"W	Standard Oil Co. of California	See 14
16	Platform A	34° 20'20"N 119° 36'30"W	Lewis & Lewis (surveyors for Union Oil Co.), 1968	Discovered before platform was installed about 800 feet from platform Oil and gas from saturated shale outcrop
17	Platform Houchin	34° 20'00"N 119° 33'40"W	U.S. Geological Survey	

Table 2-1 continued

	Location	Lat. & Long.	References	Remarks
18	Platform Heidi	34° 21'00"N 119° 30'20"W	Mattei, 1929, in Santa Barbara newspaper	On crest of anticline
19	Rincon Trend	34° 21'30"N 119° 29'40"W	L. S. Fox, 1930	Gas bubbles from crest of anticline
20	Rincon Point	34° 22'50"N 119° 29'30"W	Whitney, 1865	From fractured Monterey Formation
21	Carpinteria	34° 23'30"N 119° 31'20"W	Father Crespi, 1770	Gas bubbles
22	Carpinteria	34° 23'40"N 119° 32'10"W	Father Crespi, 1770	Similar to seeps on shore
23	Montecito	34° 25'00"N 119° 38'00"W	Calif. Div. Mines Bull. 118	Probably on a fault
24	Montecito	34° 25'00"N 119° 38'30"W	Calif. Div. Mines Bull. 118	See 23
25	La Mesa	34° 21'40"N 119° 44'30"W	Standard Oil Co. of Calif.	
26	Hope Ranch	34° 22'00"N 119° 46'20"W	Standard Oil Co. of Calif.	
27	Hope Ranch	34° 24'20"N 119° 45'50"W	Standard Oil Co. of Calif.	

No.	Location	Coordinates	Source	Remarks
28	Coal Oil Point	34° 23'40"N 119° 51'40"W	Father Pedro Font, 1776 Sixth Annual Report, State of Calif. Mineralogist, 1886 Allen & Schleuter, 1969	Best known seep area Oil, gas and tar. Oil is 12.4 gravity Typically, oil is released as globules from sand patches, some of which are in line and some of which are independent Estimated discharge for area: Max. 160 bbls. oil per day Min. 11 bbls. oil per day Avg. 50–70 bbls. oil per day More than induced seeps at Platform A
29	Coal Oil Point	34° 24'00"N 119° 52'40"W	See 28	See 28
30	Coal Oil Point	34° 24'30"N 119° 52'00"W	See 28 J. Eaton, geologist, 1931	See 28
31	Platform Holly	34° 24'30"N 119° 54'40"W	Sixth Annual Report, State of Calif. Mineralogist, 1886	Probably similar to Coal Oil Point
32	Naples	34° 25'30"N 119° 57'30"V	Holder, 1910	Tar in shale
33	Capitan Beach	34° 27'20"N 120° 00'30"W	Holder, 1910 Standard Oil Co. of Calif.	At intersection of faults
34	Refugio	34° 26'00"N 120° 03'30"W	Holder, 1910	Probably from Monterey Formation Oil slicks noted on water
35	Molino	34° 28'00"N 120° 09'20"W	Holder, 1910 Standard Oil Co. of Calif.	See 34

Table 2-1 (continued)

Location	Lat. & Long.	References	Remarks
36 3Gaviota	34° 27'30"N 120° 13'30"W	Holder, 1910 Standard Oil Co. of Calif.	See 34
37 Platform Helen	34° 26'50"N 120° 17'00"W	Standard Oil Co. of Calif.	Faulted area
38 Cuarta	34° 27'30"N 120° 17'30"V	Standard Oil Co. of Calif. Emery, 1960	Faulted area
39 San Augustine	34° 26'40"N 120° 20'30"W	Union Oil Co. of Calif. Emery, 1960	
40 Platform Herman	34° 26'00"N 120° 22'20"W	Union Oil Co. of Calif. Emery, 1960	
41 Cojo	34° 27'20"N 120° 24'30"W	Standard Oil Co. of Calif. Emery, 1960	Fractured shale on faulted a line
42 Point Conception	34° 26'20"N 120° 26'50"W	Standard Oil Co. of Calif. Emery, 1960	Fractured shale on faulted anticline
43 Point Conception	34° 26'40"N 120° 28'30"V	Mattel, 1929 Standard Oil Co. of Calif.	Oil and gas seeps onshore Globs of tar more than 1 ft. in diameter Type of tar inclicates seepage from Monterey Formation
44 Point Conception	34° 28'00"N 120° 29'00"V	See 43	See 43

45	Santa Barbara Channel	34° 06'40"N 120° 23'04"W	J. Stearns, 1969	Oil slick sighted from air Location approximate
46	Platform Holly	34° 23'30"N 119° 55'00"W	Atlantic Richfield Co.	
47	Platform Holly	34° 23'30"N 119° 56'10"	Atlantic Richfield Co.	First observed in 1969 Discharge less than 1 bbl. oil per day at present
48	Platform A	34° 20'20"N 119° 36'30"W	U.S. Geol. Survey Prof. Paper 679	
49	Point Dume	33() 59'10"N 118° 46'20"W	Standard Oil Co. of Calif.	At 50 fm. contour on line between Pt. Dume & Pt. Vicente
50	White's Point	33° 42'50"N 118° 19'10"W	Emery, 1960 Standard Oil Co. of Calif.	Analyzed by Atlantic Richfield Co.
51	Hermosa Beach	33° 51'27"N 118° 28'10"W	Standard Oil Co. of Calif.	Oil seep
52	El.Segundo	33° 53'35"N 118° 32'55'V	Standard Oil Co. of Calif.	Oil seep
53	Venice	33° 55'05"N 118° 35'05"W	Standard Oil Co. of Calif.	Oil seep
54	Platform Heidi	34° 20'30"N 119° 30'45"W	Division of Oil and Gas	Gas Bubbles

This table includes several later corrections of the California Divsion of Oil and Gas. Modified after Wilkinson, 1971, table 1, pp. 22–25.

(text continued from page 27)

seep in California was in 1776, when Padre Pedro Font, while near Goleta, noted that a great deal of tar had been washed up on shore. We now know that this tar, found at the surface, represents the final stage of decomposition and largely represents the heavy end hydrocarbons (residuals), similar to the asphalt deposits previously described.

In southern California, many underwater seeps are found along the Murray Fracture Zone (Figure 1-8) and the seaward extension of the oil-rich Ventura basin in the Coal Oil Point area. It is estimated that 50 to 70 barrels of oil per day may leak into the ocean from these seepage locations. Although there are several coastal basins north of the Santa Barbara basin that have the potential of submarine oil and gas seepage, only the Ventura-Santa Barbara and Los Angeles basins are definitely known to have seepage from the sea floor. These locations correspond to the known fault zones that emerge on the ocean floor.

The volume of oil and gas produced by an individual seep has not been well documented. Allen and Schlueter (1969) made attempts to estimate the quantity of oil flowing to the surface near Union Oil's platform "A" near Coal Oil Point. Blocker (1964) found that as much as 20% of a typical crude oil could be evaporated in the presence of strong winds and turbulent waters. Taking this into consideration, he calculated the volume of oil in an oil slick that covered the surface area of the ocean (per square foot) for various thicknesses of the oil film (Figure 2-8). He related the oil slick thickness to the film appearance on the water. These techniques are similar to those used to analyze the consequences of a tanker spill and the associated clean-up operations.

Most of the active seeps that are known to occur offshore of California have been found between Point Conception in Santa Barbara County (in the north) and Huntington Beach in Orange County (in the south). The largest concentration of seeps appears to be in the Santa Barbara Channel area, which is also the seaward extension of the oil-rich Ventura basin. The remaining documented seeps lie in the Los Angeles basin including those along the Palos Verdes Fault as it extends along the coast of Redondo Beach and Manhattan Beach. Evidence of earlier seeps has been noted in San Diego County and the northern provinces of Baja California (Wilkinson, 1971). An example of a "seep tent" for collecting migrating gases offshore California is presented in Figure 2-9.

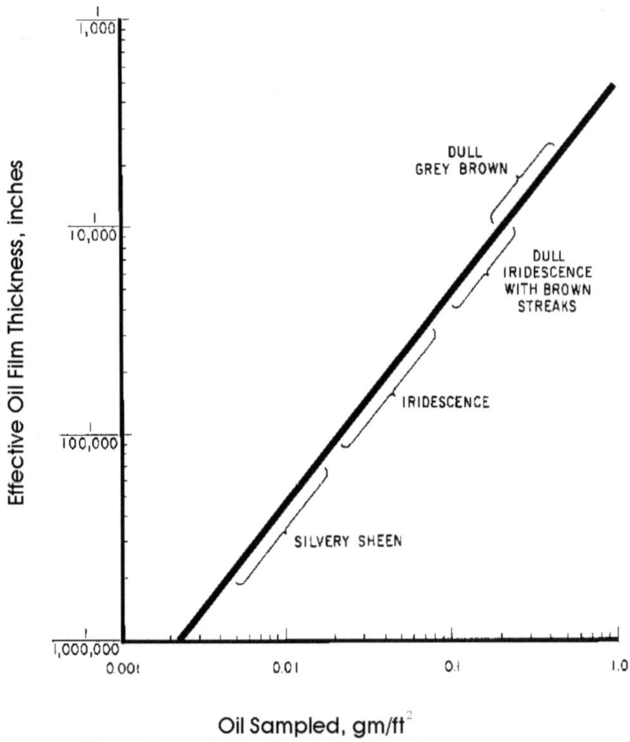

Figure 2-8. The effective thickness of a film of oil on water related to its appearance on the surface of the water. Also the quantity of oil in grams (per square foot) of the oil slick related to its appearance. An oil density of 0.9 g/cc was assumed for the oil slick. *(Data taken from technical memorandum 1230, General Research Corporation; in E. R. Wilkinson, 1971, fig. 8, p. 20.)*

The Ventura-Santa Barbara basin is a westerly trending depositional trough that includes much of the Santa Barbara Channel area. This area includes a number of oil and tar seeps (Figure 2-10). Vernon and Slater (1963) identified *Asphalt Mounds,* on the seafloor near Point Conception, to be one source area for hydrocarbon seeps. These Asphalt Mounds range up to 100 feet (31 meters) in diameter and 8 feet (2.4 meters) in height. They are distributed along an east-west trend of faulted anticlines. Tar mounds have been observed on the Southern California seafloor at three locations: the Point Conception

Figure 2-9. Example of a seep containment device ("seep tent") before being placed on the sea floor in the Santa Barbara Channel, California, in an area of intense natural hydrocarbon seepage. *(Courtesy of Paul Mount and Marina Voskanian, California State Lands Commission, Long Beach, California.)*

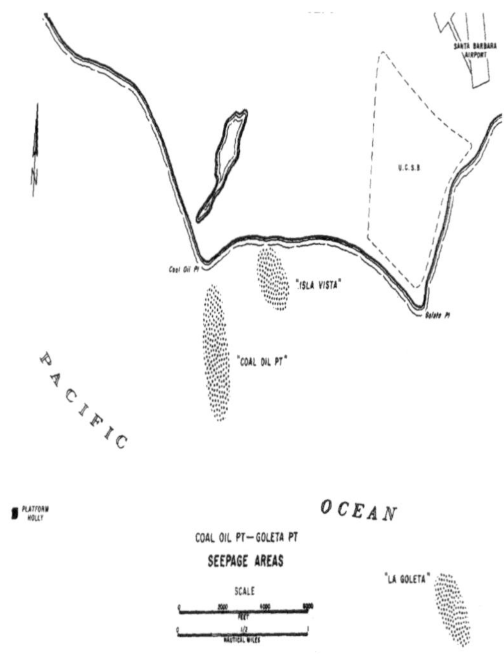

Figure 2-10. Clustering of oil seeps offshore Santa Barbara, California, at Coal Oil and Goleta points. *(Modified after Wilkinson, 1971, fig. 2, p. 10.)*

area, Coal Oil Point near Goleta, and off Carpinteria. Tar is most abundant near Point Conception where a sheet of tar covers an area at least one-quarter square mile and forms a 10- to 12-foot scarp at the seaward edge (Vernon and Slater, 1963). The migration of oil along this faulted anticline was undoubtedly accompanied by gas migration. The Asphalt Mound was formed through volatilization and loss of the light-end hydrocarbons, leaving behind heavy tar-like material. The loss of light-end hydrocarbons occurs in the form of gas bubbles rising to the ocean surface. This is direct evidence of the destruction of hydrocarbon accumulations that has been observed by Link (1952).

Figure 2-11 illustrates the location of several seeps offshore in the Santa Monica Bay. The trace of these seeps is found along the Palos

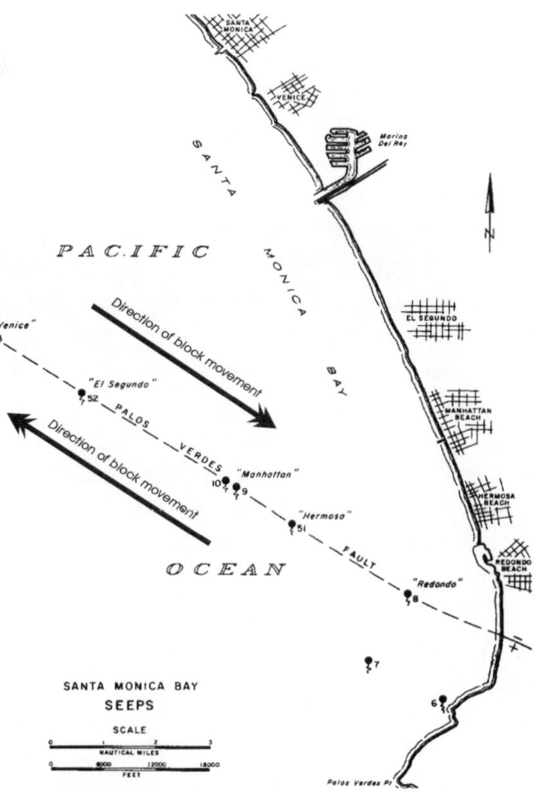

Figure 2-11. Oil seeps in Santa Monica Bay, California. Trace of seeps is along the Palos Verdes Fault. *(Modified after Wilkinson, 1971, fig. 6, p. 16.)*

Verdes Fault. The La Brea Tar Pits, an onshore seep, is one of the more famous tar seeps and is located in the northern portion of the Los Angeles basin.

Divers or remote-control underwater cameras can locate and identify the location of underwater oil and gas seeps. Samples of seep oil and gas can also be obtained. Location of underwater seeps helps identify fault planes that intercept the ocean floor and can be used as powerful tools in the location of faults hidden under surface sediments. In addition to direct visual observation, clusters of gas bubbles venting from the seafloor can be detected and monitored by acoustical reflection techniques. The bubbles reflect the location of the seep and appear as a dark vertical line against a clear background of seawater. Sieh and LeVay (1973) was able to demonstrate a direct correlation of bubble clusters with discharged gas volumes and rates from sediment cones on the seafloor. These sediment cones can form volcano-shaped mud lumps on the seafloor.

Analyses of sediment samples on the seafloor for contained gases are a direct method of determining the probable source of an underwater seep. Carlisle et al. (1975) analyzed seafloor sediment samples in an area where the geophysical data showed the presence of gas bubbles in the water and a subsurface fault. Every gas seep in the vicinity of the fault indicated that the quantities of ethane (C_2) through propane (C_4) exceeded that of methane (C_1). The presence of heavier hydrocarbons in gases migrating along the fault demonstrates the presence of an oil and gas accumulation along the fault at depth.

An isotopic analysis can show whether the gas has a biogenic or a thermogenic origin (see Part III, Chapter 17). Biogenic methane supplies little information about fault activity due to its relatively shallow origin resulting from the decomposition of organic matter at shallow depths. Thermogenic methane, on the other hand, is an excellent source of tectonic information because of its deep origin. Furthermore, the presence of ethane, propane, and/or butane provides confirming evidence that the origin of the gas is thermogenic and originates from petroleum deposits. The presence of helium, carbon dioxide, hydrogen sulfide, nitrogen, and hydrogen can also be very useful in determining the depth of origin of gas. Inasmuch as the source of helium can be from a deep source, its migration characteristics can provide useful information regarding changes in the tectonic forces acting deep within plate boundaries.

Landes (1973) prepared a detailed summary of offshore seeps covering such areas as the Gaspe Peninsula of Quebec, the United States Gulf Coast, Gulf of Paria near Trinidad, Gulf of Suez, Red Sea, Arctic Coast of Alaska and Canada, and the South China Sea.

As tectonic forces change along the fault plane, with changes in the compressive loads on formations adjoining the fault blocks, porosity and permeability of rocks also change. Reduction in porosity is reflected in the volume of gas expelled from rocks to the fault plane. The changes in porosity and permeably are reflected in the rate of gas migration along the fault plane. Accordingly, measurement of the changes in rate of gas migration along the fault provides a quantitative assessment of the change in tectonic forces. The changes in tectonic activity, along with other geologic information, can serve as a precursor of an earthquake.

Summary

Gases of deep origin, migrating along faults, can yield valuable information on the tectonic activity of the lithosphere/mantle. Biogenic gases originate in the shallower depths of the earth's crust and so are of less value. Changes in the rate of seepage (gas flow rate) can be used as a monitor of changes in tectonic activity.

RELATIONSHIP OF FAULTS TO SEEPS

Link (1952) observed that most seeps are found in younger, recently uplifted sediments. In the young and small basin areas of California, where uplift has been recent and violent and many structures that once held oil and gas have been ruptured, hundreds of seeps are known.

Accordingly, the tectonic activity occurring within the earth's crust can be viewed as a cause of destruction of many pools of oil and gas. As previously discussed, this can be related to the change in permeability along fault planes that results from the tectonic activity. The extensive number of seeps worldwide that were identified by Link (1952) in his landmark study provides important information in utilizing seeps as a predictor of tectonic activity.

Many of these seeps were identified as occurring in areas of the world that are highly prone to severe and devastating earthquakes. Examples include the mobile zone of the Mesopotamian Geosyncline;

the north side of the Monagas Basin in Eastern Venezuela; and the seepage along the eastern foothills of the Andean Chain from Columbia to Cape Horn. Especially important are the numerous oil and gas pools within California that are being subjected to active tectonic activity from both strike-slip and blind thrust faults.

Many seeps that were identified by Link (1952) are directly related to the folding and uplifting that occur as a result of the tectonic activity along blind thrust faults. The destruction of oil and gas pools by way of tectonic activity could be viewed as an overall negative; however, for the purposes considered here it can be used as a powerful predictor of tectonic activity. This is illustrated in the following discussion.

Extensive field work performed by Jones and Drozd (1983), through research funding from Gulf Research and Development Company, conclusively demonstrated the correlation of gas seeps to the earth's surface and the location of deep faults (viz., the San Andreas Fault in Central California). This finding, along with others reported by Jones and Drozd (1983), provide new insights into the potential for using geochemical signatures over fault planes as a predictor of tectonic activity.

The important findings of Jones and Drozd (1983) can be summarized as follows:

1. There is no doubt that faults provide the most permeable avenues for gas migration to the surface and they must be carefully considered whenever making a geochemical interpretation.
2. Both magnitude and composition of seep gases are important.
3. Geochemical evaluation of surface seeps provides an excellent tool for the regional evaluation of gas migration paths.
4. The compositional relationship of the seep gases proves that the soil-gas surveys can directly detect the subsurface hydrocarbons.
5. The geological characterization of the region, especially fault planes, dictated the lateral versus vertical extent of gas migration as measured at the surface of the earth.
6. Faults can serve both as migration paths and as seals. Faults can be compared to a critical orifice that acts as a partial seal, allowing gases to build-up behind the fault zones, but allow leakage to occur when a critical pressure is exceeded.
7. A lower gas pressure can cause subsurface gases to migrate into the trap (secondary collector zone), rather than out of it, creating

a "shadow" (lower magnitude) gas seep at the surface because the upward migration of light hydrocarbon gases is interrupted.

8. The soil-gas surveys constituted direct or positive geochemical characterization and not negative-type as associated, for example, with halos.
9. Many of the major faults can be mapped at the surface, allowing verification of the physical relations between the faults and the gas seeps.

CHAPTER 3

Surface Soil-Gas Surveys

INTRODUCTION

Gas from seeps passes directly through alluvium and other surface sediments that could hide the surface traces of a fault (blind faults). Soil-gas surveys help identify the type and magnitude of gas moving along the migration path (fault plane) into the soil. This volume of migrating gas is particularly high when the gas is migrating from the hydrocarbon-bearing formations. The highest soil-gas concentration readings will occur when the survey is made directly over the path and will rapidly drop off as one moves away from this avenue of migration.

The physical requirements for gas-soil analysis are shown in Figure 3-1. Important considerations for the selection of soil-gas probe sites and depths of sample points include the following: 1) water table depth; 2) unusual or nonuniform soil conditions near the surface, such as impermeable or low-permeability soil (e.g., a clay lense or aquitard); and 3) conditions that would give rise to seasonal variations in gas migration. The depth from which the sample is taken is critical. The magnitude of total gas volume and the percentage composition will often vary depending on the depth at which the sample is taken. This is particularly true for volatile hydrocarbons such as benzene. The effects of atmosphere (contamination with air) at very shallow sampling depths (less than 3 feet) and selective volatilization of the lighter components can greatly distort the results of the survey. Sampling should always be done below an impermeable layer as deeply as possible (Sokolov et al., 1959). Preferably, samples should be collected from multiple depths and then correlated with soil permeability, water depths (near surface), and other potential localized entrapment zones.

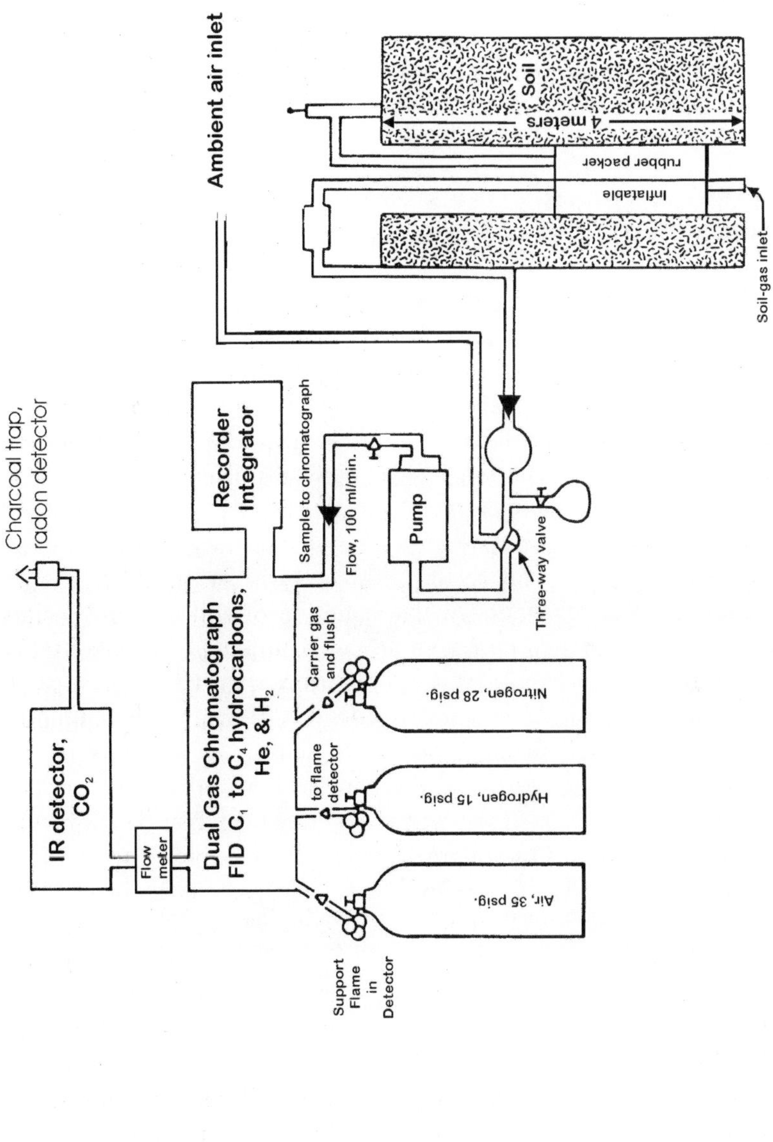

Figure 3-1. Schematic showing a method for soil-gas sampling and analysis. *(Modified after Jones and Drozd, 1983, fig. 1, p. 934.)*

The maximum usefulness of a soil-gas survey is obtained by carefully designing and instrumenting the probes to serve as permanent measuring instruments where sampling can be repeated. Plotting of this data over time is important in identifying long-term trends in the rate changes of gas migration. The location of sampling points should be carefully selected based on detailed geological studies of the individual sites. In evaluating gas migration along faults, the probes need to be placed as close as possible to the fault plane, and on either side of the fault.

MIGRATION OF METHANE AND OTHER HYDROCARBONS

Seepage of methane and other hydrocarbons along faults, fissures, fractures, and outcrops from ruptured hydrocarbon-bearing formations are prevalent throughout the world. A typical range of compositions for the migrating gases is given in Table 3-1. The geochemical distinction between the gas of gas- and oil-bearing basins is shown in Table 3-2. Examples of hydrocarbon composition for petroleum reservoirs are presented in Table 3-3.

Methane is commonly found at the earth's surface. The source of methane can be biogenic (shallow bacterial decomposition of organic matter) and thermogenic (hydrocarbon deposits formed by deep burial) (Figure 3-2). Methane can be found in great abundance in association with oil and gas fields. The major portion of natural gas is composed of methane (typical range of 80% to 90%). Nikonov (1971) demonstrated the abundance of methane in a review of many types of hydrocarbon gas sources (Figure 3-3).

Table 3-1
Empirically Determined Composition Range for Soil-Gas Hydrocarbons Over Different Reservoir Types

	$C_1/\sum C_n$	C_1/C_2	$(C_3/C_1) \times 1{,}000$
Dry gas	100–95	100–20	2–20
Gas condensate or oil and gas	95–75	20–10	20–60
Oil	5–50	10–4	60–500

After Jones and Drozd, 1983, table 1, p. 936.

Table 3-2
Average Compositions and Ratios of Light Hydrocarbons from Selected Soil-Gas Surveys

	Date	$C_1/\sum C_n$, %	C_1/C_2	$(C_3/C_1) \times 1{,}000$
Sacramento basin,	1972	95	55	6
dry gas, California	1974	95	49	8
	1975	94	55	11
San Joaquin basin,	1972	82	8	46
oil and gas, California	1974	84	7	61
	1975	82	8	56
Gas-condensate surveys				
Southwest Texas	1975	89	12	33
Western Overthrust,	1976	90	11	30
Alberta foothills, Canada	1978	88	12	30
Oil surveys				
Uvalde, Texas	1975	77	5	77
Permian basin	1976	75	5	64
Utah Overthrust, Pineview	1976	77	5	83
Appalachian folded belt, Rosehill	1978	73	4	141
Uinta basin, Duchesne	1976	68	4	171

Modified after Jones and Drozd, 1983, table 2, p. 936.

Table 3-3
Typical Compositions of C_1 to C_4 (All Data in Mole Fractions) as Found in Hydrocarbon Reservoirs

Reservoir Hydrocarbon	Dry gas	High-pressure Gas	High-pressure Oil	Low-pressure Oil
Methane	0.91	0.81	0.77	0.37
Ethane	0.05	0.07	0.08	0.21
Propane	0.03	0.07	0.08	0.21
Butanes	0.01	0.05	0.07	0.21

Modified after Katz and Williams, 1952; in Jones and Drozd, 1983, table 3, p. 937.

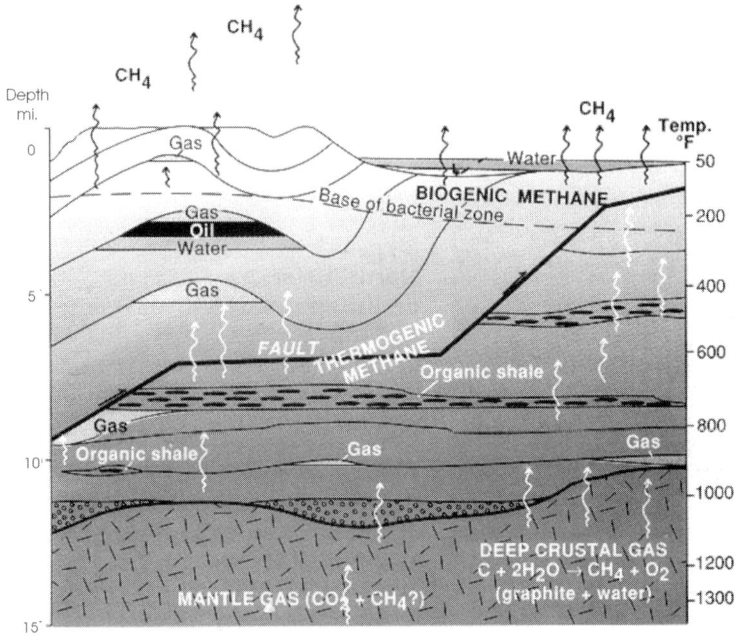

Figure 3-2. Three processes generate methane (CH_4), the main component in natural gas. Biogenic methane is produced by microorganisms during metabolism. Thermogenic methane forms when heat and pressure decompose deeply buried organic matter. Chemical reactions deep inside the earth can also generate methane. *(Modified after Howell et al., 1993.)*

Most of the methane is found in hydrocarbon traps formed by tectonic activities resulting from folding and faulting. These traps are subject to continuous, as well as intermittent, leakage of gas along faults that may extend to the earth's surface.

The work by many investigators demonstrates that the spikes of gas concentration in surface soil-gas studies can identify gas migration paths (faults). The magnitude of methane concentration (ppm) over the soil-gas pathway must be sufficient to allow statistically valid and independent populations to be defined. Bernard (1980) has hypothesized that the presence of fairly large ethane-propane-butane anomalies also suggests that the origin of the soil-gas is from oil-related sources. Gases derived from biogenic and coal deposits typically have methane-ethane ratios (C_1/C_2) that exceed 10^3 or 10^4 (Janezic, 1979; Coleman,

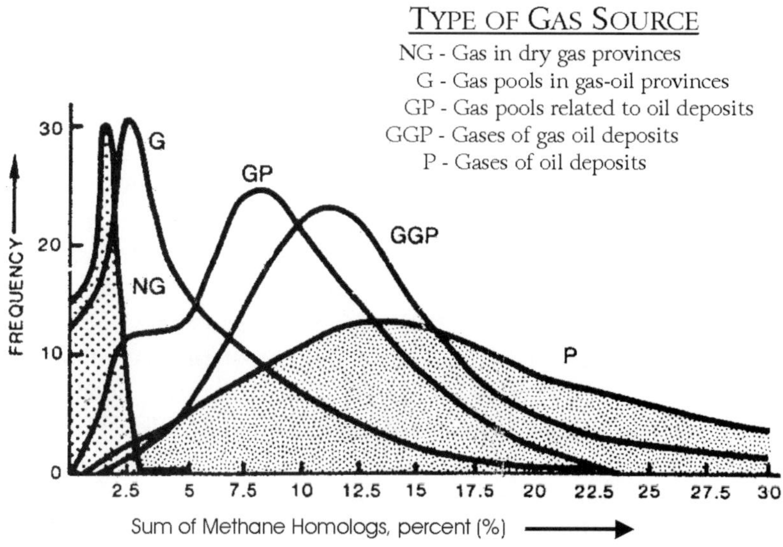

Figure 3-3. Frequency distribution of sum of methane gas homologs in different types of deposits. Figure based on analysis and classification of 3,500 worldwide reservoir gases. *(Modified after Nikonov, 1971; in Jones and Drozd, 1983, fig. 4, p. 937.)*

1976; Bernard, 1980). Oremland (1981) reported small quantities of biogenic C_2 to C_4 hydrocarbons from marine muds, where the lowest (C_1/C_4) ratio was 149:1 and the ratios of heavier hydrocarbons to methane were very small. One method used to determine the source of the methane is measurement of the stable carbon isotope of methane. Sokolov and Cheremisinov (1971) reported that the ratio of methane to heavier gaseous hydrocarbons could be used to predict if the gas is from a petroleum source or not. A ratio of 150 to 200 corresponds to gas pools formed by distant lateral migration and to gases from mud volcanoes. Lower ratios correspond to oil fields with associated gas.

The concentrations of migrating gas almost always decrease in the following order: methane>ethane>propane>butane. Leythaeuser et al. (1980) have noted that compositional changes to the migrating gas occurred along the path of migration due to diffusion. This behavior has been observed when comparing soil-gas survey data from samples gathered at very shallow depths of 0.3 to 0.6 m (1 to 2 ft) to samples gathered at depths of 4 m (13 ft) (Jones and Drozd, 1983). If diffusion were the dominant migration mechanism, a chromatographic effect would be expected for a gas that migrated through the earth. The fact that the primary effect on gas-composition is near the earth's surface indicates that the major migration of gas to the near surface is along faults and fractures rather than by diffusion. If the soil-gas samples show little chromatographic effect, it implies that the migration must be almost continuous geologically. In general, the chromatographic effect will delete the lighter, more volatile gases (methane, benzene, etc.) from the remaining less volatile soil-gases near the surface.

Soil-gas surveys by Jones and Drozd (1983) and Follansbee (1943) showed how gas concentrations can help to identify faults. Both surveys, run at different times but in the same area, identify the presence of a deep-seated fault in the San Joaquin Valley, California. The deep basement-fault near the Lost Hills Oil Field shows an increase in concentration of several magnitudes in methane and propane content of soil-gas (Figure 3-4). Figure 3-5 shows a high degree of correlation of hydrogen and methane concentrations along the deep seated San Andreas Fault. Jones and Drozd (1983) conducted another soil-gas survey many miles to the northeast of the San Joaquin Valley study area in the vicinity of Paso Robles, California, over the San Andreas Fault (Figure 3-6). In this area, the fault was clearly defined and also demonstrated a direct correlation with the gas migration.

In 1976, Veezay Geoservice, Inc. conducted a similar study showing a clear association of methane anomalies with faults as mapped by Conner and Colvin (1977) for the Pineview Field, Summit County, Utah (Figure 3-7). In this case there is a close correlation between the high spike of methane concentration and thrust faulting. The peaks of the heavier hydrocarbons are of lower magnitude than that of methane. This is consistent with both the lower concentrations of heavier hydrocarbons found in natural gas, and the lower mobility associated with the high-molecular-weight gases.

Powers (1977) ran a soil-gas survey to investigate the thrust faulting of the Ryckman Creek Field, Wyoming (Figure 3-8). Again, the hydrocarbon seeps occur directly over the thrust faults, with the gases migrating along the thrust fault and then vertically through the Tertiary deposits. An important observation to be made from these soil-gas surveys is that the soil-gas anomalies are smaller for deeper basins with longer migration paths. Conversely, the much larger magnitude seeps, as

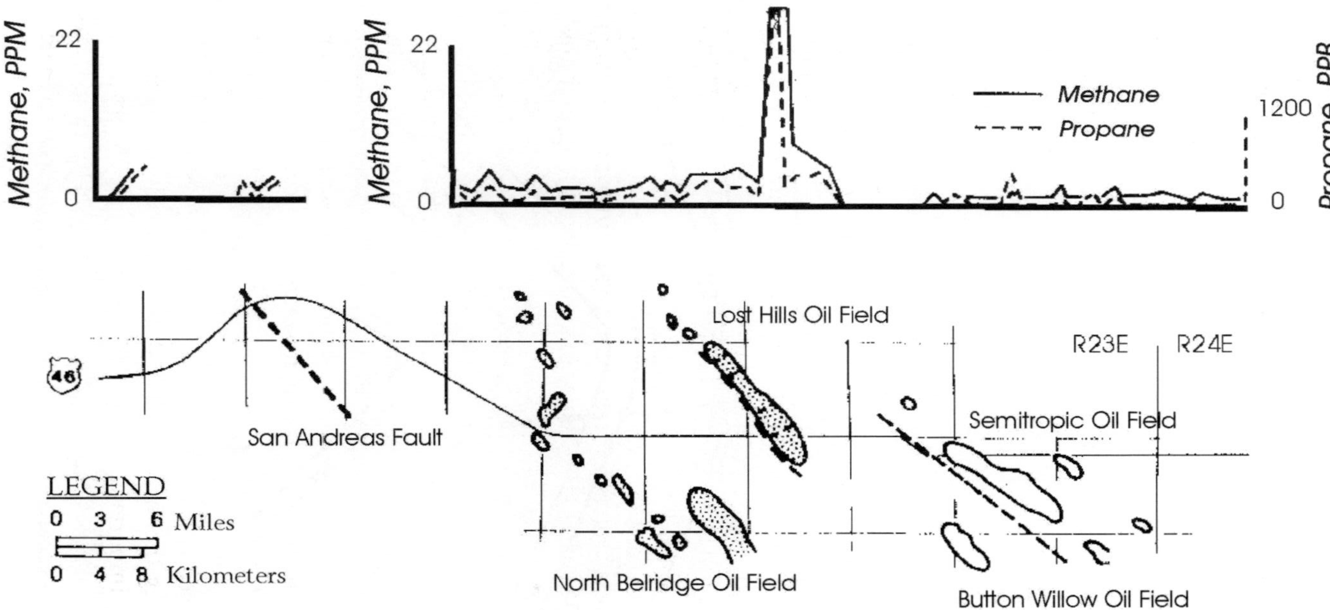

Figure 3-4. Methane and propane soil-gas survey in San Joaquin basin, California, crossing several faults and oil fields. *(Modified after Jones and Drozd, 1983, fig. 12, p. 942)*

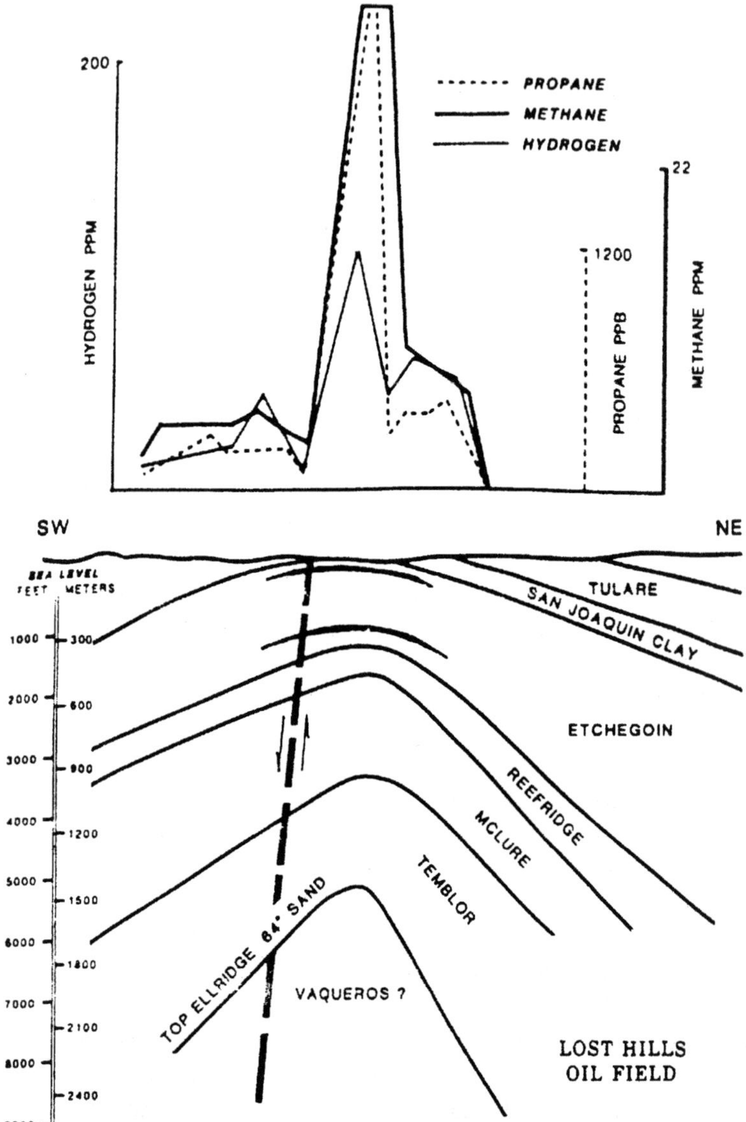

Figure 3-5. Idealized cross-section illustrating an anomaly that represents a deep fault adjacent to the Lost Hills Field, San Joaquin basin, California. *(Modified after Follansbee, 1943; in Jones and Drozd, 1983, fig. 13, p. 943.)*

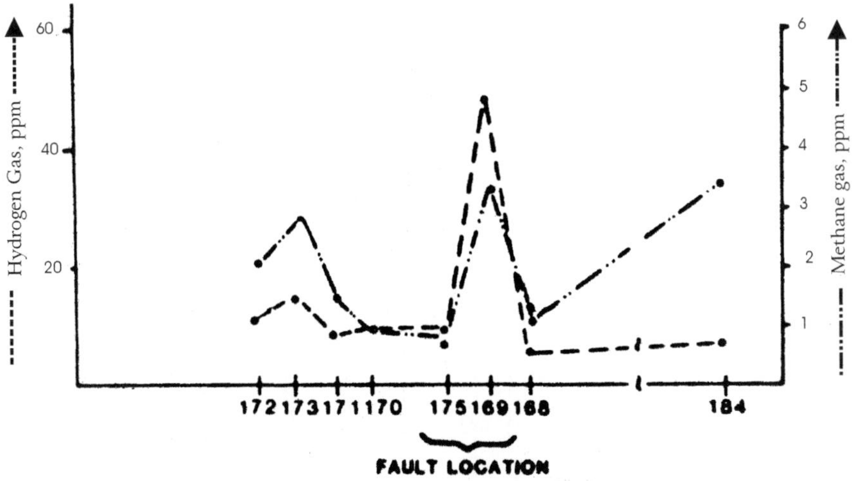

Figure 3-6. Soil-gas survey over the San Andreas Fault near Paso Robles (methane and hydrogen) (modified after Jones and Drozd, 1983, figure 14, p. 943).

observed at Pineview, Utah, for example, imply either more active tectonics or a source or trap closer to the surface (Jones and Drozd, 1983).

Summary

Soil-gas surveys can be used to define subsurface hydrocarbon deposits as well as avenues of gas migration (faults). A large portion of methane gas originates from subsurface hydrocarbon deposits. An understanding of lateral versus vertical gas migration is very important in interpreting geochemical data. The extent of either depends on knowledge of the structural scenario of rocks. Experimental data indicate that in most places there is sufficient vertical permeability (faults, fractures, etc.) for elevated soil-gas concentrations to exist directly over the hydrocarbon deposits. However, some lateral migration (particularly near the surface) generally occurs, so that the shape and location of the surface anomalies do not always match those of the prospective reservoir. Finally, formation evaluation (electric and seismic logs, etc.) and drilling data can be used to further reveal the geologic structures likely to "give rise to" gas migration.

(text continued on page 54)

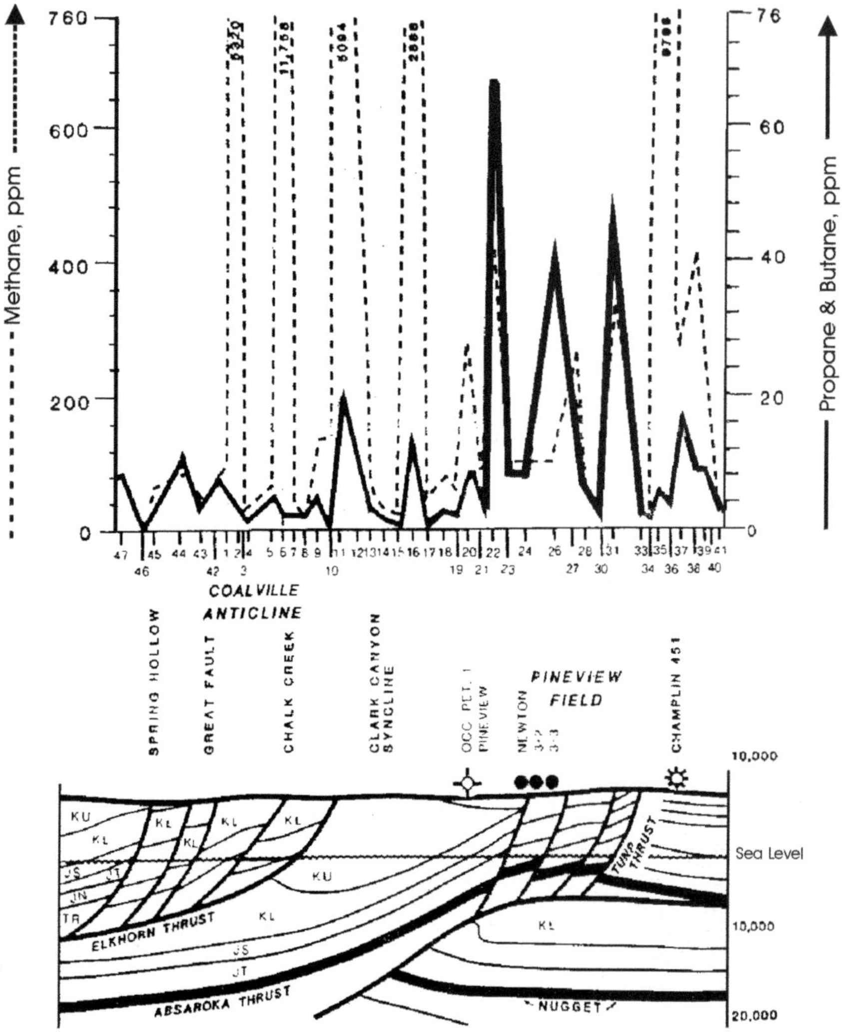

Figure 3-7. Soil-gas survey of Pineview field, Utah, utilizing methane and propane-butane concentrations of gas to identify faults. *(Modified after Jones and Drozd, 1983, fig. 16, p. 945.)* Depth in feet.

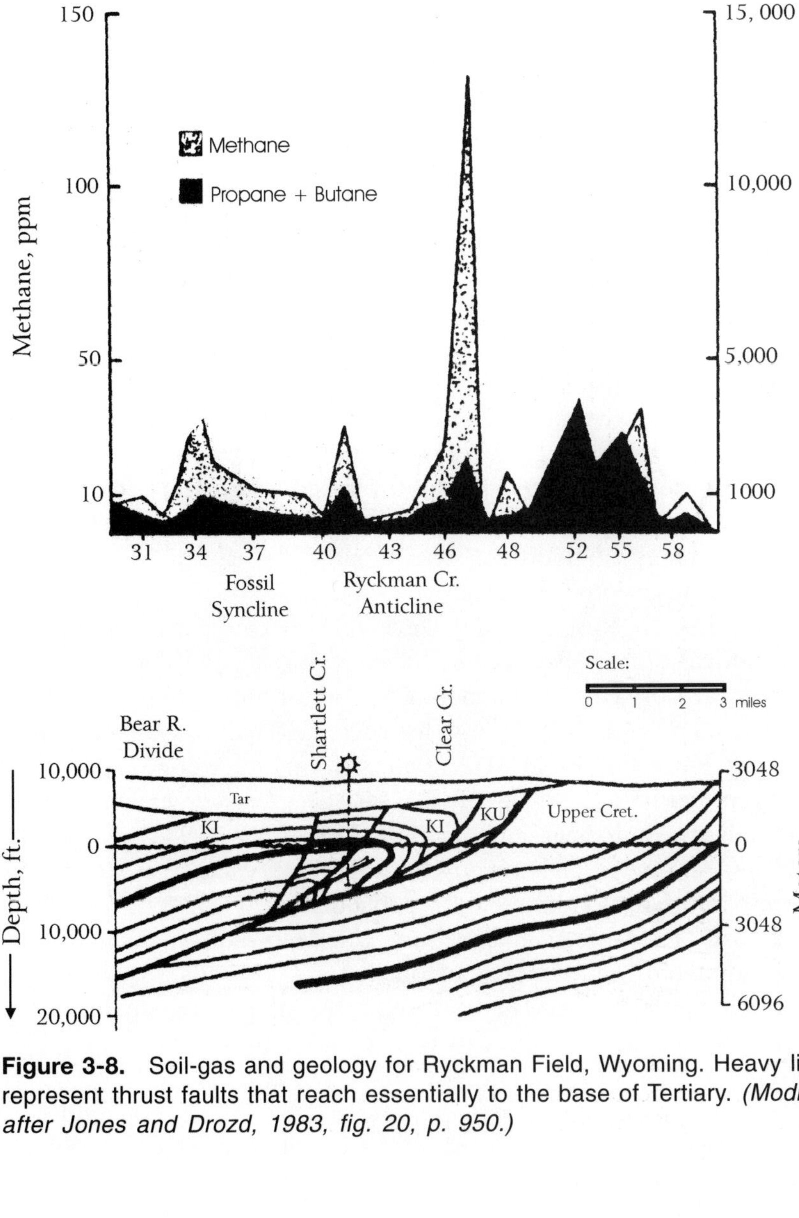

Figure 3-8. Soil-gas and geology for Ryckman Field, Wyoming. Heavy lines represent thrust faults that reach essentially to the base of Tertiary. *(Modified after Jones and Drozd, 1983, fig. 20, p. 950.)*

(text continued from page 51)

HELIUM AND HYDROGEN GAS MIGRATION

Introduction

The presence of helium measured over faults can be used as an important indicator of deep tectonic activity. According to Jones and Drozd (1983), large helium anomalies are generally correlated with deep basement faults and commonly indicate the presence of deep faults. Considerable caution must be exercised with helium anomalies because helium can also be associated with oil and gas fields, uranium deposits, and geothermal fields.

Based on examining numerous oil reservoirs in western Siberia, Nechayeva (1968) noted that hydrogen could be used as a petroleum indicator in soil-gas surveys. This study showed that hydrogen was observed in the soil-gas for 18% of all gas fields, 27% for gas-condensate fields, and 46% for all oil fields. The peak percentage of hydrogen was 0.9% for western and northern Siberian gas fields. The peak percentage of hydrogen was 6% for the eastern Siberian gas condensate reservoirs and 11.2% for the oil fields.

In 1975, Gulf Research and Development Company conducted a geochemical survey across the San Joaquin basin. The transverse started near Lost Hills and ended near Famosa in the east-central part of the basin (Figure 3-9). A large hydrogen anomaly was detected over the Lost Hills Oil Field. These measurements were repeated for hydrogen and helium when soil-gas surveys were conducted over the San Andreas Fault near Paso Robles. Follansbee (1943) observed a hydrogen anomaly and proposed that it represented a deep fault adjacent to the Lost Hills Field, San Joaquin basin (Figure 3-10). Jones and Drozd (1983) noted that helium is a deep-fault or tectonic indicator, commonly independent of oil and gas deposits. Hydrogen, on the other hand, appears to be directly related to the hydrocarbon anomalies and, thus, appears to be a petroleum predictor.

Eremeev et al. (1973) and Ovchinnikov et al. (1972) have reported the presence of helium and hydrogen in association with faults. In soil-gas studies, Jones and Drozd (1983) found helium soil-gas concentrations as high as 430 ppm over the San Andreas Fault (Figure 3-11). At adjacent sites 20 to 40 m (65 to 130 ft) from the mapped fault,

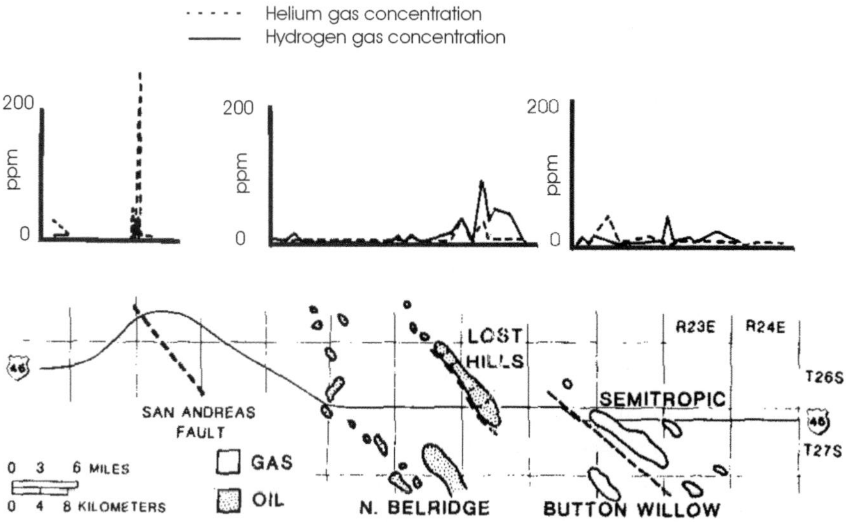

Figure 3-9. Helium and hydrogen soil gas survey in San Joaquin basin, California, crossing several faults and oil fields. *(Modified after Jones and Drozd, 1983, fig. 12, p. 942.)*

the helium soil-gas values ranged from 40 to 98 ppm. A similar relationship for helium soil-gas values was found at Gold Hill, which is about 8 km (5 mi) north of the first soil test. Jones and Drozd (1983) noted that their values obtained for soil-gas helium were much higher than those of other authors. Roberts (1981) and Reimer et al. (1979) obtained soil-gas values for helium of 5.24 ppm. Their sampling, however, was very shallow (less than one meter in depth) as compared to that of Jones and Drozd (depth of 4 meters). Reasons for low values of helium soil-gas concentration are the effects of atmospheric contamination at very shallow sampling depths. Sokolov et al. (1959) stressed that sampling must be done as deeply as possible.

Evaluation of Tectonic Activity

Helium has many unique geochemical properties that make it an ideal gas for monitoring tectonic activity of deep origin:

1. Helium is extremely light and has the smallest molecular size of any gas. For this reason, it is very mobile and will readily

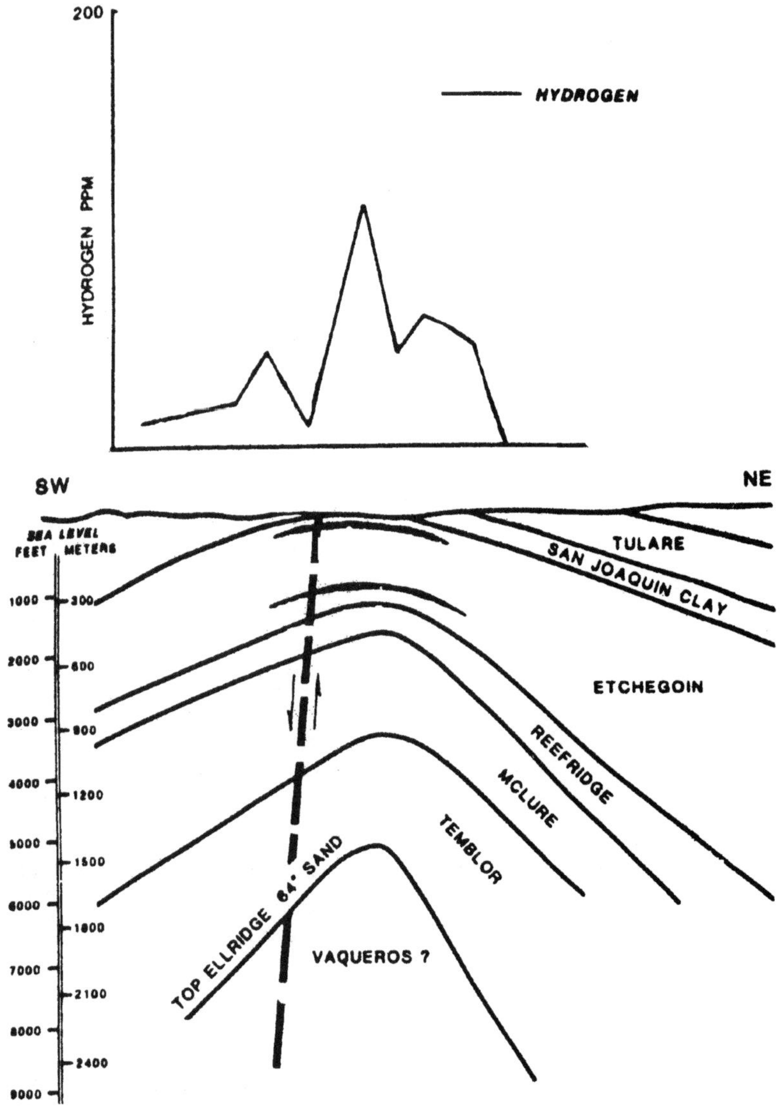

Figure 3-10. Idealized cross-section illustrating a proposed deep fault adjacent to Lost Hills Oil Field, San Joaquin basin, California. *(Modified after Follansbee, 1943; in Jones and Drozd, 1983, fig. 13, p. 943.)*

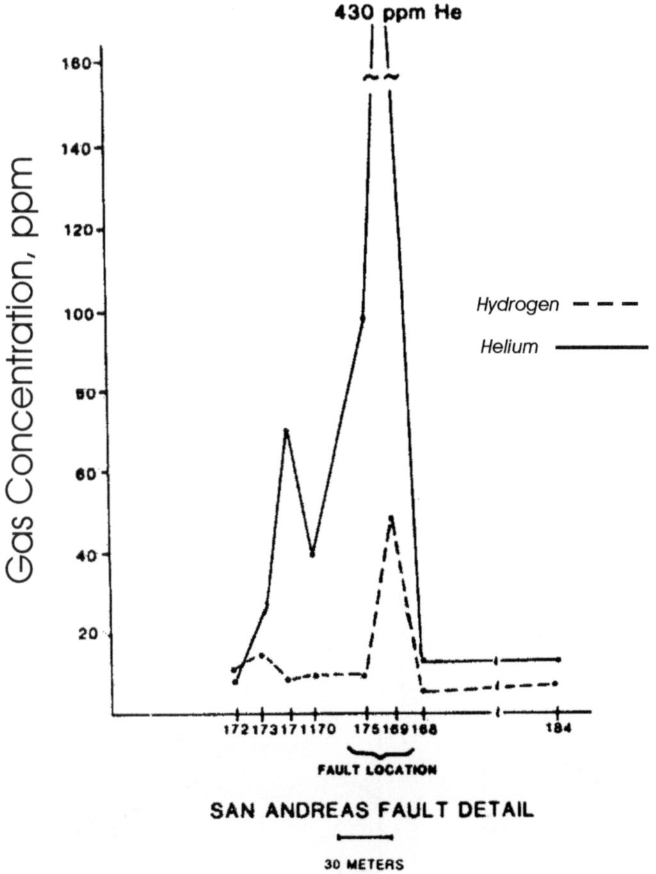

Figure 3-11. Helium and hydrogen soil-gas concentrations for a transient crossing of the San Andreas Fault. *(Modified after Jones and Drozd, 1983, fig. 14, p. 943.)*

move through a geological column (including water), propelled upward by buoyancy forces. Helium has a cross-sectional molecular diameter one-half that of the methane molecule.

2. Helium is an inert gas that does not react with other chemicals or minerals during its long migration to the surface. It also preserves its isotopic characterization and, thus, reveals its origin.
3. Helium (and its isotopes) is a common product of mantle reactions.

Depth of Origin

Heavy helium, ^4He, is created by the decay of radioactive elements such as uranium and thorium. It is also called an "alpha particle," after the discovery made by Rutherford that radiation in the form of alpha particles results from the decay of ^{235}uranium and ^{238}uranium.

Because the mantle is enriched in uranium and thorium, it is constantly producing heavy helium as the uranium breaks down into more stable elements. On average, the continental crust contains about 100 million times more heavy helium, ^4He, than light helium, ^3He. Figure 3-12 presents a range of isotope values for various sources of helium.

It has been found that helium gas escaping from the fumaroles on the high Andean volcanoes is extremely rich in ^3He. This gas contained up to 500 times as much ^3He gas as that found in the earth's crust.

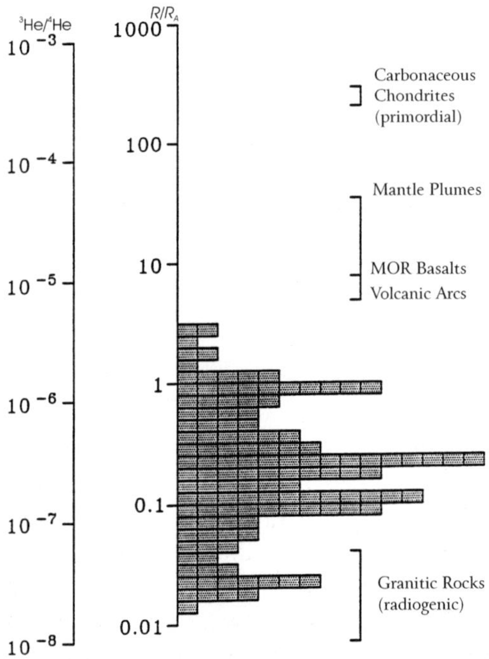

Figure 3-12. Helium isotope ratio for various types of rocks; R_A = atmospheric isotope ratio. *(Modified after Jenden, P. D. and Kaplan, I. R., 1988; courtesy of Gas Research Institute, 8600 West Bryn Mawr Ave., Chicago, IL 60631, Fig. 17, p. 34.)*

This demonstrates that at least some of the helium gas is coming from a mantle source far below the crust.

Helium Isotope Ratios

$^3He/^4He$ ratios, R, were measured by Poreda et al. (1988) in gases associated with active continental margins (principally regions of subduction) in Taiwan, New Zealand, Philippines, Thailand, Indonesia, Barbados (West Indies), and Alaska (U.S.A.). These gases contain a magmatic He component, based on measured $^3He/^4He$ ratios, R, ranging from 0.5 to 3.8 times the atmospheric value (R_A). The methane-helium data can be interpreted as the result of a mixture between a low-He/^{4}He (0.01 R_A), high–$CH_4/^3He$ crustal end member and a high-$^3He/^4He$ (0.6–3.9 R_A), low-$CH_4/^3He$ ($<10^6$) magmatic component. The presence of mantle He does not require significant amounts of mantle methane (Poreda et al., 1988). Many of the gases studied by Poreda et al. (1988) contain large amounts of CO_2. Low-$^3He/^4He$ gases (0.01 R_A, 10^{11}) owe their origin to decarbonation reactions in the crust, whereas the high-$^3He/^4He$, low-$CO_2/^3He$ gases (2–4 R_A, $<10^9$) may have a mantle CO_2 contribution. The R/R_A ratio for various types of rocks is presented in Figure 3-12. This study demonstrates that the R/R_A ratio is a sensitive detector of mantle volatiles. In tectonic regions where magmatic processes transport mantle volatiles to the earth's surface (mid-ocean ridges, oceanic islands, and volcanic arcs), the ratio of $^3He/^4He$ is enriched relative to the atmospheric ratio. Figure 3-13 maps the range for the $^3He/^4He$ ratio found in gases associated with oil and gas throughout California.

In contrast, fluids (including natural gases), which derive their volatiles exclusively from the earth's crust, contain $^3He/^4He$ ratios of $R < 0.1\ R_A$. This low "crustal" ratio is the result of production of ^4He derived from uranium and thorium decay with very minor ^3He production from the nuclear reaction on lithium. Thus, the 100-fold contrast in the $^3He/^4He$ ratio between crustal and mantle signatures provides a very sensitive test for mantle helium.

Significant mantle helium components have been reported in Kamchatka, Russia (Kamenskiy et al., 1976) and in Japan (Wakita and Sano, 1983). Wakita and Sano (1983) reported $^3He/^4He$ ratios as high as six times the atmospheric ratio (6 R_A).

Figure 3-13. Helium isotope ratios in California oil and gas fields. *(Modified after Jenden and Kaplan, 1988.)*

Summary

Soil-gas analysis based on the use of helium offers considerable promise in the examination of deep tectonic activity. Helium isotope analysis, whereby a determination is made of the relative proportions of light helium (^3He) and heavy helium (^4He), allows identification of the deep origin of the gas. Namely, a gas of deep origin would be rich in ^3He, as confirmed by helium gas escaping from volcanoes. Hydrogen can be used to augment the informational value of the gas analysis.

CARBON-DIOXIDE MIGRATION

Several investigators (e.g., Taylor, 1979; Matthews et al., 1987; and Irwin and Barnes, 1980) have identified magmatic carbon dioxide, CO_2 (from deep-seated faults) as the source for many bicarbonate springs throughout the world (Figure 3-14). They identified mantle-derived carbon dioxide in hot springs, volcanic and geothermal gases, and carbonatite magmas. This natural discharge of magmatic carbon dioxide from the earth's crust is recognized most readily in soda springs, where it has chemically combined with water to yield bicarbonate (HCO_3^{-1}) ions. Distribution of springs with an HCO_3^{-1} ion concentration of 1,000 ppm or more is restricted to areas of seismic activity at the edges of active tectonic plates. Occurrence of these springs coincides with zones of major seismic activity throughout the world (Figure 1-4). Published detailed maps showing carbon dioxide discharge for the United States (Figure 3-15), Russia (Figure 3-16), and China (Figure 3-17) show the interrelationship of magmatic CO_2 with active seismic zones. Several of the studies pertain to the discharge of carbon dioxide from vents on the ocean floor. These data reveal that the discharge of magmatic carbon dioxide is prevalent along the edges of seismically active plates. Almost all of the known carbon dioxide gas discharge is along faults of young orogenic belts located toward the margins of crustal continental plates. The presence of calcium-silicate-bearing metasedimentary minerals in geologic fold and thrust structures of older orogenic belts attests to prior production of carbon dioxide gases at an earlier age.

There are several potential sources for carbon dioxide gases in active and/or dormant volcanic areas. The venting of carbon dioxide gas from

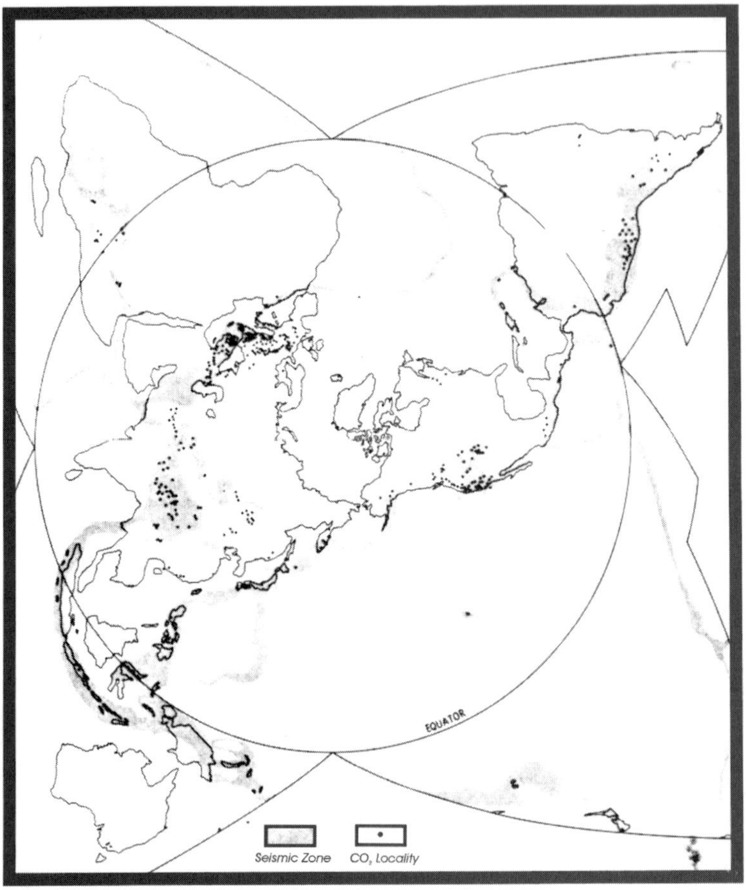

Figure 3-14. Worldwide distribution of seismic zones and areas of CO_2 discharge ($HCO_3^{-1} \geq 1,000$ ppm). Localities in China from Shi Huixin (1979); in Afghanistan from Koltov et al. (1977); and in Peru from Zapata Valle (1973). Zones of seismicity are mainly from Tarr (1974). *(Modified after Irwin and Barnes, 1980, p. 3116.)*

volcanoes and submarine hydrothermal vents is evidence of a mantle gas origin. Whereas some CO_2 may be derived directly from the mantle reactions, some CO_2 samples tested appear to be a product of thermal metamorphism of carbonate rocks that are penetrated by hot magma. Jenden and Kaplan (1989) analyzed 140 commercial CO_2 gases and plotted the CO_2 concentration versus $\delta^{13}C$ (Figure 3-18).

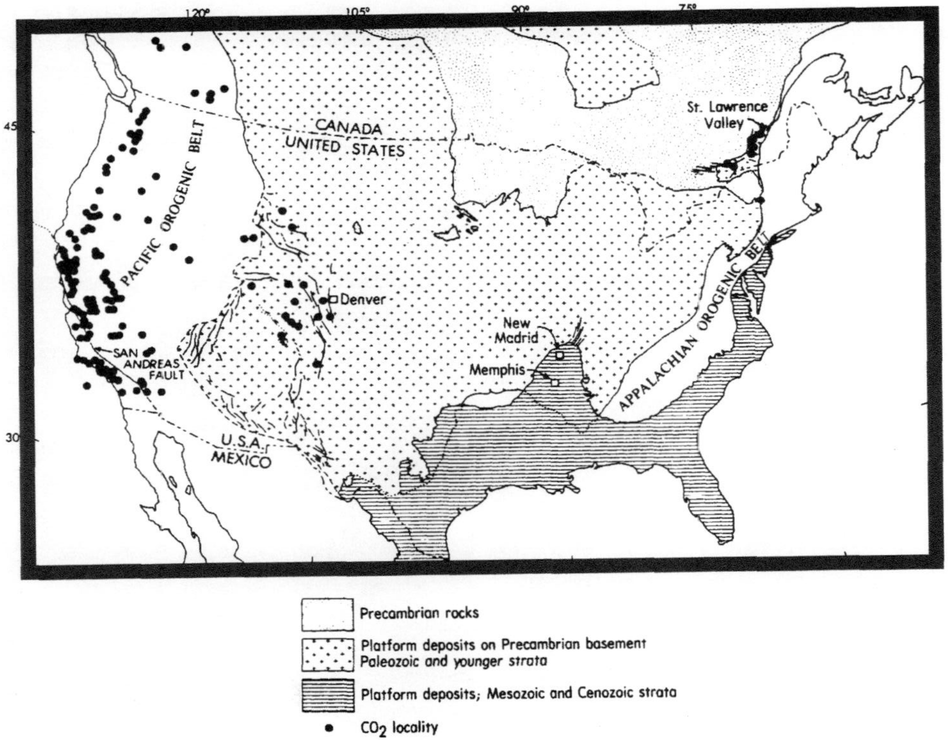

Figure 3-15. Distribution of CO_2 discharges in water ($HCO_3^{-1} \geq 1{,}000$ ppm) in the continental United States and adjacent parts of Canada. Locations of CO_2 discharges are from Barnes et al. (1975) and U.S.G.S. files; generalized geology from King (1969). *(Modified after Irwin and Barnes, 1980, p. 3117.)*

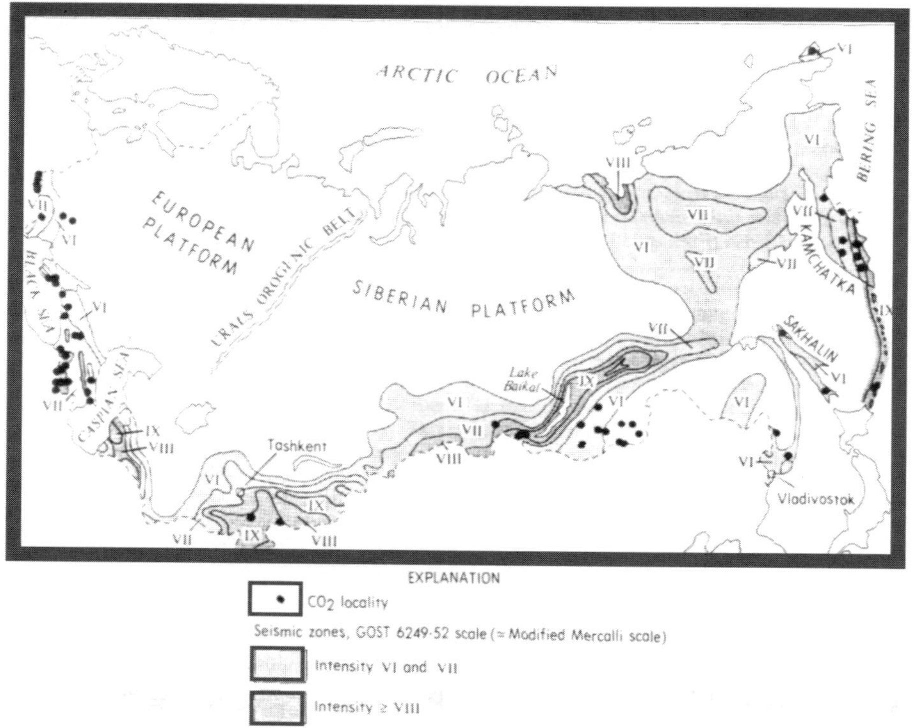

Figure 3-16. Seismic zoning and distribution of CO_2 discharge ($HCO_3^{-1} \geq 1,000$ ppm) in the USSR. The seismic zones are after Medvedev (1968) and the distribution of CO_2 localities are from Ivanov et al. *(1960) (Modified after Irwin and Barnes, 1980, p. 3118.)*

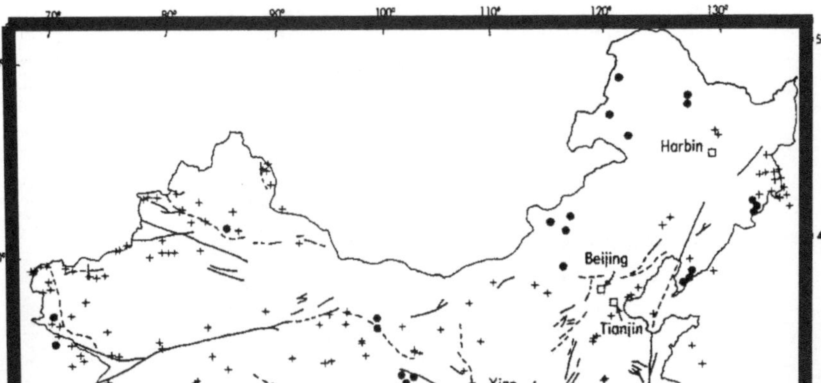

Figure 3-17. Distribution of earthquake epicenters (1901–1976; magnitude ≥6) and locations of CO_2 discharges in China. Earthquake epicenters and base map showing major faults are from Lee et al. (1978). Localities of CO_2 (gas content > 90% CO_2; for water with dissolved CO_2 (as bicarbonate) >1 g/l) are from Shi Huixin (1979). *(Modified after Irwin and Barnes, 1980, p. 3120.)*

Gases with 40% to 100% carbon dioxide compositions had $\delta^{13}C$ values of $-4 \pm 2\%$, which is outside the range of most marine carbonates. This might indicate mixing between organic and inorganic carbon dioxide gas sources; however, in tectonically active basins, values near –4% could also indicate a contribution of magmatic carbon dioxide. The isotopic characterization of associated gases (i.e., helium, hydrogen

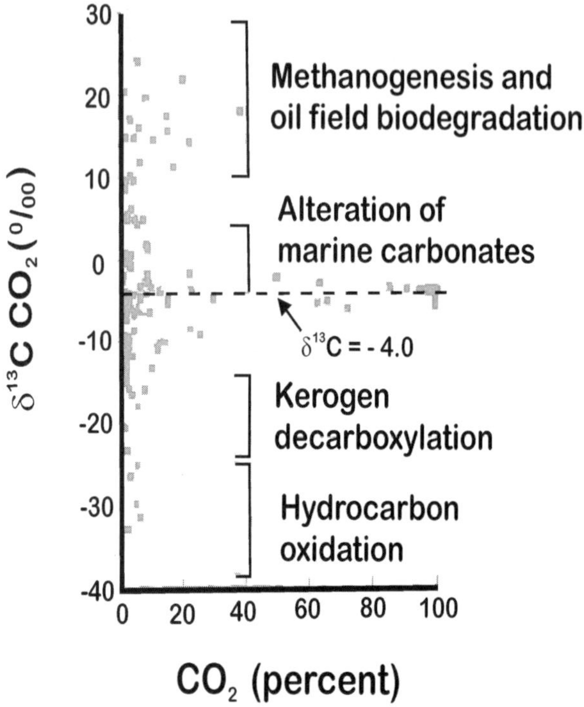

Figure 3-18. Carbon isotope characterization for carbon dioxide gas from 140 commercial natural gases. *(Modified after Jenden and Kaplan, 1989, in Howell, 1993.)*

and methane) is a useful tool to further quantify deep-origin versus shallow-origin for carbon dioxide.

On October 4, 1978, a magnitude M_R=5.7 earthquake occurred that was centered between Mammoth Lakes and Bishop, located in the high Sierras of the east-central area of California. In May 1980, four additional earthquakes occurred with magnitude M_R=6. On May 25, 1982, the U.S.G.S. first issued a volcanic hazard warning for the Mammoth Lakes area. By 1990, 100 to 300 earthquakes were occurring daily. In 1990, the United States Forest Service noted that trees on the shore of Horseshoe Lake, on the southeast flank of the Mammoth Mountain area, were dying (Figure 3-19). Sieh and LeVay (1998) identified carbon dioxide (seepage) in the soil (20% to 90%) as the gas that was killing the trees. The soils outside of this area had a carbon dioxide concentration of about 1%. Since the

Figure 3-19. Tree die-off caused by carbon dioxide seepage at the southeast flank of Mammoth Mountain, California. (Modified after Sieh and LeVay, 1998, fig. 17.)

presence of the gas was first recognized in 1990, the area of carbon dioxide contamination has increased. It is now recognized that the carbon dioxide gas is coming from a magma source and does not owe its origin to a biological process. It is anticipated that this area is similar to that of Mt. St. Helen.

Summary

Discharge of carbon dioxide gas to the earth's surface has been measured in areas corresponding to the edges of active tectonic zones located throughout the world. Most of the carbon dioxide has a mantle origin. The natural discharge of magmatic carbon dioxide is recognized readily in soda springs where it has chemically combined with water to yield bicarbonate (HCO_3^{-1}) ions. Accordingly, carbon dioxide could be a useful predictor of tectonic activity, especially when used in combination with other gases such as helium, hydrogen, and methane.

RADON GAS MIGRATION

Introduction

The migration of ^{222}radon (^{222}Rn) gas to the earth's surface has been documented in many parts of the world. Measurements have been made of the extent and distribution of radon gas in soil-gas emanations as well as radon gases dissolved in water. Japan has been a leader in developing instrumentation to measure the radon level in groundwater. This has been developed primarily to serve as a predictor of earthquake activity. Previous research has demonstrated a correlation between the increased levels of radon in groundwater and seismic activity.

^{222}Rn concentration anomalies, as well as other gases discussed earlier in this chapter (such as methane, helium, hydrogen, and carbon dioxide), are commonly associated with active fault zones and volcanic areas. Faults and fractures provide permeable paths along which these gases migrate to the earth's surface. Tectonic movement along fault zones leading up to earthquakes, may change the permeability of these zones and, thus, change the rate of migration of the gases.

Variations in ^{222}Rn concentration have been correlated to subsurface tectonic activity. Increased levels of radon gas in groundwater were detected days and weeks preceding the Tashkent, Uzbekistan, earthquake of April 26, 1966. Preceding this earthquake, Ulomov and Mavashev (1967) noted changes in concentration of ^{222}Rn gas in groundwater. The University of Tokyo (Japan) began a systematic study of radon gas migration for earthquake prediction. In cooperation with the Geological Survey of Japan, over 30 observation sites were set up throughout Japan. To date, over 400 sites have been monitoring the changes in ^{222}Rn concentrations in groundwater in the Tokyo area alone.

^{222}Rn anomalies in sampling stations located at distances of tens and up to several hundred kilometers from the epicenter, were noted prior to the Kobe (1995, Japan) earthquake (Igarashi et al., 1995). The Kobe (January 17, 1995) earthquake of magnitude M_R=6.9 was preceded by significant changes in the radon gas concentration in the groundwater. This quake resulted from the collision of the subducting Philippine Sea Plate and the Eurasian Plate. The Hanshin corridor spans the industrial zone along Osaka Bay and is located over a portion of this plate interaction area. Nishinomiya, a city immediately north of Kobe, is also located over this subduction zone. The chemistry of water in this area is carefully monitored (Miyamizu water); it is known throughout Japan for its use in sake-breweries. In the weeks before the Kobe quake, a strange pattern emerged. Levels of radon gas in the water samples rose sharply toward the end of 1994, and then declined sharply after January 8, 1995, prior to the earthquake. Other data gathered by Japan's Coordinating Committee for Earthquake Prediction showed a similar up-and-down pattern of radon levels in groundwater prior to three other earthquakes.

Systematic monitoring of ^{222}Rn concentrations in soil-gas on a regional scale was performed in the years 1975–1985 in the western USA for an active transform fault system. Large-scale variations, both in the geographic and temporal sense, were detected but could not be clearly associated with tectonic activity (King et al., 1992).

The discovery of extensive radon flux anomalies in the Dead Sea (DS) area (Steinitz et al., 1992) suggested the use of ^{222}Rn soil-gas concentration in this area as an earthquake precursor. A systematic correlation was found between ^{222}Rn fluctuations and earthquakes in the nearby sector of the Dead Sea Rift as reported by Steinitz et al. (1996). The Dead Sea Rift is a seismically active section of a major transform plate boundary (Figure 3-20A). The ^{222}Rn fluctuations exhibit large temporal changes, including influences from climatic conditions and seasonal variations. A major source of radon exists in the substratum of the studied area and technological development in instrumentation now available allows hourly measurements of radon flux variations to be accurately made.

Additional data are needed, however, to confirm whether or not variations in ^{222}Rn can be recognized as a precursor of seismic events (King, 1986). It is apparent that ^{222}Rn content provides important information on the dynamics of migration paths (Mogro-Campero and Fleischer, 1980). Whereas the significance of temporal variations of ^{222}Rn have often been disputed, Wyss (1991) has stated that of all earthquake precursors suggested so far, the measurement of radon fluctuations in groundwater offers the greatest promise.

Origin of Radon

Radon is a colorless and odorless gas found in soils worldwide. Radon gas (^{222}Rn) is continuously formed during the radioactive decay

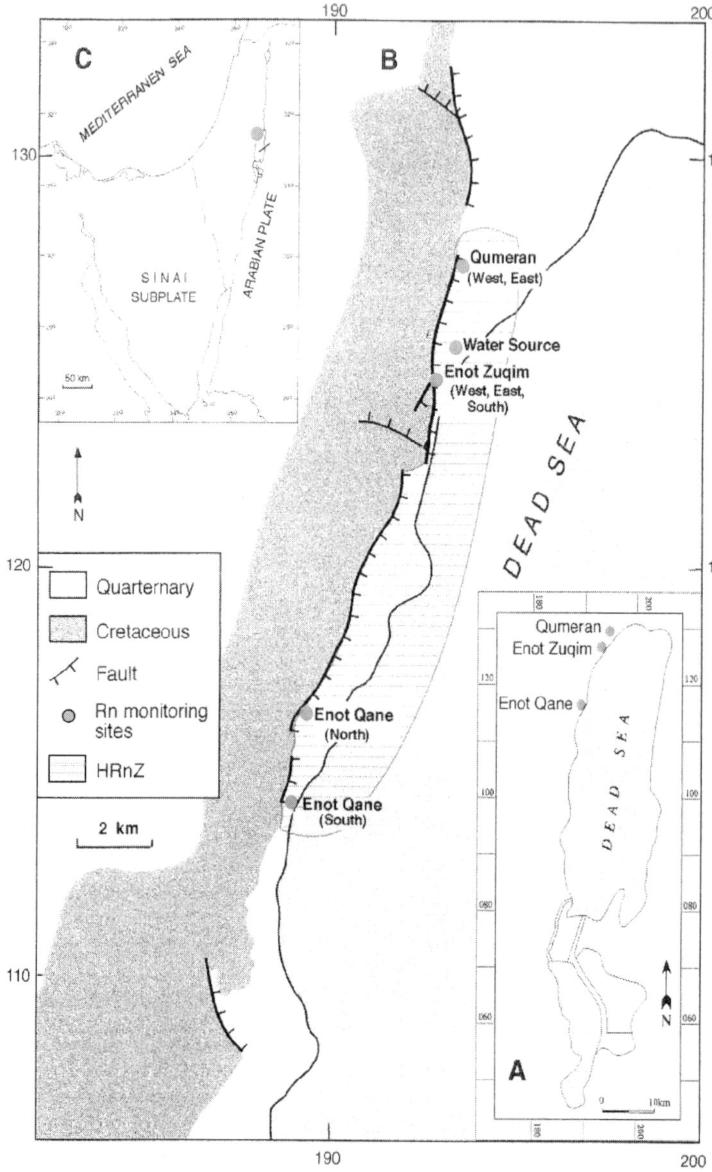

Figure 3-20. Part A displays the area studied by Steinitz at the northwest sector of the Dead Sea. Part B is a geological sketch of the northwestern area of the Dead Sea. Part C shows the detailed location of the monitoring stations of the study area and the boundary of the Sinai and Arabian Plates. *(Modified after Steinitz, 1999, personal communication.)*

of uranium (^{238}U), which is contained within the earth's substrata. The specific characteristics of radon production and decay can be summarized as follows: (1) about half of the primordial (time when the earth was formed) uranium still remains in the earth's rocks; the half-life of ^{238}U is 4.5 billion years; (2) uranium (^{238}U) decays to form radium (^{226}Rn), a radioactive and unstable mineral that has a half-life of 1622 years; (3) radium (^{226}Rn), then decays to radon (^{222}Rn) an unstable and radioactive gas that has a half-life of only 3.8 days and the ability to dissolve in groundwater. The gas migrates mainly through water along fractures and fissures; and (4) the entire decay chain of the ^{222}radon daughters, up to lead-210, occurs over a period of about 50 minutes (Table 3-4).

The chemical properties of ^{222}radon gas render it a unique tracer. Radioactive nuclear techniques provide precise measurement for the determination of radon concentrations. Radon can be measured in minute amounts and has the advantage of being relatively non-reactive with other substances. Methane, helium, hydrogen, and carbon dioxide can serve as carrier gases for radon. Transport by a carrier gas rising

Table 3-4
Decay of ^{222}Radon to ^{206}Lead

Element	Symbol	Half-Life	Particle Emitted during Formation
^{226}Radon	^{226}Rn	1622 years	
^{222}Radon	^{222}Rn	3.8 days	
^{218}Polonium	^{218}Po	3 minutes	Alpha particle (helium-4)
^{214}Lead	^{214}Pb	27 minutes	Alpha particle (helium-4)
^{214} Bismuth	^{214}Bi	20 minutes	Beta particle (high-speed electron)
^{214}Polonium	^{214}Po	.00016 seconds	Beta particle (high-speed electron)
^{210}Lead	^{210}Pb	22 years	Alpha particle (helium-4)
^{206}Lead	^{206}Pb	stable	(non-radioactive)

from depth, must be assumed in order to explain the high flux and the intense temporal variations in the concentrations of radon. Radon can serve as a unique geochemical tracer gas.

Radon as Related to Earthquakes

Radon anomalies in soil gas have been monitored since 1992 in the northwest sector of the Dead Sea, in the areas of Enot Zuqim and Enot Qane (Figure 3-20A). Geographic localities and zones with anomalous radon content are termed "High-Rn Zones" (HRnZ). Monitoring at HRnZ has shown that radon concentrations may vary with time (temporal variation). The location of the radon flux is along a 15–20 km segment oriented in a north-south direction (Figure 3-20B). Steinitz et al. (1995a) documented radon flux fluctuations with subsurface geophysical effects as well as atmospheric influences along the western boundary fault of the Dead Sea Rift. Radon anomalies occur between the scarp of the main western Rift fault and up to 1 km perpendicular to it (to the shore of the Dead Sea). The radon anomalies, found in both gravel and water sources, reach several thousand pCi/l and more. These high readings of ^{222}radon are not supported by local uranium (or radium) concentrations. ^{222}Radon background levels in the gravel were measured at less than 100 pCi/l. The spatial extent of these readings combined with their intensity suggests that there is a large vertical transport system of ^{222}radon gas from depth, probably along the fault. Down-faulted structural blocks with Senonian phosphorite (uranium-bearing) beds are believed to underlie this area at an unknown depth, and serve as the source for the ^{222}radon gas. Recent investigations of helium concentrations in the Dead Sea Rift water sources indicate that the helium gas serves as a carrier for the ^{222}radon. The Israel Science Foundation supports the research on the understanding of the radon gas migration along the Dead Sea Rift. The utilization of radon gas migration data holds promise of becoming a predictor of earthquake activity (see Marco and Agnon, 1995).

Summary

The research efforts underway in the Dead Sea Rift area offer considerable promise in the advancement of scientific knowledge regarding the use of radon gas migration as a predictor of earthquake activity. The radon gas is believed to originate from a deep source, with helium gas probably serving as a carrier gas to transport it to the surface. Other gases (e.g., CH_4 and CO_2) probably may also serve as carrier gases. Both the depth of origin and the radioactive properties of the radon gas facilitate the ability to measure and establish the changes in the rate of gas migration related to tectonic activity.

CHAPTER 4

Southern California Faults and Petroleum Reservoirs

ACTIVITY ALONG THE SAN ANDREAS FAULT SYSTEM

The Ring of Fire, a zone of frequent earthquakes and volcanic eruptions, encircles the Pacific basin (Figure 4-1). The western edge of the United States coastline lies along this ring. The coastline is the edge of two tectonic plates: the Pacific (oceanic) and North American (crustal). The northwestern portion of this coastline is a subduction zone with the Pacific plate moving under the North American Continental plate (Figure 1-8). The southern California portion of the coastline is a major transform fault, commonly known as the San Andreas Fault. This seismically active area of California presents a wide range of studied faults. Scattered among these faults are oil and gas reservoirs whose location has been controlled by faulting.

The southern California area deserves special discussion because it is one of the most intensely studied regions of the world pertaining to earthquakes. The faults have been monitored in great detail. The oil and gas accumulations along the San Andreas Fault system provide information on the gas and oil migration and accumulation along faults.

Thousands of earthquakes occur each year in California along the San Andreas Fault system, but only several hundred of these are large enough to be felt (Table 4-1). Only a few of the larger quakes have a significant effect on California (Figure 4-2). Southern California has more than 200 faults long enough to produce earthquakes with magnitude of $M_R \geq 6$. Many of these faults can be detected at the

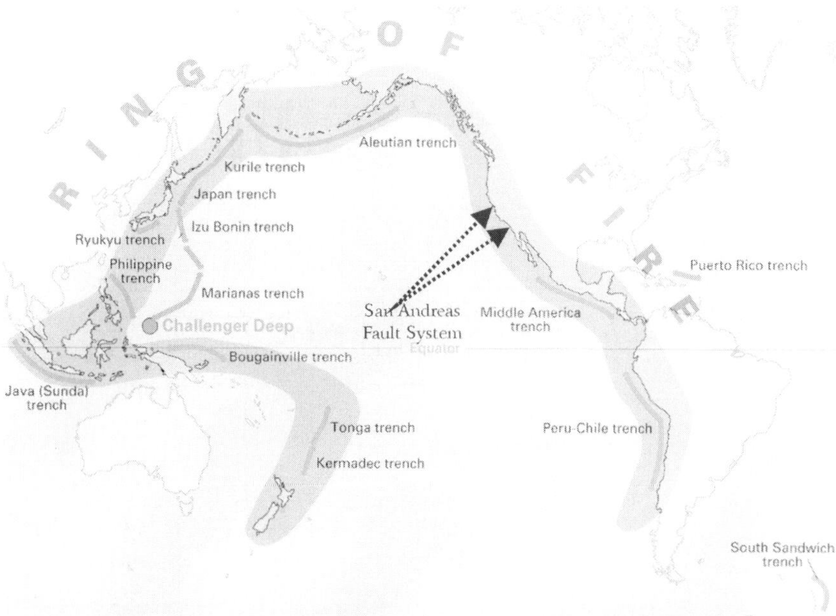

Figure 4-1. Volcanic arcs and oceanic trenches partly encircling the Pacific basin form the so-called Ring of Fire. This is a zone of frequent earthquakes and volcanic eruptions. The volcanic island arcs, although not labeled, are parallel to and always landward of, the trenches. The arrow marks the region of the California United States coastline under discussion. *(Modified after Kious and Tilling, 1997, p. 38.)*

Table 4-1
Classification of Earthquakes by Magnitude or Size

Magnitude	Size
Under 5	small
5 to 6	moderate
6–7	large
7–7.8	major
7.8 or above	great

Modified after Jones, 1995, p. 23.

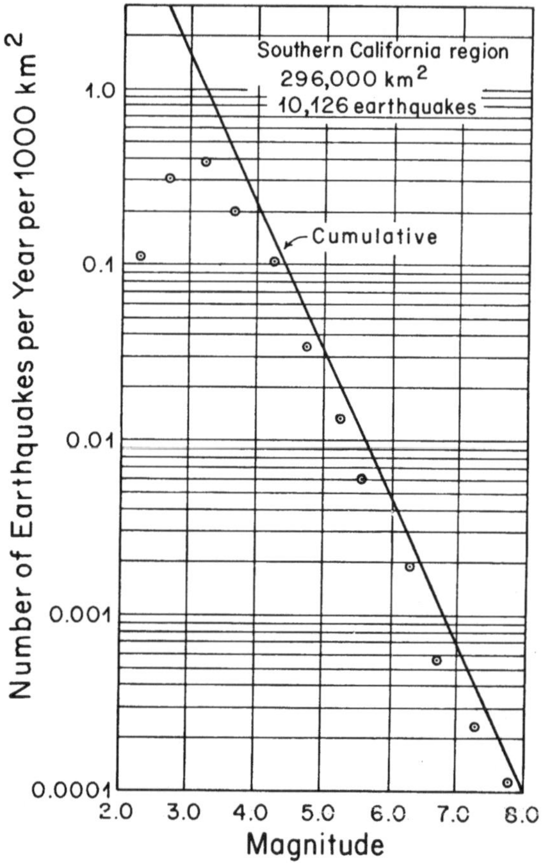

Figure 4-2. Histogram showing the number of earthquakes per year per 1,000 square kilometers in half-magnitude intervals. Cumulative curve indicates the number of earthquakes equal to or greater than any given magnitude. *(Modified after Allen, written communication; in Albee and Smith, 1965, fig. 5, p. 16.)*

earth's surface; others are buried under alluvial sediments of the Los Angeles basin or the inland valleys (blind faults). A map illustrating the general pattern of major faults in southern California is presented in Figure 4-3.

Faults vary in length. The fault plane can be vertical, horizontal, or at any angle to the surface of the earth. A person living anywhere in Southern California, is within 50 kilometers (30 miles) of a major fault. The greatest concentration of faults is in the mountains along

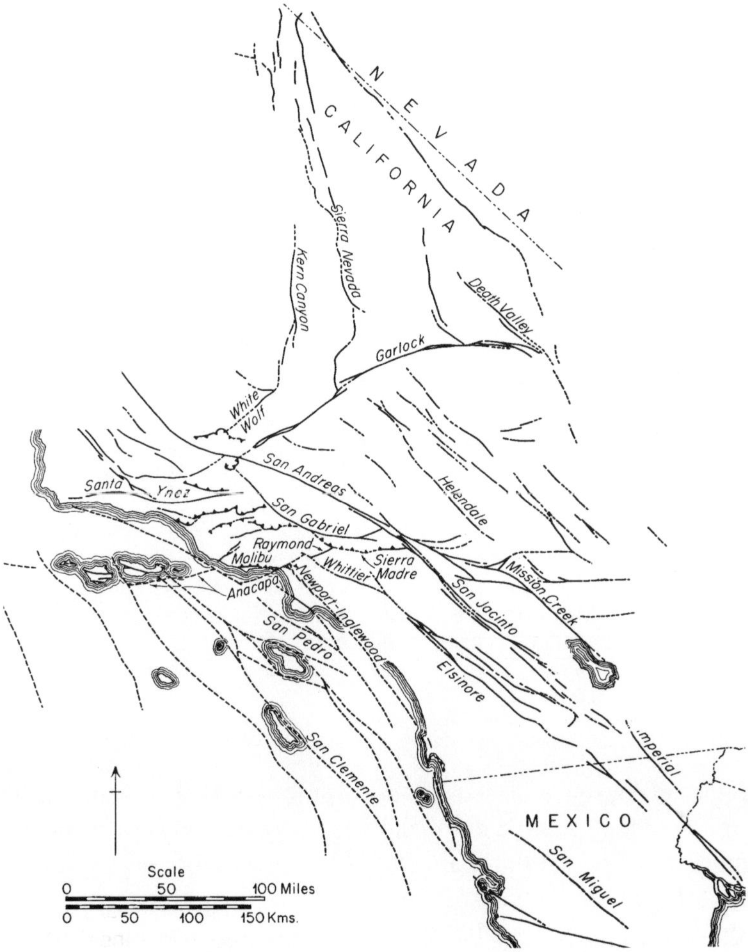

Figure 4-3. Map of major faults in southern California. *(Modified from Allen et al., 1965; California Department of Water Resources, 1964; Emery, 1960; Hill, 1916; in Albee and Smith, 1966.)*

the San Andreas Fault system (i.e., San Bernardino, San Gabriel, Santa Susana, and Santa Ynez mountains). Those faults that produce the greatest number of earthquakes are also those that accumulate stress at the fastest rates (Table 4-2). As a rule, faults of greater length produce a larger magnitude earthquake, which lasts for a longer period of time. The magnitude of an earthquake is proportional to the log of

Table 4-2
Magnitude vs. Accumulated Stress*

	Fault	Probable Magnitudes	Length, kilometers	Slip rate, millimeters per year	Average Time between Earthquakes
San Andreas Fault System	San Andreas	6.8–8.0	1,200[1]	30 ± 5	140–210
	San Jacinto	6.5–7.5	230	12 ± 5	85–250
	Whittier-Elsinore	6.5–7.5	200	5 ± 2	250–700
	Imperial	6.5–7.0	50	30 ± 5	40–60
Other Faults	Garlock	7.0–7.5	250	7 ± 4	200–3,000
	Oakridge	6.5–7.5	90	5 ± 1	200–700
	Sierra Madre	6.5–7.5	85	4 ± 2	200–1,000
	Santa Susana	6.5–7.3	40	6 ± 1	100–300
	San Cayetano	6.5–7.3	40	4.5 ± 2	200–600
	Cucamonga	6.0–7.0	25	4 ± 2	200–1,000

* *These 10 faults are the fastest-slipping faults in southern California, moving at least 4 millimeters per year. The Newport-Inglewood fault, although well known, moves less than 0.5 millimeter per year, slower than many other faults in the same area.*

[1] *550 kilometers of the San Andreas fault are in Southern California, south of Parkfield, California.*
Modified after Jones, 1995, p.7.

the amount of energy released in the rupture (Figure 4-4 and Table 4-3). A magnitude M_R=6.0 earthquake has about 32 times more energy than a magnitude M_R=5.0 and almost 1,000 times more energy than a magnitude M_R=4.0 earthquake. The amount of displacement as a result of the quake, along the fault, increases with the magnitude of the quake (Figure 4-5 and Table 4-4). Movement along the San Andreas Fault, California, can occur either in sudden jolts or in a slow, steady motion called "creep." Fault segments that are actively creeping, experience many small to moderate earthquakes, which cause little or no damage to urban property. It is the locked segments of the San Andreas Fault that store tremendous amounts of energy, which can buildup for decades or even centuries prior to being released in devastating earthquakes (i.e., the San Francisco earthquake (M_R=8.3) in 1906 ruptured along a previously locked 430 km-long segment of the San Andreas Fault).

An earthquake is the result of a sudden slip along the fault plane. In general, friction along the San Andreas Fault plane prevents move-

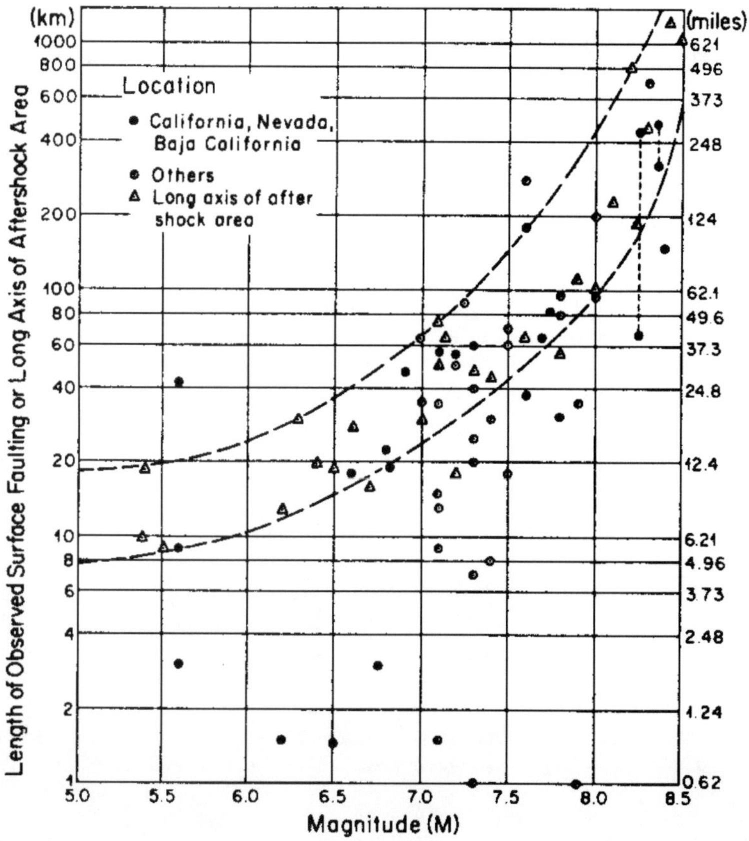

Figure 4-4. Relationship of earthquake magnitude with length of associated surface faulting and the long axis of aftershock area. *(Modified after Albee and Smith, 1965, fig. 1, p. 14.)*

ment. When the stresses along the fault plane exceed the friction forces preventing motion, an earthquake occurs. Southern California has had more than 200,000 earthquakes between 1984 and 1994—few were damaging and most were never felt (Jones, 1995). Since 1857, few quakes have been located on the San Andreas Fault itself. In the Southern California portion of San Andreas Fault system, the last quake located on this fault was in 1857. Figures 4-6 and 4-7 illustrate historic locations for large earthquakes (M_R>6). There are three major factors that affect the movement (or shaking) felt in an earthquake:

Table 4-3
Listing of Major California Earthquakes by Magnitude*

Magnitude	Date	Location	Length, km	Duration, sec
7.8	January 9, 1857	Fort Tejon	160	130
7.7	April 18, 1906	San Francisco	400	110
7.5	July 21, 1952	Kern County	75	27
7.3	June 28, 1992	Launders	70	24
7.0	October 17, 1989	Loma Prieta	40	7
6.9	May 19, 1940	Imperial Valley	50	15
6.7	February 9, 1971	San Fernando	16	8
6.7	January 17, 1994	Northridge	14	7
6.6	November 24, 1987	Superstition Hills	23	15
6.5	April 9, 1968	Borrego Mountain	25	6
6.4	October 15, 1979	Imperial Valley	30	13
6.4	March 10, 1933	Long Beach	15	5
6.1	April 22, 1992	Joshua Tree	15	5
5.9	July 8, 1986	North Palm Springs	20	4
5.9	October 1, 1987	Whittier Narrows	6	3
5.8	June 28, 1991	Sierra Madre	5	2

* *For comparison, the largest earthquake ever recorded was a moment magnitude 9.5 in Chile on May 18, 1960. The largest earthquake ever recorded in the United States was in Alaska on March 27, 1964, with a magnitude of 9.2. Prior to recorded history, the Missouri earthquake and early California earthquakes were of similar magnitude or greater.*

Modified after Jones, 1995, p. 23.

(1) magnitude of the earthquake, (2) distance from the epicenter of the earthquake (or where the fault ruptured), and (3) local soil conditions.

The primary cause of earthquakes along the San Andreas system is the relative movement of the Pacific and North American plates. In general, the Pacific plate is moving northwest, scraping horizontally past the North American plate at a relative rate of 45 millimeters ($1\frac{3}{4}$ inches) per year. The movement along the San Andreas Fault system in southern California is not always parallel to the plate movement. Between the Southern end of the San Joaquin Valley and the San Bernardino Mountains (in the so-called "Big Bend"), the San Andreas Fault system bends in a more westerly direction. Where the

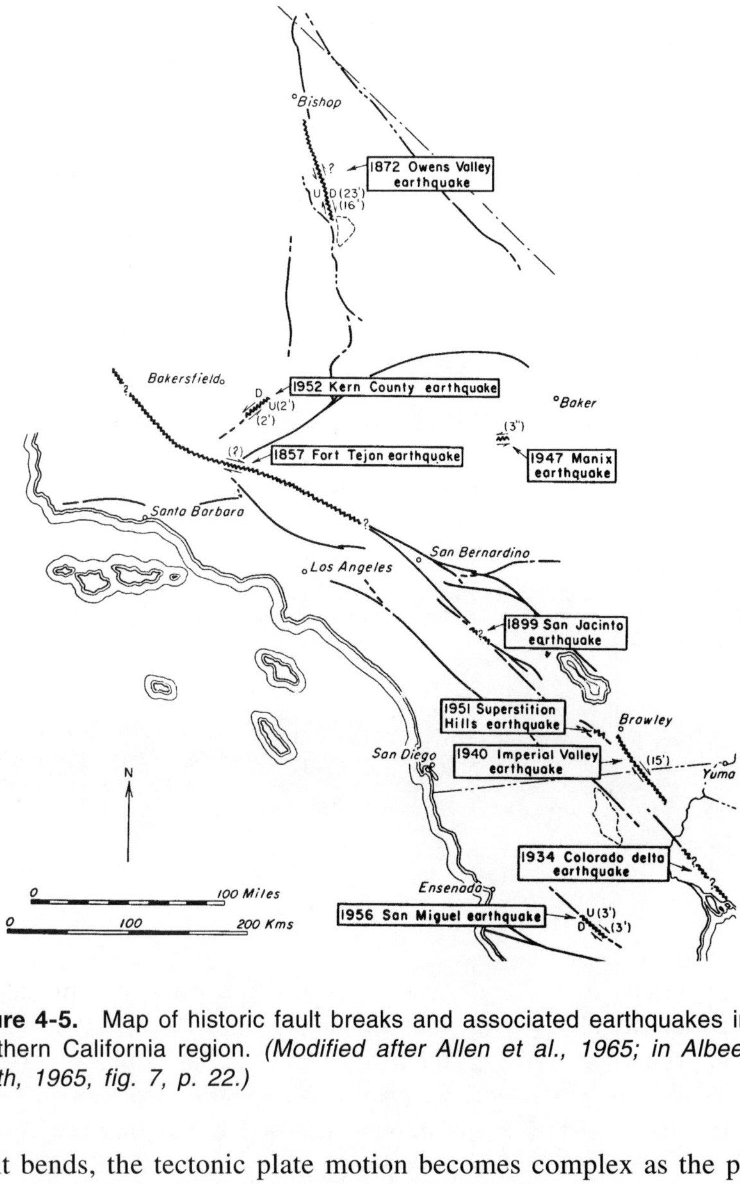

Figure 4-5. Map of historic fault breaks and associated earthquakes in the Southern California region. *(Modified after Allen et al., 1965; in Albee and Smith, 1965, fig. 7, p. 22.)*

fault bends, the tectonic plate motion becomes complex as the plates push against each other creating a compression zone. Approximate areas of high earthquake risk are shown in Figure 4-8, which takes into consideration the soil conditions. In general, the more common smaller earthquakes will produce greater shaking in softer soil, old stream or lakebeds, or a landfill that was not properly compacted

Table 4-4
Length and Amount of Observed Displacement Related to Earthquakes in California, Nevada, and Baja California

Date	Location	Magnitude	Length, miles	Observed Displacement: Max. Vert. Component	Max. Horiz. Component	Max. Total across Zone	Displacement by Resurveying
1857	Fort Tejon, Calif.	$7\frac{3}{4}$	40–250	?	Large		
1868	Hayward, Calif.	7	?	?	Some		
1872	Owens Valley, Calif.	$7\frac{3}{4}$	ca. 50	13'	18'		
1899	San Jacinto, Calif.	ca. $6\frac{3}{4}$	ca. 2	?	?		
1906,	San Francisco, Calif.	8.3	190 or 270	3'	21'	16'	
1915	Pleasant Valley, Nev.	7.6	24	15'	—	15'	
1932	Cedar Mtns., Nev.	7.3	38	24"	34"	35"	
1934	Excelsior Mtns., Nev.	6.5	0.85	5"	—	5"	
1934	Colorado River delta, Baja Calif.	7.1	?	?	?	?	
1940	Imperial Valley, Calif.	7.1	40	—	19'	19'	ca. 9' Horizontal
1947	Manix, Calif.	6.2	1	—	3"	3"	
1950	Fort Sage Mtns., Calif.	5.6	$5\frac{1}{2}$	5"–8"	—	8"	
1951	Superstition Hills, Calif.	5.6	1.9	—	Slight	Slight	
1952	Kern County, Calif.	7.7	40	4'	2'–3'	3.6'	1–2' Horizontal 1–2' Vertical
1954	Rainbow Mtns., Nev.	6.6	11	12"	—	12"	
1954	Fallon, Nev.	6.8	14	30"	—	30"	
1954	Fairview Peak, Nev.	7. 2	35	14'	12'	$18\frac{1}{2}$'	ca. 8' Horizontal ca. 4' Vertical
1954	Dixie Valley, Nev.	6.9	31	7'	7'	7'	ca. 7' Vertical
1956	San Miguel, Baja, Calif.	6.8	12	36"	31"	36"	
1966	Parkfield, Calif.	5.6	ca. 25	—	4"	4"	

Major sources of data in Tocher, 1958; Richter, 1958; Allen et al., 1965. Modified after Albee and Smith, in Lung and Proctor, 1969.

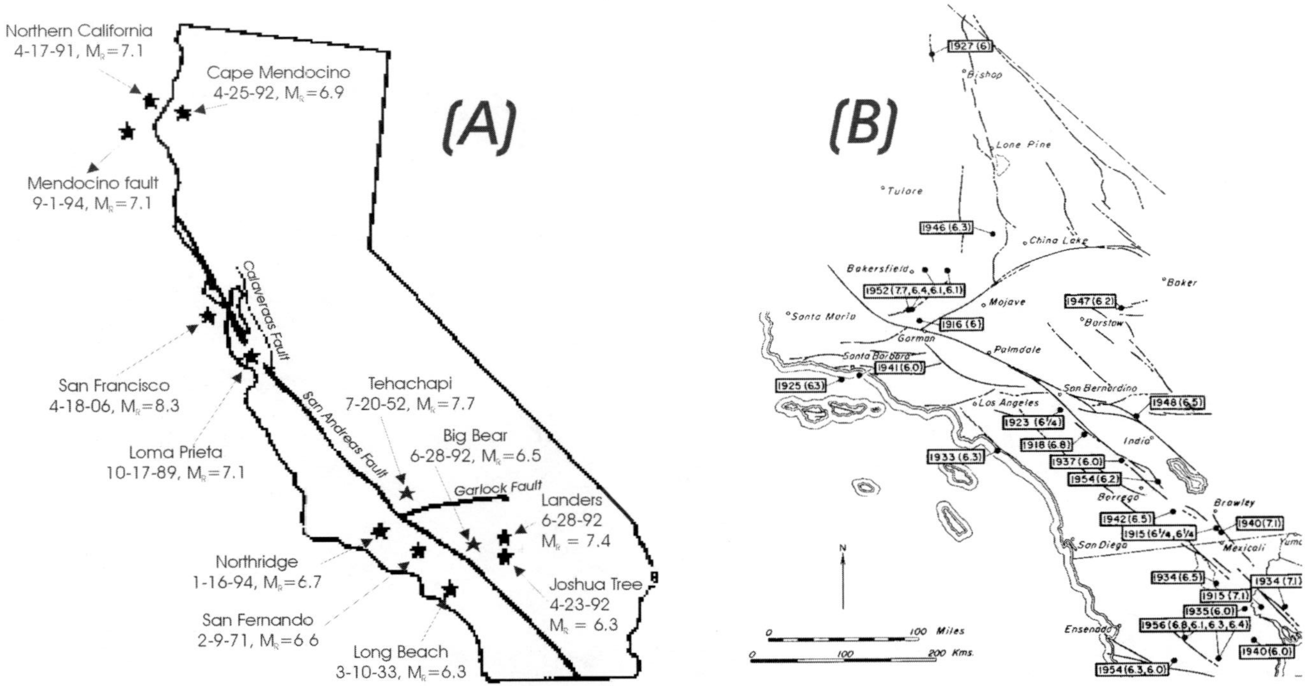

Figure 4-6. Earthquakes of magnitude 6.0 and greater in the Southern California region. *(Modified after Harden, 1997, fig. 14-4, p. 320, updated from Geotimes, 1993; and Allen et al., 1965; in Albee and Smith, 1965, fig. 6, p. 18.)*

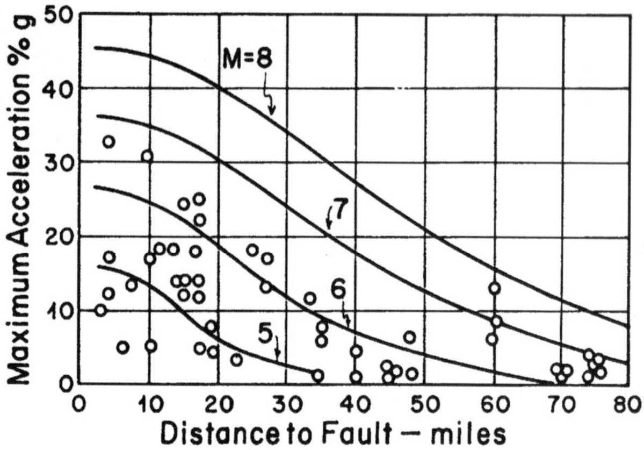

Figure 4-7. The maximum acceleration of an earthquake versus the distance from the epicenter. *(Modified after Albee and Smith, 1965, fig. 4, p. 15.)*

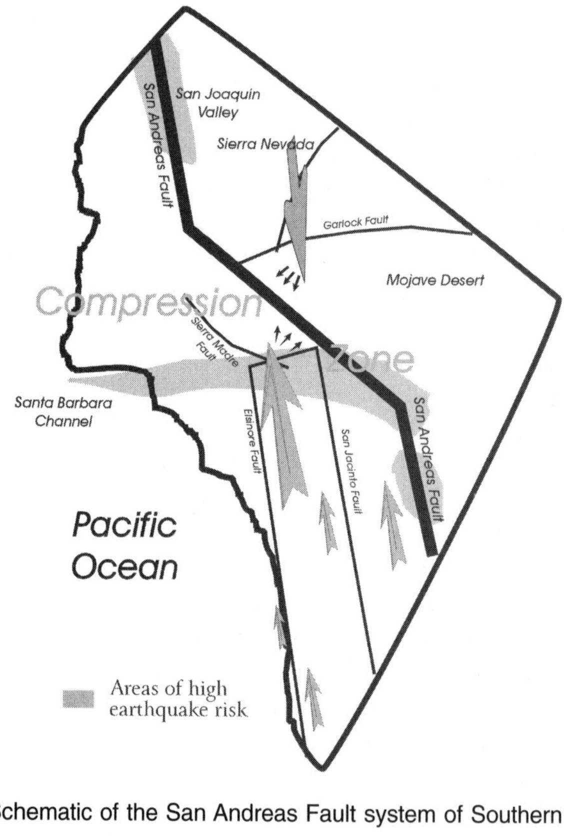

Figure 4-8. Schematic of the San Andreas Fault system of Southern California.

(Jones, 1995) because the softer soil amplifies the earthquake waves. The area shaded light gray in Figure 4-8 is likely to have three or more severe earthquakes in the coming century. In general, the Southern California area is expected to experience a M_R>7.0 earthquake seven times per century (Jones, 1995). It should be pointed out that an earthquake occurring in one portion of the shaded area affects only a limited area about its epicenter.

SOUTHERN CALIFORNIA FAULTS

Palos Verdes Fault

The Palos Verdes Fault is located in the coastal area of Southern California, forming an important part of the complex fault system of the Los Angeles Basin. The Palos Verdes Fault can be traced by separate gas and oil offshore seeps trending in a northwesterly direction into the Pacific Ocean from Redondo Beach to Santa Monica (Figure 2-11). Oil and gas seeps have been found at six separate locations. The offshore seep nearest land is located about two miles off Redondo Beach and originates near the head of the Redondo Submarine Canyon at a water depth of about 800 feet. Five additional seeps occur along the 12-mile trace of this fault, including the "Manhattan" seep located about four to five miles off the shoreline. The "Venice" seep is the most seaward, nearly eight miles off the Venice beach. The activity of these seeps varies and they often remain dormant for extended periods of time.

The length of this fault can be traced for about 80 miles. This is a "strike-slip" fault, with more horizontal than vertical movement. This fault has produced uplifting, especially in the area of the Palos Verdes peninsula. A recent study conducted by McNeilan et al. (1996) has established that the slip rates along this fault are among the highest for any fault in the Los Angeles Basin. The peninsula side of the Palos Verdes Fault is moving northwest into the Santa Monica Bay at a rate of 2.7 to 3 mm/year. At this rate of movement, it is the most active fault within the Los Angeles Basin. By comparison, the nearby Newport-Inglewood Fault (responsible for the devastating 1933 Long Beach earthquake) has a slip rate of about 2 to 10 mm/year. The length of this fault indicates that it has the potential of generating a M_R>7.0 earthquake.

Because much of this fault is offshore, the threat of a "tsunami" generated by an earthquake along this fault must be considered. A "tsunami" can be created when an earthquake shakes the ocean floor. A series of waves radiates out from the epicenter. Without protection, coastal cities will be seriously exposed. For example, following the 1964 Alaskan earthquake, the harbor at Crescent City, California, was destroyed by a seven-meter high tsunami and 11 people were killed. Japan has been hit by several tsunamis resulting in both human death and destruction of property. One of the worst tsunamis in history occurred on June 15, 1896, when it hit the eastern coast of Honshu, Japan, following an earthquake produced by a wide arc of ocean floor thrusting upward along a submarine fault off the Japan Trench. More than 10,000 homes were destroyed and 26,000 people were killed.

Northridge Earthquake

On January 17, 1994, the Northridge earthquake (M_R=6.7) awakened southern Californians. Figures 4-9 and 4-10 present the basic cross-

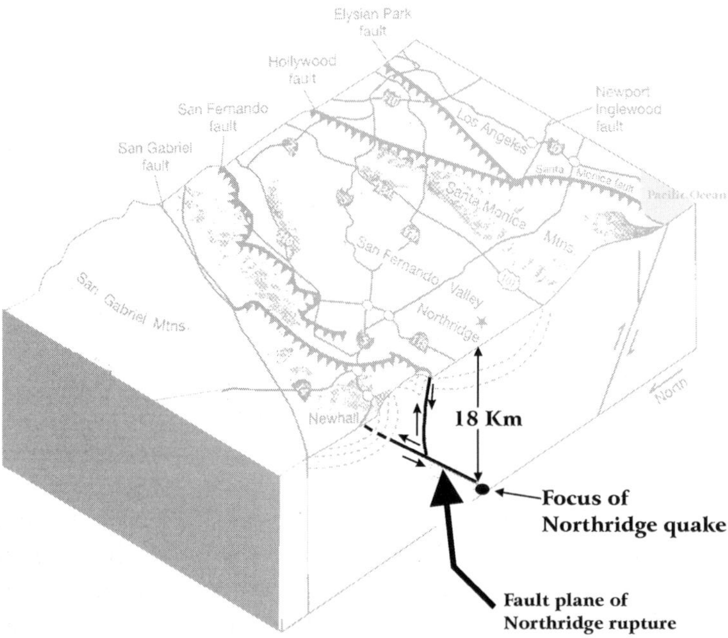

Figure 4-9. Cross-section of the area where the Northridge earthquake occurred in January 17, 1994. The reverse faults below the San Fernando valley and the northern Los Angeles basin are shown. *(Modified from W. P. Lawrence Berkeley Laboratory; in* Earth Magazine, *Learning from Los Angeles, September 1994.)*

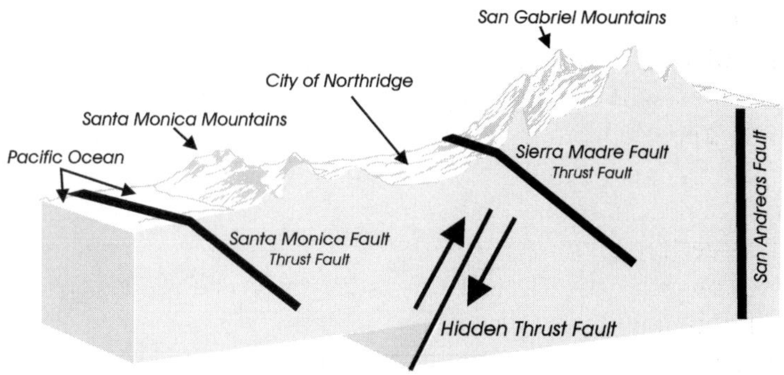

Figure 4-10. Hidden faults under Northridge. *(After Levy and Salvadori, 1997, fig. 13.2, p. 154.)*

section of the area surrounding Northridge, California. The quake occurred along a fault that was not exposed at the surface (blind fault).

Following the Northridge earthquake of 1994, about 15 miles to the west, there was an increase in gas bubbling up from the ocean bottom near the Malibu pier in California. This seepage occurred close to the Malibu Fault, which is an active fault that extends along the coastline of Southern California. Additional information on this fault and associated seismic events are discussed in greater detail in later chapters.

LOS ANGELES BASIN PETROLEUM DEPOSITS

Inasmuch as the migration of natural gas from the Los Angeles basin oil and gas deposits is prevalent, they are discussed in greater detail here. There are four major oil- and gas-producing regions in California (Figure 4-11): (1) San Joaquin-Sacramento (Great Valley), (2) Los Angeles, (3) Ventura, and (4) Santa Maria. Two smaller valleys (Cuyama and Salinas) produce oil and gas in lesser quantities. Oil in California occurs primarily in formations of Pliocene and Miocene age with lesser quantities in the Oligocene, Pleistocene, Eocene, Jurassic, and Cretaceous. Although most of gas in these regions is associated with oil, there are relatively few fields with dry gas. All the dry gas and a very large part of the oil (with its associated natural gas) reside in sands, sandstones, or sandy shales. A minor amount of oil occurs in fractured shales and fractured schist. The petroleum in

(text continued on page 90)

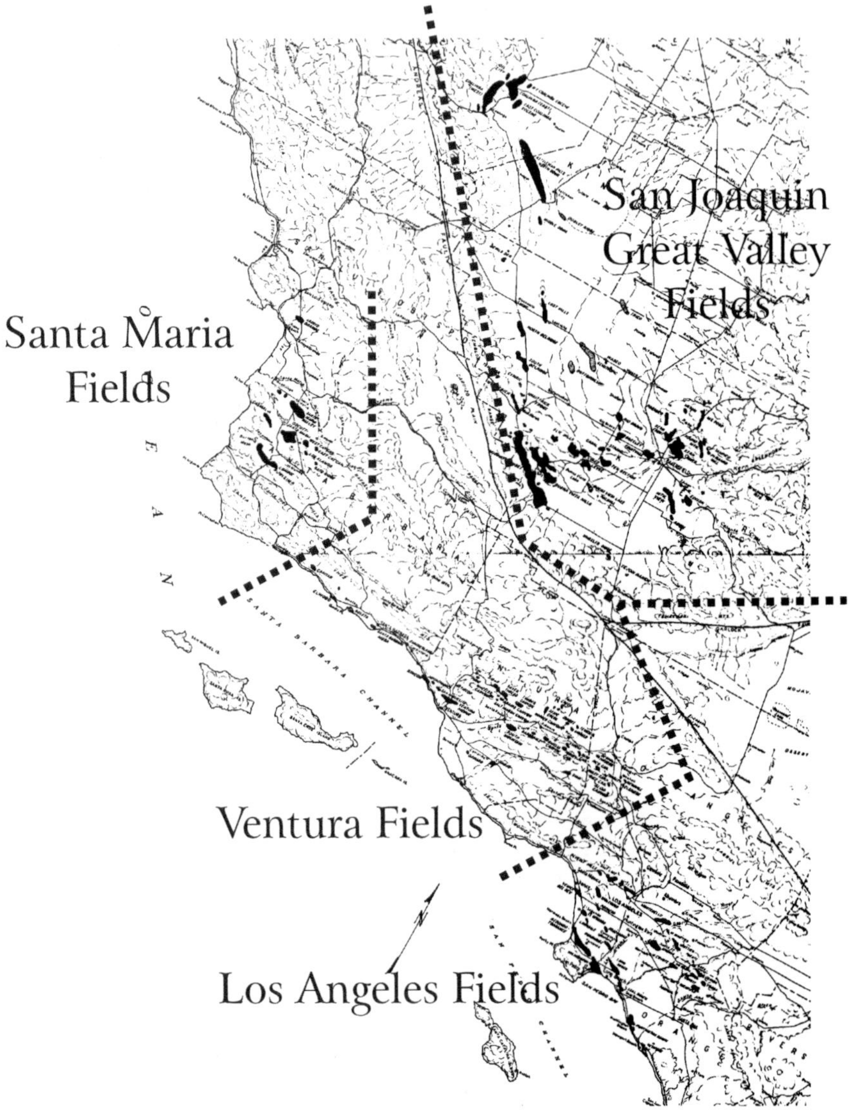

Figure 4-11. Map of central and southern California showing California's oil and gas fields. *(Modified after Richfield Oil Corp. map; in Hoots, 1948, Plate VI, p. 275.)*

Table 4-5
Types of Oil Accumulations Found in California

Types of Accumulation	Examples
A. Accumulations in normal closed anticlines; faulting minor or not of such type as to greatly affect accumulations; unconformities supposed to be absent or of slight importance	1. North Dome of Kettleman Hills 2. Middle Dome of Kettleman Hills 3. North Belridge (middle and lower Miocene oil) 4. Ten Sections (faulting possibly influential) 5. Rio Bravo (faulting possibly influential) 6. Wheeler Ridge 7. Elwood (marginal faulting important?) 8. Ventura Avenue 9. Dominguez (Pliocene oil) 10. Santa Fe Springs 11. West Coyote 12. Seal Beach
B. Accumulations in anticlines, with faulting important secondary control	1. South Mountain (data not very conclusive) 2. Inglewood 3. Dominguez (Miocene oil) 4. Long Beach 5. Santa Maria or Orcutt 6. McKittrick 7. Huntington Beach 8. Wilmington 9. Richfield
C. Accumulations in anticlines, with overlap or unconformity as essential secondary control	1. Elk Hills (particularly eastern development) 2. Buena Vista Hills (particularly northwestern area) 3. North Belridge (Pliocene oil) 4. Belridge (Pliocene oil) 5. Playa del Rey
D. Accumulations in sand lens on plunging anticline, possibly without effective unconformity	1. Coalinga Eastside (Eocene oil)

Table 4-5 (continued)

Types of Accumulation	Examples
E. Accumulations due to unconformity; type of fold incidental or secondary	1. Coalinga Westside 2. Midway-Sunset fields 3. Santa Maria Valley 4. Edison
F. Fault accumulations	1. Mount Poso 2. Round Mountain 3. Mountain View 4. Whittier 5. Brea-Olinda

Modified after Reed, 1948, table 1, p. 97.

(text continued from page 87)

the fractured shales and schist migrated into these formations from nearby source beds.

In California, the major oil sources are only found in a small part of the state, in a portion of the coastal mountain-and-valley province. Oil has not been found in commercial quantities anywhere in the northern coastal ranges or in approximately half of the southern coastal ranges. Much of the greater part of the coast ranges appears to be entirely barren of commercial quantities of petroleum. Reed (1948) and others have classified the oil accumulations of California structurally and stratigraphically (Table 4-5 and Figure 4-12). A high percentage of the oil and gas occurs in anticlines or domes; however, nearly every known type of geologic structure is present. Faults are present in nearly all oil accumulations and are an important factor in the entrapment of oil and gas.

The Los Angeles basin is approximately 22 miles wide and 46 miles long (Figure 4-13). The Santa Monica Mountains and a line of hills extending southeastward to the Santa Ana Mountains form the boundary of the Los Angeles Basin on the north. The foothills of the Santa Ana mountains bound it on the east. The San Joaquin hills, the Pacific Ocean, and the Palos Verdes hills bound it on the south and west. Structurally, the major oil fields within this basin are primarily anticlinal (Figure 4-14). Some of the lower oil-producing fields (i.e., Whittier, Brea-Olinda, and those within the Los Angeles central area)

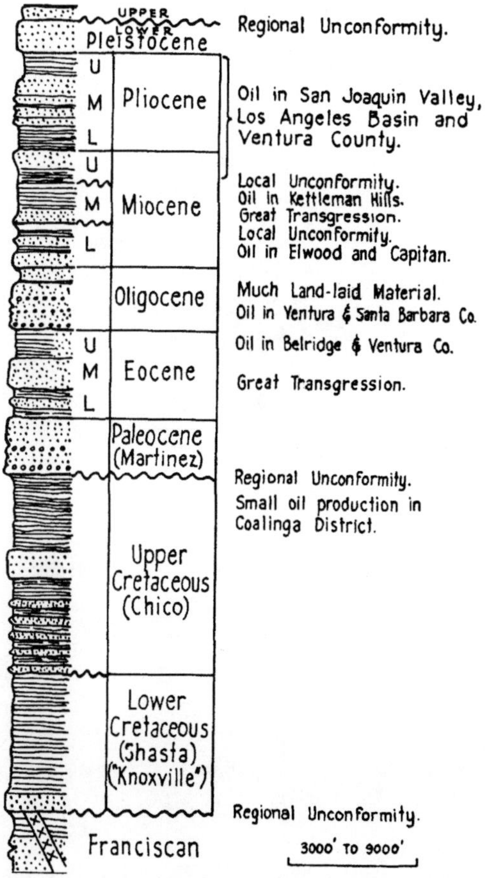

Figure 4-12. Generalized stratigraphic column of the coast ranges showing positions of major oil zones. *(Modified after Reed, 1948, fig. 44, p. 95.)*

are faulted monoclines. The oil fields within the Los Angeles Basin are all within a radius of 30 miles from the central area of Los Angeles. These faulted monoclines are not strong structural traps and are, thereby, especially prone to gas migration to the surface during seismic events.

As early as 1921, it was recognized that the proven oil fields within the Los Angeles Basin lay roughly along three parallel lines, which have a northwest–southeast trend. Leck (1921) described these lines of structure as a series of domes, anticlines, and faults along which the oil has accumulated (Figure 4-15). The first and most northerly

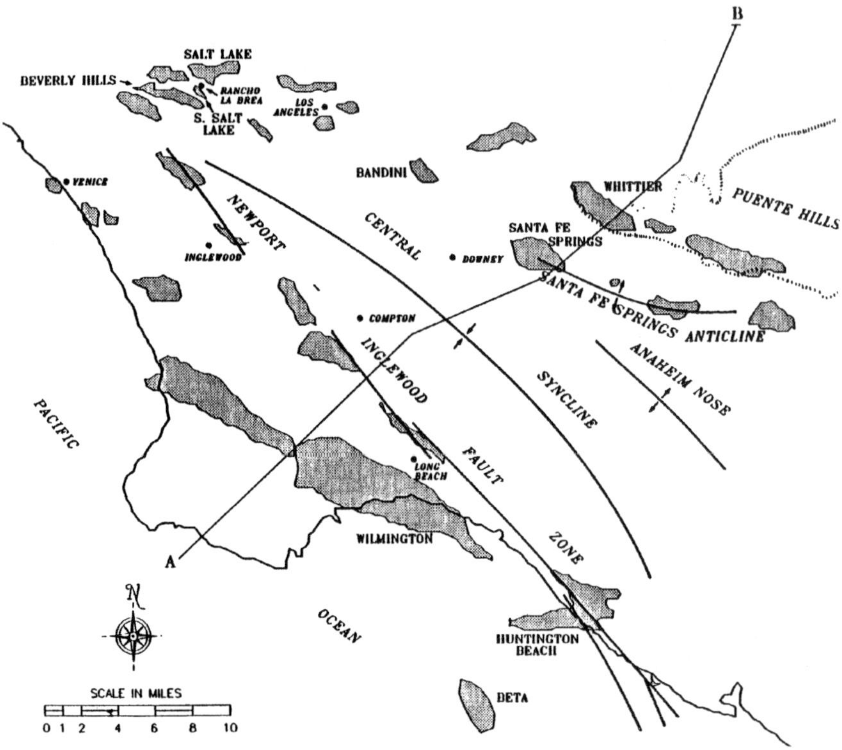

Figure 4-13. Los Angeles basin oil fields. Cross-section A-B is shown in Figure 4-15. *(Modified after Yerkes et al., 1965; in Jeffrey et al., 1974.)*

of these begins at the Salt Lake field, near Hollywood, and runs east to the Los Angeles field and then eastward through the Montebello, Whittier Hills, Brea Canyon, and Brea Olinda fields. The second line of structure lies about three miles south, and parallel to, the first and is separated from it by a narrow syncline. Beginning at the southeast, this general structure starts in the Richfield district and runs northwest to the Coyote Hills field and then to the Santa Fe Springs area. The third line is the longest in extent and is separated from the second by a broad, flat area of about 12 miles in width, which is apparently a syncline, but which might contain buried folds (Leck, 1921). Beginning in the southeast, this general line of structure starts in the Newport area and runs northwest to the Huntington Beach field and then to the Signal Hill dome. At this point, the strike changes more to the north and the general line on uplift continues through Dominguez,

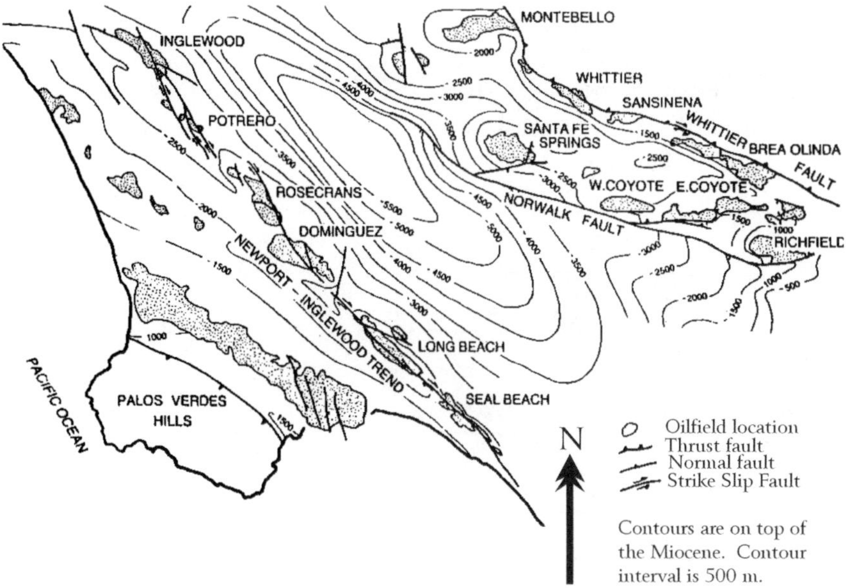

Figure 4-14. Location of productive trends along major faults for the Los Angeles basin. *(Modified after Harding, 1973, and Harding, 1974.)*

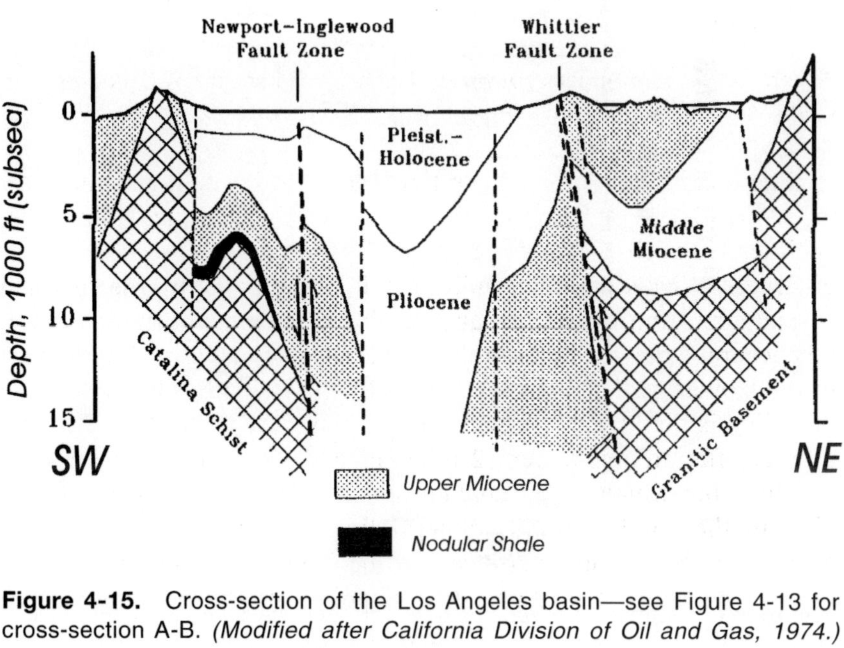

Figure 4-15. Cross-section of the Los Angeles basin—see Figure 4-13 for cross-section A-B. *(Modified after California Division of Oil and Gas, 1974.)*

Athens on the Hill, Inglewood, Culver City and finally ends in the area around the Wolfskill lease of the West Oil Company about one mile southwest of Beverly Hills. Figure 4-13 illustrates this general trend of northwest–southeast faulting within the Los Angeles basin and how faulting has controlled the location of the oil fields for this area.

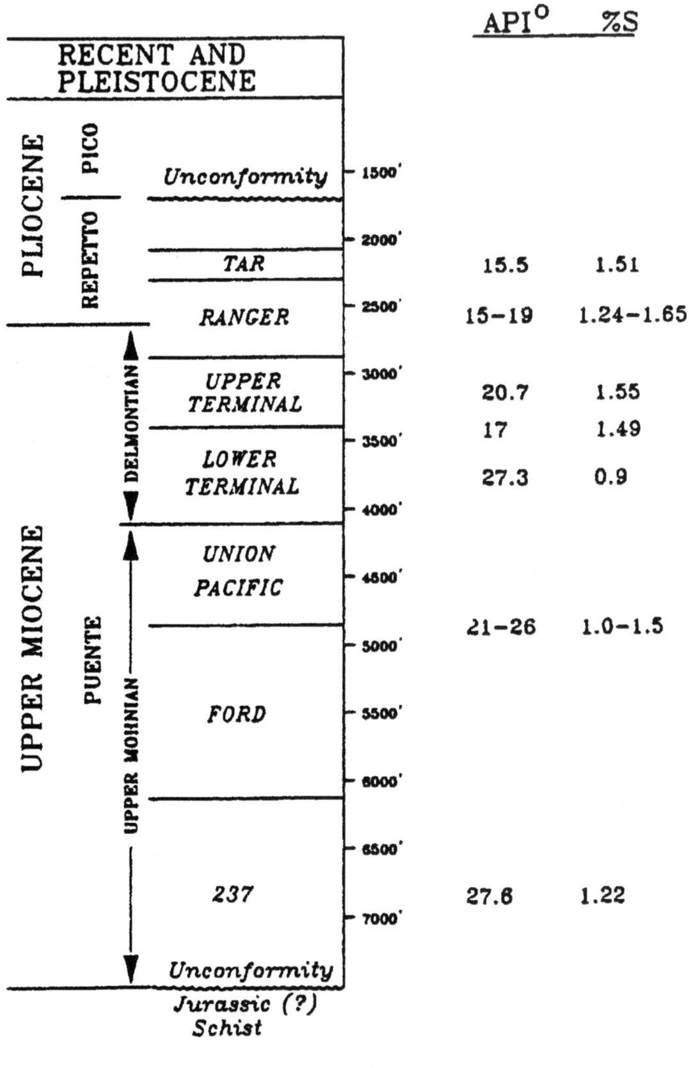

Figure 4-16. Straigraphic records and typical oil gravity from each producing horizon for the Los Angeles basin. *(Modified after Wissler, 1948.)*

Wissler (1948) proposed a stratigraphic column for the Los Angeles basin based on foraminifera succession found in numerous core samples. The foraminifera offer a feasible basis for reliable inter-field as well as intra-field oil well correlation. Within the basin, petroleum is produced from the Pliocene and upper Miocene sediments (Figure 4-16). The rocks penetrated by wells of the Los Angeles basin range from Recent Alluvium to weathered and fractured Jurassic schist of the Franciscan formation. The Franciscan serves as a basement rock for the area west of the Inglewood-Newport uplift. In general, the sand or sandstone varies in hardness, texture, and permeability and represents 98% of the Los Angeles basin petroleum reservoir rocks. The sands are thick, highly permeable, and have high rates of production and unusually large recoveries per acre (Hoots, 1948). Most of these prolific sands are of the blanket type covering extensive areas, but some are lenses having only limited distribution.

Summary

Examining the discussion above and the publications of the California Division of Oil and Gas on California oil and gas fields, it can be concluded that avenues for natural gas migration to the earth's surface from petroleum deposits along faults are plentiful. The problem of migration of natural gas from some oil and gas fields is more severe than others. Thus, hazards associated with each oil or gas field must be evaluated separately.

PART II

Events Preceding Earthquakes

Hospital destroyed by the Spitak earthquake. (modified after Agbabian and Chilingarian, 1991).

CHAPTER 5

Earthquake Precursors

Devastating earthquakes are commonplace events in the earth's history. During the 20th century alone, 22 large earthquakes occurred that have caused over 1.25 million casualties and multibillion-dollar damage around the world. To reduce catastrophic (human and financial) losses, scientists and engineers are trying to develop techniques of predicting future earthquakes.

Danger and fear of earthquakes are so great that nations worldwide are spending millions of dollars trying to forecast quake occurrence in spite of the failure of previous attempts. Inasmuch as the origin of the seismic events is hidden deep underground and quakes last only seconds, it is very difficult to investigate earthquakes directly. People, however, attempt to recognize an oncoming event based on indirect information, such as tremors, smells, and sounds preceding and accompanying earthquakes.

The primary idea of all past, present, and future developments in short-term seismic forecasting is the concept of precursors—the events that supposedly precede and indicate oncoming earthquakes. Scientists have spent years of research trying to understand their causes, describe their development, and discover these precursors. They have attempted to apply both natural and artificial anomalies that can be correlated with earthquakes as possible precursors, from current seismic activity to the behavior of animals. Tangible results, however, have not been attained.

Enthusiastic amateurs have also attempted forecasting earthquakes. For example, one promising source of information may be the "lost pets" section of the newspapers. Some "investigators" correlate the pronounced increase in the number of lost cats and dogs with the occurrence of big earthquakes. It seems plausible that on the eve of major seismic events, sensitive pets feel some changes in the environment and try to escape by fleeing the home.

It is important, however, to link earthquake precursors with a probable direct cause (causes) of the event. A great deal of understanding of the causes of earthquakes was attained after the devastating San Francisco earthquake of 1906 in California. Due to continuous field observations, it was discovered that the primary cause of any shaking of the earth is the release of accumulated rock stresses. This release, in the form of crustal rupture followed by the earth's vibrations, is what is felt as an earthquake. After this discovery, it became clear that reliable causal precursors can be linked to accumulated stresses in the rocks.

There is a consensus in the World Scientific Community (e.g., Bolt, 1988; Sadovsky et al., 1984; Mogi, 1967) that major earthquakes occur most often and repeatedly in the vicinity of existing faults because they represent the so-called "weakened zones." For these zones, the chances of the earth's crust rupturing are considerably higher and, for this reason, they represent an excellent area for monitoring precursory events. (For details, see Chapters 1 and 2.)

It is commonly believed that an earthquake occurs because the accumulated stresses exceed a certain frictional resistance limit. Therefore, if for a particular point (points) an accumulation of stresses at the fault is detected, it can be inferred that there is a possibility of earthquake occurrence. The problem is that the point measurement, which represents local changes in stresses, can vary and change considerably for the points in close proximity.

The processes preceding an earthquake are difficult to monitor because they may be hidden from observation deep underground. Under these circumstances, researchers are forced to use all kinds of indirect information that can be related to earthquake preparation.

A short-term forecast (on the order of days, hours, minutes) is important to reduce and mitigate potential damage caused by a large earthquake. The possibility of short-term prediction of earthquakes has been studied by many scientists worldwide. They recommended using various possible precursors such as:

1. Cosmic events
2. Abnormal animal behavior
3. Electrical phenomena associated with earthquakes (lightening and lighting of the lower layer of air atmosphere).
4. Crustal movement, stresses, and strains.
5. Seismic activity

6. Changes in the ratio V_p/V_s, where V_p and V_s are velocities of propagation of primary and secondary seismic waves, respectively
7. Hydraulic resistivity.
8. Hydrodynamics anomalies (groundwater level, discharge rates, etc.).
9. Acidity (pH) and chemical composition of groundwater (pH = negative logarithm of H^+ ion concentration in water).
10. Radon concentration in the groundwater and soil air
11. Changes in the concentrations of various gases

We have found that the most promising precursor is the variation in the rate of upward gas migration. Changes in the rate are most significant in the vicinity of existing faults and fractures. The changes in stresses preceding a major earthquake may either increase or decrease the permeability of faults and fractured zones. Thus, pronounced changes in the rate of migration of gases to the earth's surface in the vicinity of faults and fractured zones can be used as short-term precursors of seismic events. This and other possible precursors are discussed in subsequent chapters.

CHAPTER 6

Occurrence and Consequences of Earthquakes

INTRODUCTION

As previously stated, large earthquakes can cause catastrophic damage with numerous casualties around the world. The worst death toll (over 650,000 people killed) in the 20th century was reported in 1976 after a strong earthquake (M_R = 7.9 on the Richter scale) in the city of Tangshan in the People's Republic of China. Afterward, the Chinese government reduced this death toll estimate to 242,000. The strongest recorded earthquake (M_R = 8.9 on the Richter scale) was the Alaskan earthquake, which occurred on Good Friday, March 27, 1964. Even in the unpopulated Alaskan territory it caused 750 million dollars damage. Natural disasters triggered by this quake were enormous. Five hours after the first tremor, a giant tsunami wave totally demolished the Alaskan port of Valdez. At the Seward port, a 33-foot high wave instantly swept away all the port's facilities. The Spitak earthquake of 1988 in Armenia killed more than 25 thousand people and completely demolished the town structures (Karapetyan, 1991). Spitak ruins and surface ruptures are presented in Figures 6-1 through 6-4.

Although the recent Northridge earthquake of 1994 in California struck in a well-prepared area, it killed 61 people and caused damages totaling billions of dollars. The final estimate of damage reported at the Pasadena Seismic Conference in March of 1997 was 42 billion dollars. The record for financial damage is held by Japan's Kobe earthquake of January 17, 1995—about 100 billion dollars.

Figure 6-1. A hospital destroyed in the Spitak earthquake of 1988. *(Courtesy of Agbabian and Chilingarian, 1991.)*

Figure 6-2. A villager's house destroyed in the Spitak earthquake of 1988. Hundreds of masonry homes collapsed during the earthquake. *(Courtesy of Agbabian and Chilingarian, 1991.)*

Figure 6-3. Faults produced by the Spitak earthquake of December 7, 1988. *(Courtesy of Agbabian and Chilingarian, 1991.)*

Figure 6-4. Surface rupture on Armenian highlands after the Spitak earthquake of December 7, 1998, which was 8 km long. *(Courtesy of Agbabian and Chilingarian, 1991.)*

Lioning Earthquake of 1974

There is only one documented example of successful short-term prediction of a large earthquake in modern history. On the afternoon of February 4, 1974, a radio broadcast to the 3 million inhabitants of Lioning Province, People's Republic of China, commanded them to leave their homes immediately because of an impending earthquake. At 7:30 P.M., an earthquake with magnitude of $M_R = 7.3$ destroyed almost all the buildings in the city of Haicheng with a population of 90,000. The death toll was surprisingly small, only 300 people, because most of the people obeyed the order and left their houses. For the first (and probably the only) time anywhere in the world, an earthquake of this magnitude was predicted, and this prediction saved thousands of lives.

Choluca Earthquake

Artistic pictures of earthquakes are horrifying. In the book entitled *Earthquake* by Berle and Roeburt (1959), the feelings of survivors of the Choluca earthquake in Mexico are described:

> Earth trembled under their feet. The sea swelled and foamed in the harbor, and beat to pieces the vessels riding at anchor. Whirlwinds of fire and ashes covered the streets and public places; houses fell, roofs were flung upon the pavements, and the pavements were scattered. Thirty thousand inhabitants of all ages and sexes were crushed under the ruins.

The cause of this earthquake was a fault plane more than 100 miles long and three feet deep. The sudden slip and rupture of rocks along the fault were telegraphed to the observatories. The recorded seismic waves showed that primary waves (*P* waves) traveled outward in straight lines in all directions from the hypocenter, at a velocity of six miles per second. Secondary waves (*S* waves), vibrating at right angles to the direction of propagation, had lower speed, but with more than twice the destructive power (Berle and Roeburt, 1959).

The earthquake struck the highland slopes on which the picturesque village of Choluca was situated. The surface of the ground rose in low, very fast moving waves. Small cracks opened in the soil on the crests of waves, closing in wave troughs.

Small wooden houses and poorly constructed mortar structures collapsed at once. People and livestock disappeared into the earth as the ground opened. Water pipes and sewers broke and telephone poles fell. The waterways of Choluca were filled with soil and debris.

The earthquake lasted only 56 seconds. Fire spread rapidly due to the lack of firefighters and special equipment. In less than one minute the village of Choluka was completely destroyed (Berle and Roeburt, 1959).

Spitak Earthquake of 1988

On December 7, 1988, a devastating earthquake of $M_R = 7$ occurred in the Spitak region of Armenia. The hypocenter of the earthquake was located at the border of two huge tectonic blocks, and their mutual displacement caused enourmus damage. More than 25 thousand people were buried under the ruins of the completely destroyed city of Spitak, whereas Leninakan, Kirovakan, and many other populated areas in the northern part of Armenia were only partially destroyed.

Northridge Earthquake of 1994

At 4:31 A.M. on January 17, 1994, a powerful earthquake struck 11 miles beneath Northridge, California, northwest of downtown Los Angeles. As reported by Cal Tech in Pasadena, it measured $M_R = 6.8$ on the Richter scale and caused severe damage to structures. As a result of the earthquake, 61 people were killed and more than 10,000 were injured. Damage estimates were set at about 20 billion dollars. There is a widespread opinion that this was possibly one of the worst earthquakes in U.S. history.

The principal writer who lived in the city of Sherman Oaks, located only five miles away from the epicenter, will remember this event for the rest of his life:

> On January 17, 1994, at 4:31 A.M., I was in the bathroom brushing my teeth. The first thrust hurled me down on the floor. After crawling out of the bathroom, I stopped at the doorframe and called my wife. She joined me momentarily, and we stood in the shaking doorframe calling for our son Andrew. Pieces of glass from the broken dishes and windows flew around the room, the

walls squeaked and cracked filling in the space with unforgettable cacophonic sounds.

The vibrations stopped abruptly. I rushed to the son's bedroom and attempted to take him out. Andrew tried to get up but was thrown down by the second shock that hurled me to the floor and out of the bedroom. In a moment, the door of my son's bedroom was blocked by pieces of broken furniture.

The second shock stopped as abruptly as the first one, and I was able to reach my son. Power was out, flashlights were in the car, my son moaned in his bed in the darkness. I helped him to get up. In two minutes we were out. Our neighbors gathered around barely dressed and looked deeply shocked. It was very quiet on the street, just a remote siren signaled that something terrible had happened. Smell of gas was in the air indicating a gas leak and danger of explosion. A neighbor came in with the tools and shut down the gas pipe. Now it was the time to count our casualties.

There was a terrible mess in the house. Pieces of broken furniture and dishes were scattered around, a massive granite table was torn apart with two big pieces resting in the opposite corners of our living room.

The nose of my 16-year old son was badly swollen and bleeding. I grabbed my son and drove him to the Encino hospital. We were lucky. The emergency surgeon was a Vietnam veteran, and it took him 5 minutes to examine the nose and fix it with iodine solution and a piece of adhesive tape. The Northridge earthquake was recorded as a tiny white scar across the bridge of Andrew's nose.

For many other people the consequences were much more severe. An old three-story apartment building in the Northridge area collapsed in the garage space killing 17 people. A policeman's motorcycle fell into a crack in the freeway and the rider was instantly killed when the ramp collapsed. Many families were forced to abandon their homes because they were destroyed.

The Northridge earthquake of 1994 was composed of two main shocks separated by a five-second quiescence interval. Strong vertical motion of the earth's crust caused vertical displacements of about 15 feet at the epicenter that resulted in severe damage to the structures.

The apparent energy liberated by this earthquake considerably exceeded the normal level for earthquakes with a magnitude of M_R = 6.8 (Giusti, 1995).

The compound underground waves were extremely destructive. Complex waves generated by the two shocks rippled through many types of rock blocks interfering with each other in many ways. They traveled back and forth through the earth's crust after possibly being bounced off the dense rocks of the San Gabriel and Santa Susana mountains. As a result of superposition of traveling waves, severe damage occurred where the crests of seismic waves intersected.

The thrust fault that produced the Northridge earthquake was located at a depth of 11 miles and did not have a visible trace on the earth's surface. For faults of this kind, blocks of rock are thrust at an acute angle above other blocks of rock. This causes the vertical displacement of the crustal pieces and results in the release of large amounts of energy at the earth's surface. These kinds of earthquakes are difficult for scientists to investigate because they do not leave any visible traces at the earth's surface. Thrust faulting is caused by the compression of the earth's crust. Squeezing upward compression produces pronounced vertical displacement of the crustal blocks and results in massive destruction of surface structures. The local fault zones of the Northridge earthquake of 1994 are shown in Figures 6-5 and 6-6.

EARTHQUAKES IN GENERAL

If earthquakes could be predicted, necessary precautions could be taken to minimize the damage. Hence the importance of earthquake forecasting studies is obvious. The question is whether or not it is possible to develop a reliable forecast.

The *point of strike*—the focus of the earthquake—is most often situated within 70 km from the earth's surface. A few earthquakes occur at the surface, whereas deep-focused earthquakes (300–700 km) are very rare and occur in the subduction zones exclusively. Approximately 30% of all earthquakes occur at a depth greater than 70 km and nearly 8% of them occur at depths below 300 km (Green, 1994).

More than 90% of the world's earthquakes occur in the three major zones (of earthquake belts or seismic belts) shown in Figure 6-7. The most active zone is the Circum Pacific Belt where 80% of the total number of earthquakes occur. This belt begins on the east coast of

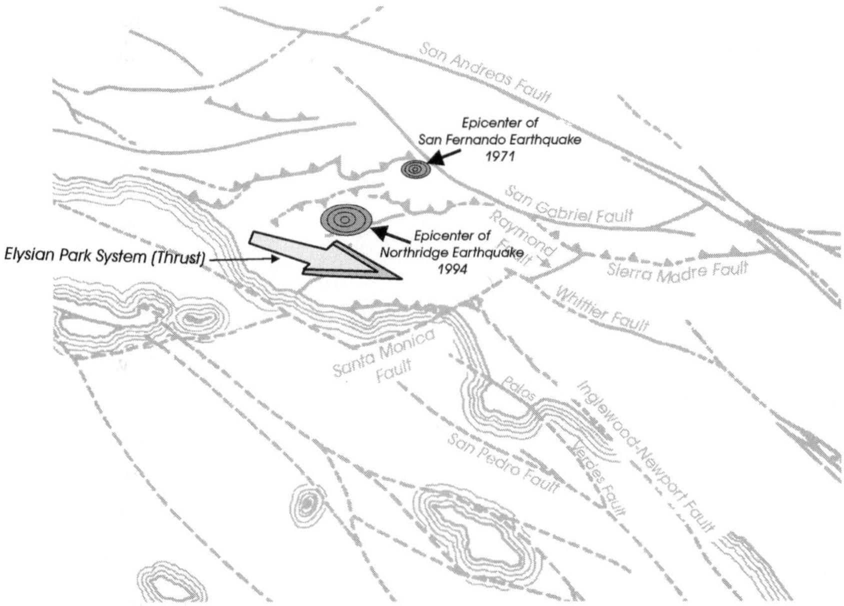

Figure 6-5. Epicenters of the 1971 San Fernando and 1994 Northridge quakes. *(Modified after Levy and Salvadori, 1995, fig. 13.3, p. 155.)*

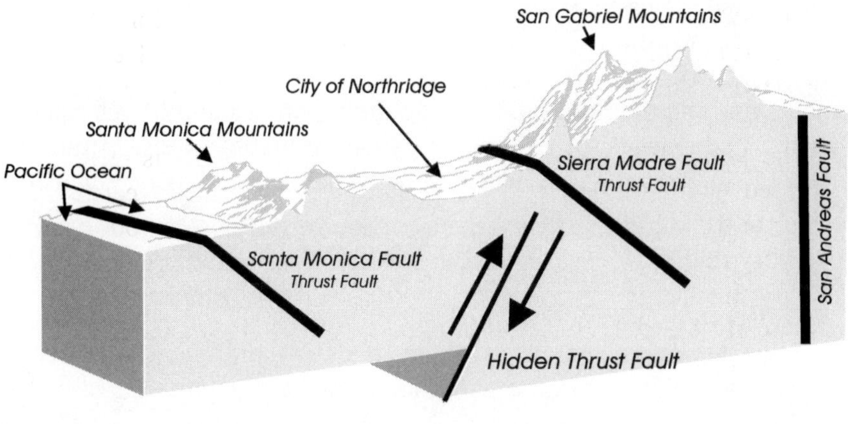

Figure 6-6. Hidden faults under the city of Northridge, California. *(Modified after Levy and Salvadori, 1995, fig. 13.3, p. 154.)*

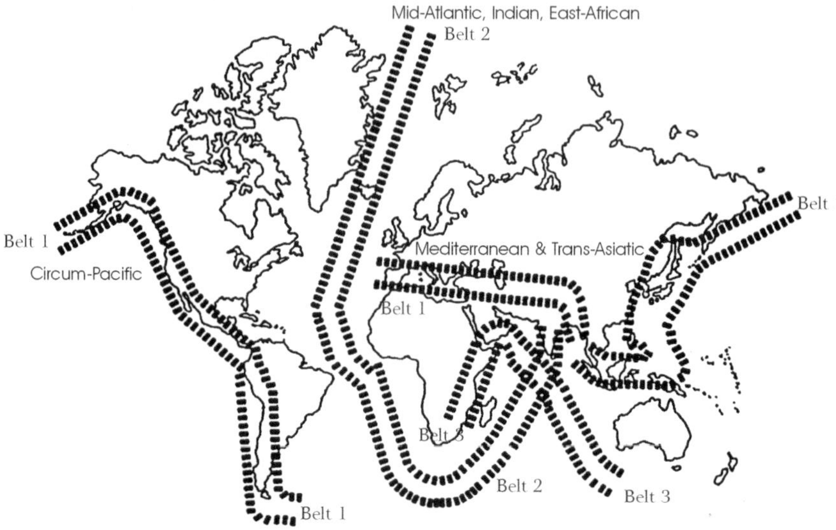

Figure 6-7. The three major delineated earthquake belts where 90% of all earthquakes occur. *(Modified after Giusti, 1995, fig. 1-5, p. 9.)*

Kamchatka and extends to Alaska through Kuril, Commander, and Aleutian Islands. It proceeds south toward Mexico along the west coast of North America. It then extends along the Pacific Coast of South America and ends at the Antilles loop, including the Falkland Islands and New Georgia. Its second branch passes along Kuril and Japanese Islands and the east coast of Asia, reaching Australia and New Zealand.

The second belt, the Mediterranean and Trans-Asiatic, is responsible for about 15% of all earthquakes. It extends from Burma to the Pamir mountains and then to Iran, covering the Black Sea and Mediterranean basins, and reaches the Atlantic Ocean at the Azor Islands. In Europe, the belt includes Cyprus Island, Crete Island, the Balkan Peninsula, and Carpathian mountains.

The Arctic-Atlantic region, together with the western part of the Indian Ocean and East-African grabens (where approximately 5% of the seismic events occur), forms the third belt.

Even this fragmental description shows the enormous scope of earthquake events in the world and, thus, the importance of earthquake prediction.

The development of an earthquake takes years and is hidden in the earth's crust. Rupture occurs underground, is localized in a small crustal area, and occurs unexpectedly. The earthquake movement lasts for only seconds. These conditions make studying the mechanism of earthquake development and its precursors in order to predict future earthquakes extremely difficult.

In the mid-1980s, there were hundreds of scientific contributions published annually on the subject of earthquake forecasting. Recently this number has decreased. Public opinion has also changed dramatically: from desperate hope to skeptical disbelief in the power of scientific prediction. In spite of this skepticism, investigations on this subject continue because of the great need for them and the driving force of scientific curiosity. In addition, the seismic forecasting is one of the most challenging research fields.

Many events presumably preceding earthquakes have been investigated as possible precursors. Here, we will analyze the available information and try to select the most appropriate phenomena (signs) associated with earthquakes for the purposes of long- and short-term prediction.

Long-term earthquake prediction is necessary to determine the safety standards of industrial and residential constructions for a region of interest. It is generally based on the accumulated statistical data about a location, the frequency and intensity of past seismic events, and their consequences. These describe the seismic environmental conditions (or seismic climate) under which people are expected to live in a given region in the future.

Short-term prediction is mostly based on assessing the pronounced anomalies in continuous observations of the related processes. Considerable efforts on earthquake prediction by scientists worldwide in the 1980s has led to understanding that precursors that reflect the general energetic state of a location of interest must be relied on to facilitate short-term prediction (Mogi, 1985). Numerous phenomena have been suggested and investigated as possible precursors; however, all practical attempts to develop a short-term forecast of seismic events has failed in the past.

We have analyzed the causes of this failure and outlined a plan for developing a methodology and technical system for the short-term prediction of seismic events. The basis behind this technique is founded in the links between (1) the patterns in the rate changes of vertical and subvertical gas migration from deep layers of the earth's crust (and from the oil and gas fields), and (2) accumulation, redistribution, and discharge of the stresses in the rocks.

The rate of vertical and subvertical migration of the natural gases as a precursor of major seismic events was mainly studied by a few Chinese and Russian scientists. Chinese seismologists stated that the rate would increase considerably on the eve of a major earthquake because of the generation and desorption of gases (mostly methane and carbon dioxide) in the very deep layers of the crust. This explanation seems unsatisfactory because, as the Russians demonstrated, the same process was observed repeatedly for shallow- and intermediate-depth earthquakes.

In a pioneering paper, Chilingar et al. (1996) pointed out that it is the changes in the rate (increase or decrease) which is of vital importance. Professor Chilingar and his associates at the University of Southern California independently developed the hypothesis that the rate changes were primarily caused by the alterations in the permeability of faults and fractured zones. They proved that natural gases do migrate along faults (Gurevich et al., 1993). The permeability is a very sensitive gauge, and monitoring the rates of the gas migration by measuring the concentration of various gases in time, presents the possibility of discovering considerable anomalies in stress changes prior to earthquakes. If the migration rates of various gases (and their concentrations) are measured continuously at many points (preferably along faults and fractured zones) of a chosen site, a gaseous portrait of the site can be constructed that in a certain way characterizes crustal distribution of stresses and their changes.

In many areas there are water wells (some abandoned). The concentration of various gases in water can be continuously measured. This allows for a convenient method of monitoring the rates of the gas influx to the surface. In addition, changing chemical composition provides important information on the gas migration. For example, influx of CO_2 will change the pH[1] (acidity) of the water:

$$CO_2 + H_2O \rightleftharpoons H_2CO_3$$

$$H_2CO_3 \rightleftharpoons H_2CO_3^- + H^+$$

---

[1]$pH = -\log_{10}[H^+]$

CHAPTER 7

Causes of Earthquakes

INTRODUCTION

What is the cosmic image of our planet? It looks as a solid body of approximately spherical shape (with flattened poles) moving in space. If one happened to live on the earth at the end of twentieth century, then one's experience and knowledge created an image of an uneven solid surface partially covered by water with possible changes in the surface elevations of approximately 10 km.

The earth, however, is not a solid body. Like a Russian doll (Matreshka) it consists of several somewhat spherical layers of matter inserted into each other. This layered structure was discovered due to experiments with the velocities of propagation of seismic waves. The velocity of seismic waves is a function of the physical properties of the materials through which waves are traveling. Abrupt changes in the seismic velocities indicate the boundaries between the earth's layers. The schematic diagram of the earth's body is presented in Figure 7-1. The upper solid layer of the earth's crust, 50–100 km thick, consists mostly of granite (the continental part of the earth's surface) and basalt (the oceanic part of the surface). This uppermost layer is called the "lithosphere." Deeper toward the center of the earth there is a layer consisting mostly of olivine (to a depth of 400 km) and spinel (to a depth of 700 km). This layer is a dense, hot, semi-solid rock approximately 2,900 km thick called the "mantle." The central part of the earth is called the "core," which is nearly twice as dense as the mantle because it is composed of iron-nickel alloy. It is divided into a liquid (2,200 km thick) outer core and inner solid core (1,250 km thick).

Estimated physical properties of the concentric earth's layers and their depths are presented in Table 7-1. The oceanic crusts are relatively

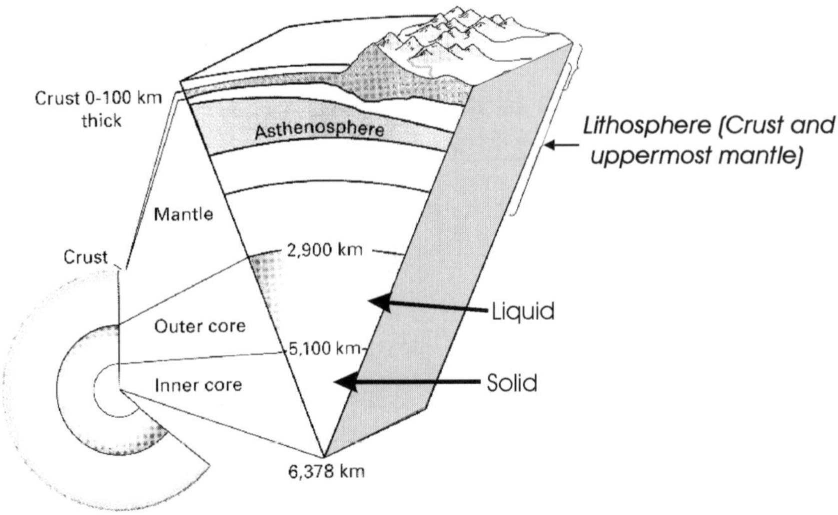

Figure 7-1. Cross-section of the earth's body showing the various layers of matter. *(Modified after Kious and Tilling, 1997, p. 4.)*

thin and seldom exceed the thickness of 10 km. Deep ocean trenches at continental margins and those associated with island arcs are the boundaries of subduction zones, places where the oceanic crust is subducted into the asthenosphere and melted.

The depth of the continental crust does not exceed 100 km as delineated by the Mohorovicic discontinuity (Moho). The Moho indicates the surface, which reflects seismic waves, where the seismic velocity suddenly increases. It is a sharp compositional boundary between the lighter crust (rich in feldspar) and denser mantle composed of pyroxene and olivine (feldspar is practically absent). The continental crust is lighter than the ocean crust. It consists primarily of granite (silica and aluminum) with an average density of 2.8 g/cm^3. The continental crust seems to float on the top of basaltic material (silica and magnesium) with an average density of 3.2 g/cm^3. This continental matter is not always consumed at the subduction zones; instead, it is uplifted into mountain ranges such as the Andes.

As previously stated, the thickness of the "solid" surface shell (the earth's crust) does not exceed 100 km. Taking into consideration that

Table 7-1
Estimated Physical Properties (Average) of the Earth Layers

Layer	Depth, km	Density, g/cm^3	P-velocity, km/s	S-velocity, km/s	Press. Grad., kbar/km	Temperature, °C
Crust	0–50	2.8–3.3	6.2–7.8	3.4–4.7	0–12	0–500
Lower lithosphere	5–50	2.8–3.3	6.4–8.0	4.0–4.8	1.2–12	125–500
Asthenosphere	100–150	3.1–3.4	8.0–7.8	4.8–4.3	28–44	1,200–1,300
Asthenosphere	150–200	3.4	7.8–8.2	4.5–4.4	44–61	1,300–1,500
Upper mantle	200–450	3–4–3.8	8.2–9.6	4.4–5.1	61–150	1,500–2,000
Upper mantle	450–650	3.8–4.1	9.6–9.9	5.1–5.4	150–223	2,000–2,200
Lower mantle	650–1,500	4.1–4.8	9.9–12.2	5.4–6.7	223–616	2,200–2,500
Lower mantle	1,500–2,900	4.8–10.0	8.1–12.2	6.7–0.0	616–1,404	2,500–3,500
Core (outer, liquid)	2,900–5,000	10.0–12.0	8.1–10.3	0.0	1,404–3,216	4,000–4,200
Core (inner, solid)	5,000–6,370	12.0–13.1	10.3–11.3	0.0–3.6	3,216–3,660	4,200-4,400

Modified after Dott and Batten, 1976; Decker and Decker, 1982, and Gubbins, 1990; in Chilingarian et al., 1995, p. 3.

the earth's diameter is approximately 12,000 km, it is easy to visualize that on a planet scale this crustal layer is relatively thin. It is difficult to maintain the crustal layer without faults and fractures (under the conditions of permanently moving matter of a hot dense core) because its thickness to the earth's diameter ratio is 1:120. As a comparison, chicken eggs have approximately the same shell thickness to the average diameter ratio (average thickness of the shell is 0.4 mm, whereas the average diameter is 50 mm).

The earth's crust, however, is not a continuous monolithic shell. Permanent matter and energy fluxes inside the hot liquid outer core create huge temperature stresses and pressure changes, which tear apart the earth's crust into large blocks called "crustal plates." Each plate consists of the crust and the more rigid part of thc upper mantle (lithosphere). The matter and energy fluxes inside the liquid core and variations in gravity force these pieces of crust to move. The direction of their movement and corresponding deformation of the crust are determined, primarily, by the direction of energy and matter fluxes. In Figure 7-2, two possible schemes of crustal plate movement are illustrated: along their boundaries and floating apart.

Distinct plates of the fractured lithosphere are carried along the surface of the mobile asthenosphere (from *asthenes*, the Greek word for "weak"). One prevalent theory is that convective currents of matter within the mantle cause the motion of the plates, which sometimes collide and sometimes move away from each other. The slow movements (about 2.5 cm per year) have been detected with seismic tomograms, which show the regions of ascending and descending currents in the asthenosphere and upper mantle. The principal source of the currents within the mantle is an uneven distribution of the temperature at the surface of the core that ranges from 4,000–5,000°C.

These considerations, supplemented with natural observations, have led to the plate tectonic theory or the theory of continental drift. This theory presents the earth's crust as twenty big pieces (tectonic or lithospheric plates) moving on the surface of the outer core.

According to this theory, there is an upwelling of red-hot molten matter from the mantle at the ocean ridges that stretches along the bottom of the ocean. These oceanic plates repeatedly collide with or subduct under the thicker continental plates. This permanent interaction of the plates produces areas with extremely high stress levels within the plates and especially in the vicinity of the plate boundaries.

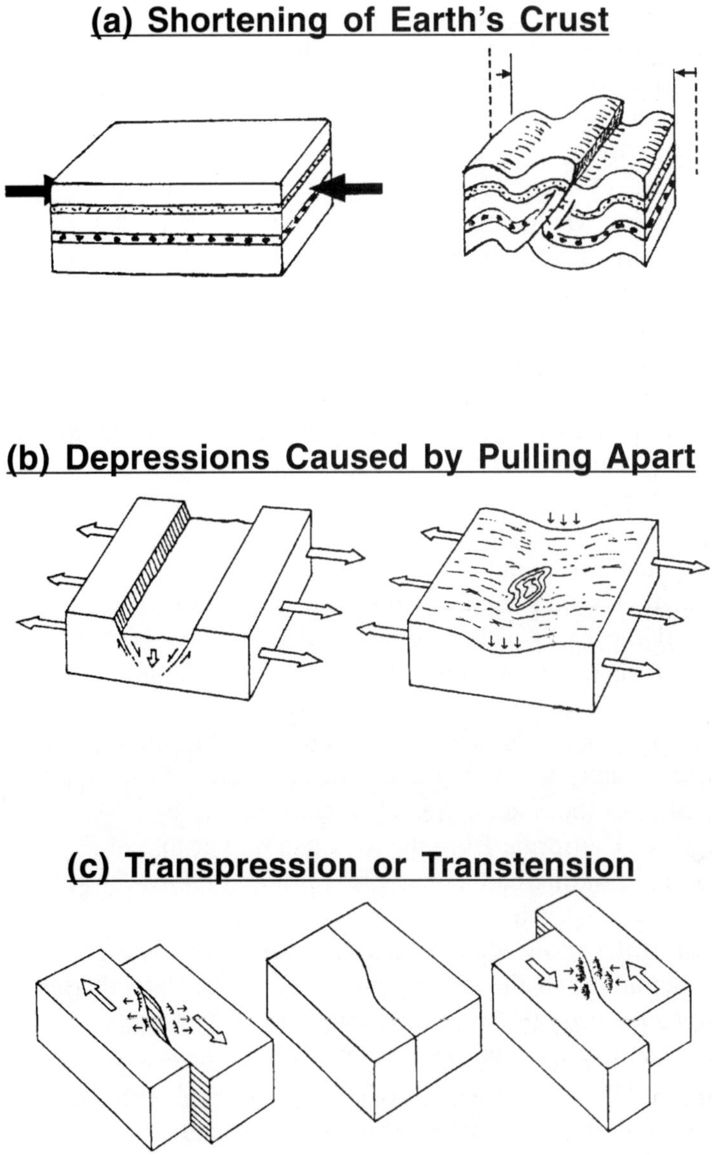

Figure 7-2. Possible types of crustal plate motion: (a) shortening of earth's crust, where rocks are folded and faulted resulting from compression; (b) depressions (grabens), where crust is pulled apart; and (c) transpression or transtension, where compression or extension is accompanied by transform motion along the plate boundary. *(Modified after R. S. Creely, 1997; in Harden, 1998, p. 13.)*

Possible prevailing directions of movement are shown in Figure 7-3. Because of this movement and frequent collisions, the plate boundaries exhibit extreme geological activity and represent the crustal areas where about 90% of tectonic events (earthquakes, mountain forming processes, and eruptions of volcanoes) occur.

Earthquakes occur (1) at the mid-ocean spreading ridges (e.g., East Pacific Rise), where ocean-floor spreading indicates that the plates are pulling apart (this occurs at shallow depths resulting in relatively small earthquakes); (2) at the subduction zones where one plate dives beneath another [deep-sea trenches form where the lithosphere converges (large percentage of the seismic energy is generated at the subduction zones)]; and (3) along the transform faults, where the plates move parallel to one another. The San Andreas Fault in California represents a transform fault where a horizontal sliding (north–south) occurs with a relative average rate of plate movement (with respect to each other) of approximately 2–5 cm per year.

Continual movement of the crustal pieces (blocks) creates extensive internal forces that squeeze and stretch rock formations of the outer layers of the earth. When, at a certain location, the stress becomes greater than the rock can withstand, the rock ruptures (fractures) producing an earthquake. These breaks in the crust are called "faults." Most of the faults lie beneath the earth's surface and are not visible. They are usually situated in the areas of intense seismic activity. Some of them (like the San Andreas Fault, extending over 700 km through the state of California from the Mexican border to San Francisco) can be observed on the earth's surface. Surface rupture can reach up to a distance of 10–15 m.

Most of the knowledge about what causes earthquakes was acquired at the beginning of the twentieth century. Since then, the basic theory has not changed. In 1921, a special meeting at the Commonwealth Club of California on earthquake prediction took place in San Francisco. Stanford's Professor of Geology, Bailey Willis, spoke on the possibility of future earthquakes (Willis, 1921):

> There will be a next earthquake. It will come in the course of accumulation of that elastic strain . . . at the time when the two great masses, that beneath the Pacific on one side and that beneath the continent on the other, which are pushing past each other, have accumulated the elastic strain sufficient to overcome the friction.

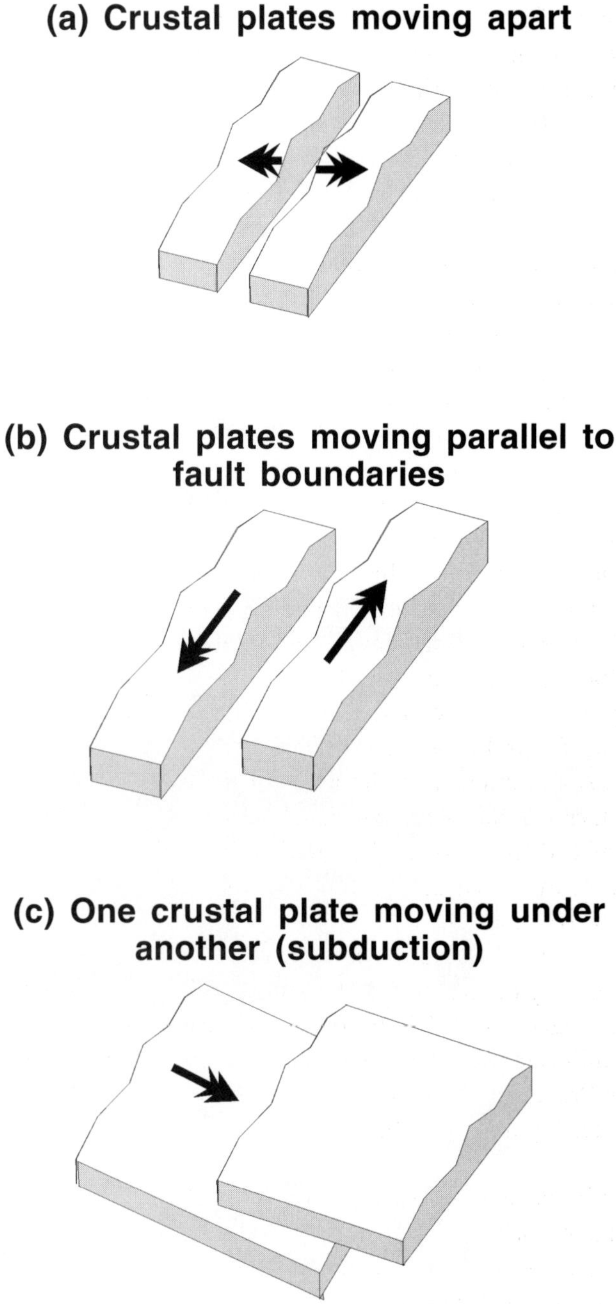

Figure 7-3. Possible directions of crustal plate movement.

It is impossible to observe and investigate the processes of earthquake development in the focal zone because it occurs unexpectedly and is of short duration. For this reason, scientists attempt to describe the sequence of events leading to an earthquake (or mechanisms of earthquake generation) by developing hypotheses or models of the focal processes. Reid made the first step in this direction in 1910. Investigating the consequences of the Great San Francisco earthquake of April 18, 1906, he formulated the theory of elastic relaxation. This theory stated that the elastic strains accumulated at certain places along the boundaries of great faults (the San Andreas Rift, in particular). When the accumulated strain exceeds the level that rock can withstand (or exceeds the friction forces between the rock blocks), the rock ruptures and movement occurs producing the blow that releases seismic energy in the form of waves (see Karapetyan and Karapetyan, 1981).

Comparing this statement with current theoretical developments, the only difference is that the word "elastic" is omitted in the contemporary theory. Actually, elastic deformation relates to the situation when the primary form and size of a rock sample are restored after the force causing the strain ceases, which is not typical for crustal rocks.

SIMPLE MODELS FOR EARTHQUAKES

The Great San Francisco earthquake of 1906 was obviously produced by the "slip-stick" sliding of the North American and Pacific tectonic plates along the San Andreas Fault. As a result of the stress release, the opposite sides of the fault shifted with respect to each other for a distance ranging from 3–6 m along the line of rift. Fortunately, geodetic measurements were made: prior to and immediately after the earthquake, which made it possible to measure the abrupt shift in the positions of the tectonic plates as a result of this earthquake.

A schematic illustrating development of the rupture during the San Francisco earthquake is shown in Figure 7-4. Suppose that line *AB* represents the San Andreas Fault. Draw the control line *CD* (a benchmark) perpendicular to *AB*. Half a plane *ABC* represents the Pacific Plate that moves gradually to the north, whereas half a plane *ABD* represents the North American Plate that slides past the Pacific Plate to the south. The arrows indicate the direction of movement of the plates.

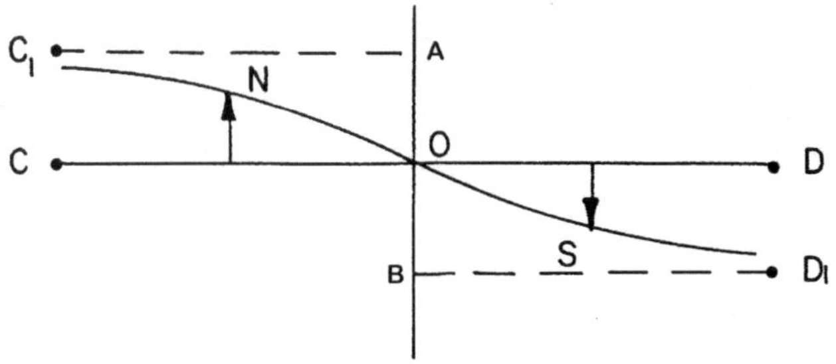

Figure 7-4. Fault development during the great earthquake of San Francisco (1906). *(Modified after Karapetyan and Karapetyan, 1981.)*

Because of the friction at their boundaries, the relative motion of plates is not constant and can slow down or even completely stop for a while ("for a while" can last for 10–20 years, which is a short time period on the geological time scale). The forces causing the motion, however, continue to be applied. This results in an accumulation of stresses at the contact surfaces of the plates. If the deformation were elastic over a broad range of deformation, then after, for example, 100 years the line CD would bend to become curve C_1OD_1.

For example, suppose that this position was reached on the morning of April 18, 1906. At this time, the accumulated strains had exceeded the level that the consolidated rocks in the area of contact could withstand. As a result, these rocks ruptured. The curve C_1OD_1 was then transformed into two-half lines: C_1A and BD_1. The accumulated energy in the area of contact was released as an abrupt shift of the rock blocks that, upon termination, produced an enormous blow, generating seismic waves and, thus, the earthquake.

The small area around the hypocenter in which primary rupture occurs may be called the "focus" of an earthquake. Sometimes the focus is simply identified with a geometric point of the hypocenter location. The accumulated energy is released at the focus of an earthquake, not only in the form of mechanical ruptures and seismic waves, but also in many other forms, such as: irreversible deformation of rocks, repositioning of rock blocks, heat generation, chemical transformations, and transport of underground fluids. Thus, if one intends to scrupulously describe a seismic rupture, all of these aspects must be taken into consideration.

The broad application of the term "mechanism" in scientific developments relates to all types of studies from physics to psychiatry to describe the essence of a phenomenon of interest. In a general scientific glossary, it refers to a succession (sequence) of essential events related to the phenomenon and the links among them. Despite the fact that the term "mechanism" has been severely criticized, it is widely used for the purposes of describing the models of a phenomenon of interest.

A possible mechanism for the formation of a crack under tensile stress is presented in Figure 7-5. Consider a possible mechanism of the crack's development in a brittle material under compression. When a compressive stress is applied to a piece of brittle material (Figure 7-6), it creates perpendicular tensile stresses, which tear the sample apart. As the brittle sample experiences increasing stress, micro-cracks open parallel to the direction of compression. If the stress continues to increase, the multiple micro-cracks appear, link up, merge, and form a single open crack (fault). Movement then occurs along this fault. This sliding is accompanied by a blow when the movement is suddenly terminated. This blow is felt as an earthquake.

Energy generated by the rupture and following the shaking is primarily released in the focal zone. There have been many attempts to compose

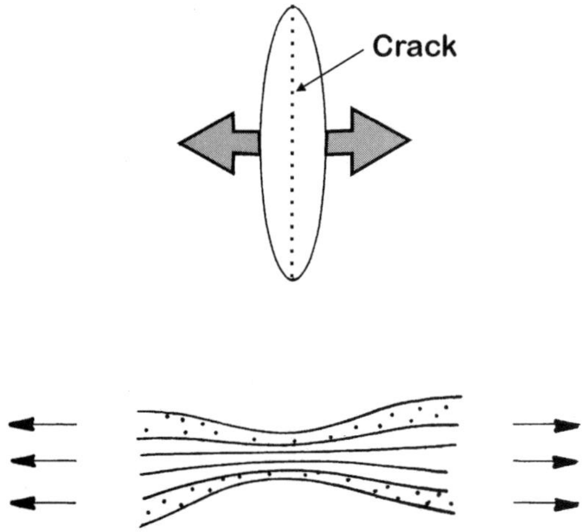

Figure 7-5. Crack development under tensile stresses. Arrows indicate the directions of applied stresses.

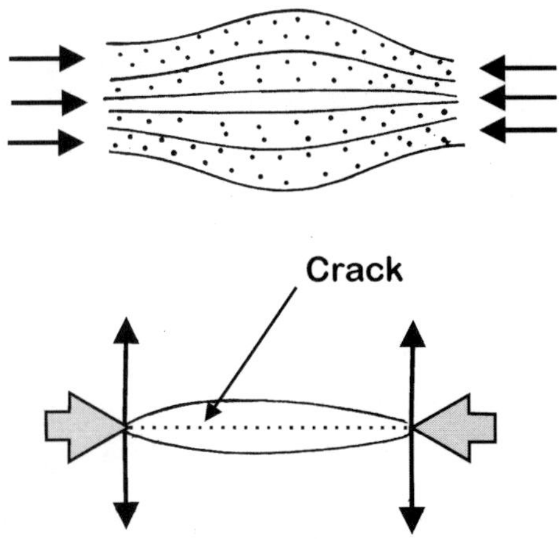

Figure 7-6. Crack development under compressive stresses. Arrows indicate the directions of applied and created stresses.

simple mechanistic models of the focal zone for describing and investigating the causes and effects of earthquakes. In 1923, Japanese seismologist Nakano suggested a basic principle for the construction of a theoretical model of the focal zone. The principle is based on producing the same-modeled dislocation fields for *P* and *S* waves as for the observed real fields (see Karapetyan and Karapetyan, 1981).

In 1932, using Nakano's principle, Japanese seismologist Honda developed a theoretical model of the focal zone based on the lucid mechanical considerations. He supposed that the focal core could be presented as some sphere of small diameter with stresses applied to its surface (Figure 7-7). He divided an imaginary sphere by the two mutually perpendicular planes going through its center, one of which is supposed to contain a plane of rupture. Thus, the sphere was divided into four parts. Alternating tensile and compressive forces were applied to all quarters of the sphere producing vibrations. For the source of vibrations, Honda composed the fields of dislocations and showed that the composed fields were similar to the fields of dislocations for *P* and *S* waves observed in real earthquakes (see Karapetyan and Karapetyan, 1981). Later, a series of generalizations and specifications were made based on the Honda model, although the basic principle of construction remained unchanged: generation of the same distribution of dislocations as for a real earthquake.

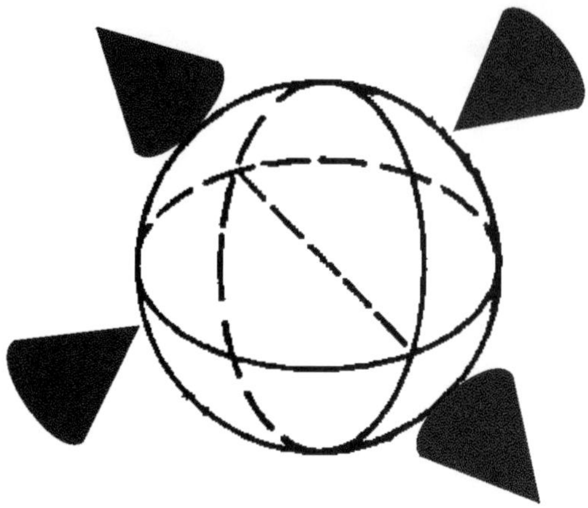

Figure 7-7. Earthquake focal model. *(Modified after Karapetyan and Karapetyan, 1981, fig. 4, p. 47.)*

Crustal Structure and Rock Mechanics

Some crustal features that help explain the development and occurrence of earthquakes are discussed here. The earth's solid shell is a complex crustal mixture of rocks and fluids experiencing permanent matter-energy transformations in geologic time-scale. More precisely, it is a set of porous rock pieces containing migrating fluids, which are kept together by gravitation and lateral stresses.

Developments in geotectonics (Sadovsky, 1979) have shown that the earth's shell is composed of separate pieces of rock of various dominant sizes (Figure 7-8). The full spectrum of sizes of rock fragments forming the earth's crust is presented in Figure 7-9. Large blocks of the earth's crust having characteristic size of 10–100 km (L_2) are usually involved in the development of strong quakes together with separate small 200–1000 m (L_1) rock blocks (tectonic blocks). The latter are probably responsible for generating the most destructive seismic waves of high frequencies when an earthquake occurs.

Under increasing stress, this mixture of rock blocks having various sizes are likely to disintegrate and rupture at the boundaries of large blocks with characteristic sizes (which are multiples of the sizes of the separate small blocks). This hypothesis can be partially justified

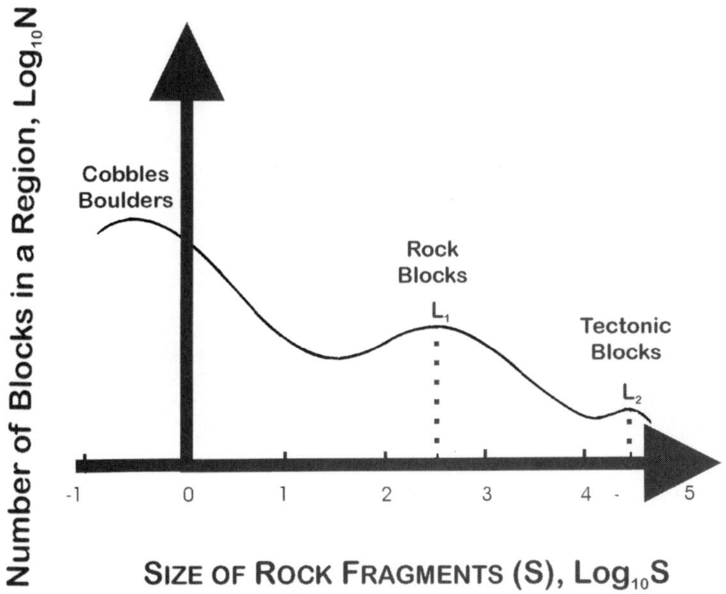

Figure 7-8. Relationship between the number of blocks versus the size of blocks in a region.

by the experiments conducted in the 1970s in the Siberian branch of the Soviet Academy of Sciences. In these experiments, the scientists applied shear stress to a uniform and isotropic layer of sand. Under the uniform shear stress, the layer of sand disintegrated into the smaller sliding layers of approximately the same thickness (multiple of the grain size), and the deformation was concentrated at the boundaries between the smaller sliding layers. If it is taken into account that the primary rock blocks have the characteristic size L_1, then it is plausible that with respect to the whole crustal block of the size L_2, these pieces will behave like the sand particles in the Siberian experiments. Consequently, the earthquake rupture will probably occur at the boundaries of these primary blocks (plates).

Myriads of silent microquakes occur continuously throughout the earth's crust. They are caused by the nonuniformity of distribution of stresses in rocks and their abrupt change. The resulting microfracturing is due primarily to disequilibrium between the total overburden pressure and formation fluids pressure.

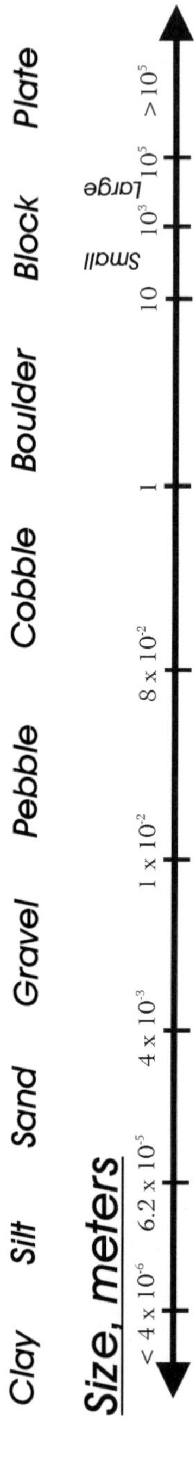

Figure 7-9. Size classification of rock fragments.

Sedimentary crustal rocks are often porous. Much like a sponge filled with water, the rock's voids contain mixtures of fluids. In the earth's crust, the rocks experience pressure due to the weight of overlying dry rocks and the weight of fluids filling the pores (pore or fluid pressure p_p). The total overburden pressure (p_t) is equal to the fluid (pore) pressure (p_p) plus the grain-to-grain stress (effective pressure p_e):

$$p_t = p_p + p_e \qquad 7\text{-}1$$

The compaction of sedimentary rocks is due to the effective stress. The degree of compaction is determined by many factors: lithology (type of rocks), porosity, initial thickness, rate of loading, type of clay minerals, particle size and shape, chemistry of interstitial fluids, temperature, etc. (for details see Rieke and Chilingarian, 1974; Chilingarian and Wolf, 1975).

The compaction process changes the distribution of stresses in the rocks: at some places they accumulate, at others they relax. Local relaxation of stresses in the rocks occurs due to fracturing.

The abnormal formation pressure of fluids (in excess of hydrostatic) may generate fractures (hydraulic fracturing). Generally, it is assumed that the total overburden pressure uniformly increases with depth. For example, average Tertiary deposits exert an overburden pressure gradient of 1 psi/ft of depth (0.231 kg $cm^{-2}m^{-1}$) (Chilingarian and Wolf, 1975). This corresponds to the pressure exerted by a formation with an average bulk density of 2.31 g/cm^3 (144 lb/ft^3).

The average depth of crustal earthquakes is about 10 km, and the frequency of quakes changes with depth. If, for example, the bulk density is 2.31×10^3 kg/m^3, then the overburden pressure at a depth of 10 km is estimated at 4067 kg/cm^2. In oil and gas production operations, the hydraulic fracture may occur at half of the total overburden pressure. Therefore, at a depth of 10 km, hydraulic fracture can occur at a formation pressure of 2033 kg/cm^2, which is a realistic estimate for the normally-pressured formations. If the rupture occurs in the vicinity of an existing fault or at the boundary between tectonic blocks (already overstressed), then this small rupture (cause) can trigger a large earthquake. A large earthquake as defined in Table 4-1 refers to earthquakes with a Richter magnitude greater than $M_R = 6$.

Under continual application of stresses to the zone of rupturing, more fractures are generated. Small fractures tend to merge, creating faults along which the tectonic blocks of the crust can move until the collision with the other tectonic blocks terminates this movement. Kinetic energy of movement dissipates at the place of collision in the form of seismic waves, rock deformation, and frictional heat generation.

In the process of block dislocation one can distinguish at least two stages of energy transformation: (1) transition of potential energy of accumulated stresses into kinetic energy of block (blocks) movement, and (2) dissipation of kinetic energy in the process of collision in the form of nonelastic deformation, heat transfer, and seismic elastic waves that are perceived at the earth's surface as an earthquake.

Thus, the underground realm is very unstable: a small rock rupture (cause) can trigger a chain reaction of rock fracturing and dislocation of the tectonic blocks, which generate large earthquakes.

Deep-Focused Earthquakes

Most earthquakes occur within 70 km of the earth's surface. As previously discussed, the process of earthquake development occurs at these depths. The mechanism of rupture of rocks illustrated in Figures 7-5 and 7-6 is frequently called "brittle-fracture" and "friction-sliding." This mechanism explains the origin of earthquakes that occur within 70 km of the earth's surface.

Almost 30% of earthquakes occur at depths greater than 70 km (Green, 1994). Eight percent of earthquakes occur below a depth of 300 km, in the molten hot medium of the mantle at a temperature of about 2,000°C and enormous pressures ranging from 10–100 gigapascals. These earthquakes are called deep earthquakes. They are very powerful and are felt over vast territories. The causes of "deep" earthquakes (below 300 km) are quite different from a "crustal" rupture. Under very high pressures and temperatures, brittle rocks do not exist at these depths. Thus, the "brittle-fracture friction-sliding" process cannot develop and earthquakes at these depths are very difficult to explain (Green, 1994).

For years, deep earthquakes have been the most challenging puzzle of seismic geology. Recently, Professor Harry W. Green and Dr. Pamela

C. Burnley (1989) of the University of California, Riverside, developed and tested a plausible hypothesis that deep-focused earthquakes occur due to matter densification, development of the so-called anticracks, and fast sliding masses of matter along these anticracks. The term "anticrack" is used by researchers to name the lenses of denser plastic material inside the mass of mantle.

Deep earthquakes occur exclusively in the subduction zones. In these zones, a block of cold uppermost layer of earth (the lithosphere) sinks into the hot mantle. This creates an upward flow of the liquid, hot mantle matter, which solidifies in the form of ridges on the ocean floor.

The distribution of earthquake frequencies based on depth is shown in Figure 7-10. The frequency of seismic events steadily declines beginning at 70 km to a depth of 300 km. This suggests that they originate by the mechanism of brittle rupture. Below a depth of about 300 km, the frequency of events increases, reaching a maximum at a depth of approximately 600 km. Below this depth, earthquake frequency rapidly declines. Earthquakes do not occur below a depth of 700 km.

Local densification and plastification of the mantle matter are two processes that, combined, create the conditions necessary for deep earthquake occurrence. The most abundant mineral in the upper part of the mantle is olivine. At a depth of about 300–400 km, under enormous pressure, olivine becomes unstable and changes into spinel, which is 6% denser than olivine. At a depth of 660 km, spinel becomes unstable and transforms into two phases, which are 8% denser.[1] The depth of stability of spinel ranges from 300 to 700 km, the interval in which all deep earthquakes occur.

---

[1]Chilingar and Terry (1954) showed that dolomitization (i.e., replacement of calcite (sp. gr. = 2.71) by dolomite (sp. gr. = 2.87) during diagenesis and catagenesis) created porosity of 13.1% and mircofractures:

$$\underset{\text{(Calcite)}}{CaCO_3} + Mg^{++} \rightarrow \underset{\text{(Dolomite)}}{CaMg(CO_3)_2} + Ca^{++}$$

The above-described findings showed that even a small degree of densification could cause significant changes in the properties of the rocks (i.e., increase in porosity (13.1%) and microfracturing).

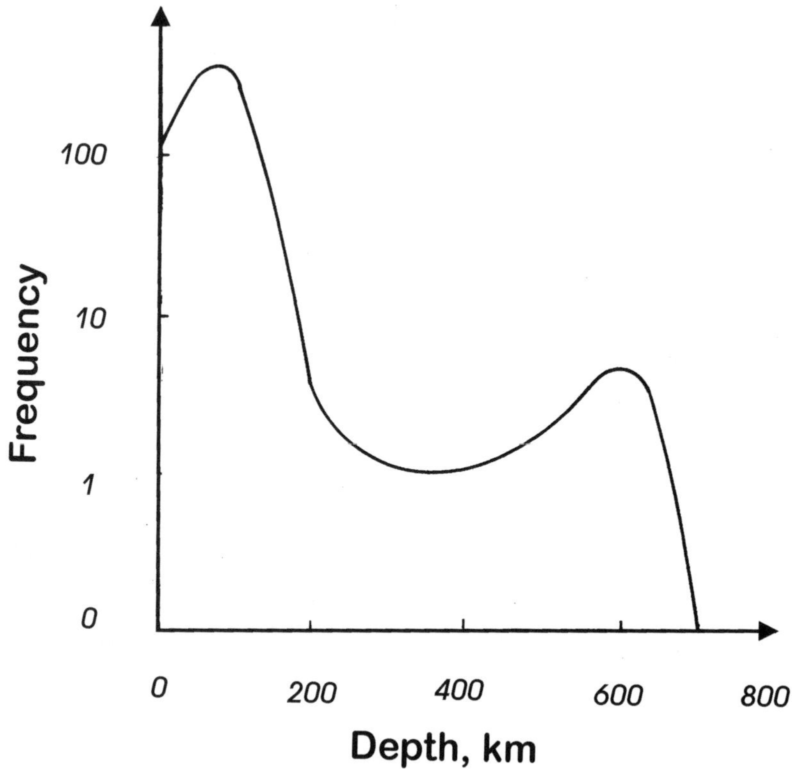

Figure 7-10. Distribution of earthquakes with respect to the earth's depth. *(Modified after Green, 1994, p. 67.)*

Because of the correlation between the degree of densification and earthquake frequency, researchers hypothesized that the above-described matter transformations (olivine to spinel) were responsible for deep earthquakes. At first, it was thought that densification of a large mass of mantle matter created implosion (a sharp shrinking of the matter) that, in turn, radiated energy waves felt as an earthquake. Later investigations, however, showed that the patterns of seismic waves were indistinguishable from those of shallow earthquakes and, consequently, were generated by the mechanism similar to the "brittle fracture and friction sliding."

A long series of experiments with olivine-spinel transformations conducted by Green and Burnley (1989), Green et al. (1990), and Iidaka and Suetsugu (1992), revealed important properties of these materials that depicted plausible picture of deep earthquake occurrence. They discovered that under very high pressure and temperature, olivine transforms into heavier spinel. In addition, the fine crystals of spinel (approximately 10^{-5} mm in diameter) pack together to form an extremely plastic anticrack fault, along which large masses of viscous matter can slide producing a great impact. Moreover, anticracks filled with the spinel crystals are oriented perpendicularly to the direction of applied stresses.

From these three discoveries (higher density, extreme plasticity, and normal orientation of the compact sets of spinel crystals with respect to the direction of applied stresses), the theory of transformation-induced faulting was developed. This process of fault development followed by shocks is similar to the brittle fracture formation; however, it differs considerably in its microphysics. When a high pressure is applied to the olivine crystals, the spinel clusters (spinel lenses) begin to grow. The spinel lenses have shapes similar to those of open tensile cracks; however, they are perpendicular to the applied stress. Thus, these features are totally opposite to the properties of the brittle cracks developed under the stresses in the upper layer of mantle. For this reason, Green and Burnley (1985) called this process "anti-crack faulting."

Brittle fracture is the primary cause for earthquakes with a depth of focus up to 70 km. Olivine-spinel transformation accounts for deep earthquakes below 300 km; above this depth, olivine is stable. Therefore, the question remains: what mechanism is responsible for intermediate earthquakes in the depth range of 70–300 km?

When olivine reacts with water under low temperature and pressure, it forms the hydrous mineral serpentine. When the latter sinks into deep layers of the mantle, it dehydrates at high pressure and temperature. The process of dehydration is accompanied by a large emission of acoustic energy. At very high temperatures and pressures, H_2O steam formed in the process of dehydration exerts an enormous force on the surfaces of the crack, which widens further with resulting shaking (see Green, 1994). In conclusion, examining the curve of the earthquake distribution with depth and the experimental results on transformation of olivine into spinel, it can be deduced that there are three mechanisms responsible for the occurrence of earthquakes at different depths: (1) Shallow earthquakes (less than 70 km in depth) are produced by brittle fracture mechanism as a response of solid fragile medium to accumulated stress. (2) When a tectonic plate subducts to a depth below 70 km, the water-containing minerals are dehydrated producing additional pressure due to the generation of steam that enhances the forces producing the brittle fractures. This amplifies the usual brittle-fracture friction-sliding mechanism. (3) Below a depth of 300 km, the unstable transformation of olivine into spinel is the primary cause for deep-focused earthquakes produced by the anticrack-sliding mechanism (Green, 1994).

CHAPTER 8

Magnitude and Intensity of Earthquakes

INTRODUCTION

The process of rupture originates, as a rule, in a small area inside the earth's crust, called the "focus" or "hypocenter." A hypocenter is usually located within 50 km of the earth's surface. There are records on deep earthquakes, however, with hypocenters located at depths ranging from 100–700 km. Projection of the hypocenter on the earth's surface is called the "epicenter," and is located on the earth's surface by its geographic coordinates. Mechanical energy released at the focus is transmitted through the crustal medium mostly in the form of seismic waves. There are three characteristic types of seismic waves: *P* (primary) waves, *S* (secondary) waves, and surface (Love and Rayleigh) waves.

THE NATURE OF WAVES

P Waves (Longitudinal Waves)

For longitudinal waves, the direction of propagation coincides with the direction of vibration of the medium. They are responsible, in particular, for the terrifying sounds accompanying large earthquakes. These waves travel at about 5 miles per second and force the crustal rock to vibrate along the direction of wave travel. These vibrations cause the rocks to push together (compress the blocks) and rip apart (dilate the blocks) (Figure 8-1). Because of their soundlike nature, when *P* waves emerge at the surface from deep inside the earth, a

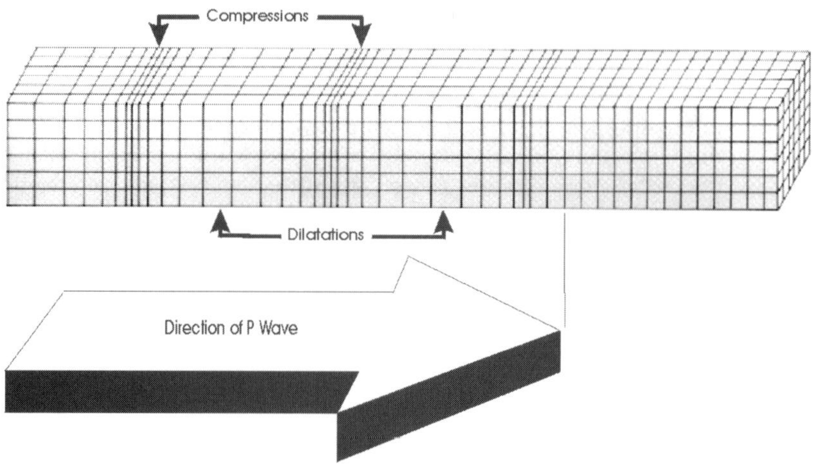

Figure 8-1. Diagram illustrating the forms of ground motion near the ground surface for *P* waves. *(Modified after Bolt, 1988, fig. 6, p. 29, and Bolt, 1976.)*

fraction of their energy may be transmitted into the atmosphere as sound waves (Bolt, 1988).

S Waves (Shear Waves)

Sometimes referred to as secondary waves, the direction of displacement of a particular point of the crust space due to *S* waves is perpendicular to the direction of propagation. They spread in the medium at a velocity of about 2.5 miles per second. Direction of vibration forms a right angle with the wave traveling direction. These vibrations cause the rocks to shake from side to side (Figure 8-2). This wave produces the wave motion (shaking) that is very damaging to structures at the earth's surface.

Surface Waves (Love and Rayleigh Waves)

Surface waves can be divided into two types. The first type is called the Love or *L* wave. *L* waves only move along the earth's surface and travel slightly slower than *S* waves. Their motion is similar to the *S* waves in that they have no vertical displacement; it moves the ground

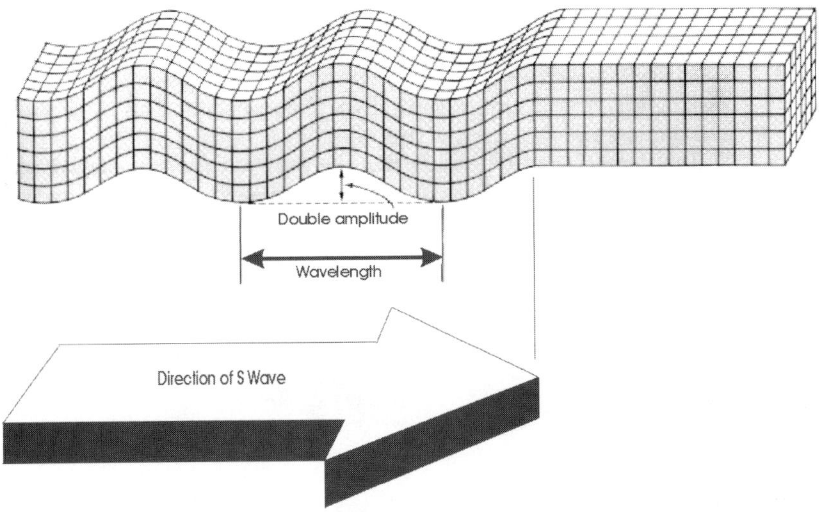

Figure 8-2. Diagram illustrating the forms of ground motion near the ground surface for *S* waves *(Modified after Bolt, 1988, fig. 6, p. 29, and Bolt, 1976.)*

from side to side in a horizontal plane, but at right angles to the direction of propagation (Figure 8-3). This type of wave is distructive to foundations of structures (Bolt, 1988), because it causes horizontal shaking of the earth's surface.

The second type of surface wave is known as a Rayleigh or *R* wave. The pieces of material disturbed by a Rayleigh wave move both vertically and horizotally in a vertical plane pointed in the direction in which the waves are traveling (Figure 8-4). The arrows in the figure illustrate the eliptical movement of a piece of rock as the wave passes.

CHARACTERISTICS OF EARTHQUAKES

Magnitude of Earthquakes

To record seismic waves and measure their intensity, scientists use special measuring devices called "seismographs." Seismographs record the vibrations of rock at the place of installation. Amplitude of the vibrations is determined by measuring the maximal height of deflections (during one period of vibration). Gathering the records of many

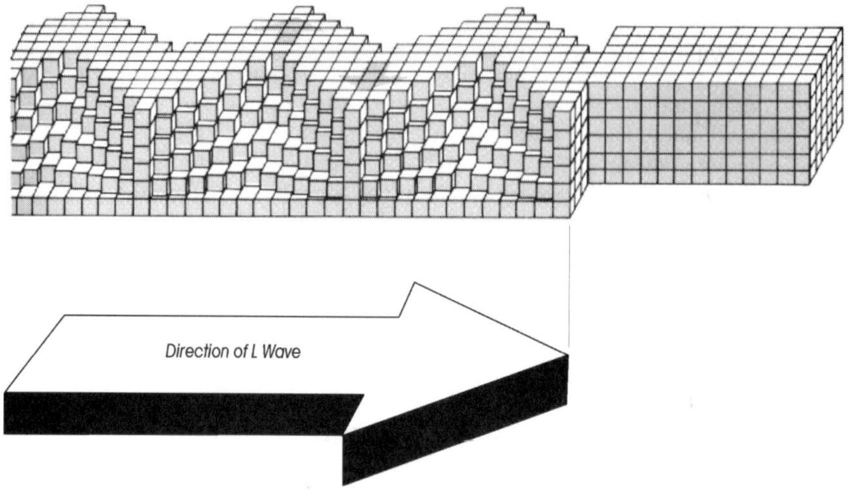

Figure 8-3. Diagram illustrating the forms of ground motion near the ground surface for *L* (Love) waves. *(Modified after Bolt, 1988, fig. 6, p. 29, and Bolt, 1976.)*

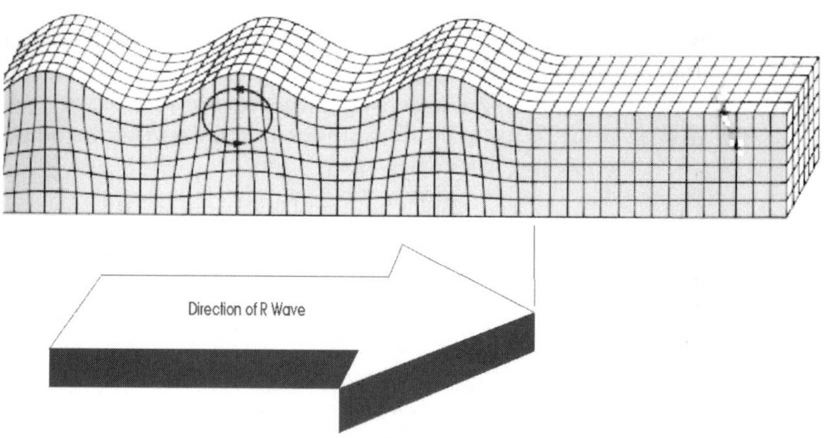

Figure 8-4. Diagram illustrating the forms of ground motion near the ground surface for *R* (Rayleigh) waves. *(Modified after Bolt, 1988, fig. 6, p. 29, and Bolt, 1976.)*

seismograph stations, scientists can investigate the propagation of seismic waves and study the distribution of seismic energy in space and time.

The main integral characteristic of an earthquake's power is "magnitude." Magnitude is determined by the amount of energy released at the hypocenter. The magnitude on the Richter scale is not limited at the top or bottom. The highest recorded magnitude was $M_R = 8.9$ and the lowest, $M_R = -3$. By using the logarithmic scale, one unit of increment in magnitude corresponds to a 30-time increase of released energy at the focus of an earthquake.

To characterize the effect of an earthquake at the epicenter, the parameter called the "intensity" is commonly used. Intensity is a relative measure of the strength of an earthquake on the earth's surface, based on the scale of destruction and human sensations. Seismologists use four relative scales to characterize the surface effect of earthquakes.

The two major scales of measuring earthquake magnitude are the Mercalli and the Richter. The Mercalli scale ranges from I (weakest) to XII (strongest) and represents the relative degree of released energy in terms of surface effect on structures and human sensations. Its description is given in Table 8-1 (Giusti, 1995). As shown, it is a qualitative rather than quantitative measuring system.

The Richter scale is a quantitative measure of earthquake magnitude. It requires, however, instrumental measurements and the use of special formulas to obtain the final result. This scale is usually applied to evaluate small- to moderate-intensity quakes. The surface-wave and body-wave magnitude scales are applied to evaluate the scope of moderate to large earthquakes (Bolt, 1988). For the strongest seismic events, the moment magnitude scale is used to evaluate their power.

Because the Richter scale is used almost everywhere it should be considered in detail. Formal definition includes two measurements at two different locations of maximum amplitudes of seismic waves (e.g., A_1 and A_2) by horizontal seismographs of the Wood-Anderson type situated 100 km from the epicenter of an earthquake. Amplitudes A_1 and A_2 must be measured in microns (μm). Average amplitude, A, of these two measurements $\left[A = \left(\frac{A_1 + A_2}{2}\right)\right]$ is calculated, and $M = \log(A)$ is taken as the Richter magnitude of a particular earthquake (see, for example, Karapetyan and Karapetyan, 1981).

Table 8-1
Mercalli Earthquake Intensity Scale

I	Usually detected only by apparatus. Not felt, or rarely felt by people. Sometimes felt when the area is located a great distance from a large quake.
II	Felt indoors by a few, especially on upper floors, or by sensitive persons. Delicately suspended objects may swing. Sometimes, dizziness or nausea is experienced.
III	Felt indoors by many, with usually a rapid vibration. The vibration is like the passing of light trucks. Hanging objects might swing slightly.
IV	Felt indoors by most, by a few outdoors. Vibration is like that of the passing of a heavy truck. Dishes and windows rattle, walls creak, hanging objects swing.
V	Felt indoors by all, outdoors by most. Buildings trembled, broken dishes and glassware, cracked windows in some cases. Occasional falling of unstable objects. Hanging objects swing considerably. Opened or closed doors and shutters. Moved small objects. Spilled liquids from filled containers. Trees and bushes slightly shaken.
VI	Felt by all. Some frightened. Person's movement is unsteady. Trees and bushes shaken moderately. Small bells, like church and school bells, ring. Slight damage in poorly constructed buildings. Small amounts of plaster fall. Some chimneys have fine cracks. Broken dishes and glassware, also some windows. Overturned furniture in some instances.
VII	All frightened. Some found it difficult to stand. Noticed by persons driving cars. Trees and bushes shaken strongly. Waves on bodies of water like ponds. Turbidity of water increased from mud stirring up. Some incaving of sand stream banks. Rang large bells. Slight damage in well-constructed buildings, considerable damage in poorly built buildings. Some chimneys cracked extensively, also walls cracked. Considerable fall of plaster. Numerous windows broken.
VIII	All frightened, some near panic. Trees shaken strongly with branches and trunks sometimes broken off. Springs and well water flow changes, altered temperatures of waters. Slight damage to buildings made to withstand quakes, considerable damage to buildings of ordinary construction. Some buildings collapsed. Chimneys twisted and collapsed.

IX	Most in panic. Cracked ground. Considerable damage to most buildings. Some buildings collapsed, damage to reservoirs. Underground pipes cracked in some instances.
X	Cracked ground, especially when loose and wet, with cracks up to several inches wide. Larger fissures run parallel to canal and stream banks. Considerable landslides. Changed water level in wells. Serious damage to dams. Some wooden buildings destroyed. Bent railroad rails, slightly torn apart pipelines underground. Open cracks in cement work and roads.
XI	Widespread ground disturbances: Fissures, cracks, land slips, etc. Sea waves, tidal waves formed. Severe damage to wood-frame houses, especially nearest the epicenter. Great damage to dams and bridges. Ruined many buried pipelines, bent railroad rails greatly.
XII	Total damage, nearly all construction works greatly damaged. Numerous shearing cracks in the ground. Landslides, rock falls are widespread. Fault slips in firm rock, geologic displacements observed. Greatly altered water channels, surface and underground, disrupting flow patterns. Dammed lakes and ponds. Waves seen on the ground, distorted levels and lines of sight. Objects thrown in the air.

Modified after Giusti, 1995.

Theoretically, the Richter scale can grade the magnitude by any real number from $-\infty$ to ∞. For example, if $A = 0.1$, then $M_R = \log(0.1) = -1$. If the magnitude was reported as $M_R = 7.0$ on the Richter scale, then the average amplitude of the seismograph dislocations should be 10,000,000 microns or 10 meters. This sets certain limitations on the use of this scale in direct form and, therefore, is only convenient for evaluating small to moderate earthquakes.

In order to avoid unnecessary confusion, seismologists and news media widely use the term Richter scale regardless of what scale is actually applied to evaluate the magnitude of the seismic event. The Landers and Big Bear earthquakes of June 28, 1992, were graded $M_{SM} = 7.6$ and $M_{SM} = 6.7$, respectively, on the surface-wave magnitude scale (Giusti, 1995), and this gradation was reported in the news media as the Richter scale magnitude. In the meantime, the evaluation on the moment magnitude scale gave $M_M = 7.3$ and $M_M = 6.2$ for these events, respectively. Table 8-2 gives a comparison between the Richter and Maximum Intensity scales.

Occurrence Time and Location of an Earthquake Hypocenter

Time of occurrence of earthquake and location of the hypocenter can be determined as explained below. Inasmuch as the *P* and *S* waves are propagated with different velocities (approximately 5 miles/sec and 2.5 miles/sec, respectively), for any particular seismic station, it is possible to determine the arrival times for each of these wave types. The difference between the times of arrival is proportional to the distance from a hypocenter to the station. The simplest method of evaluating these important parameters (time and location) is based on measurements (estimates) of the difference of travel times for

Table 8-2
Approximate Relation Between Richter Magnitude (M_R) and Maximum Intensity (M_M)

Richter Magnitude, M_R	Maximum Intensity, M_M	Typical Effects
2.0 and under	I–II	Not generally felt by people
3.0	III	Felt indoors by some people; no damage
4.0	IV–V	Felt by most people; objects disturbed; no structural damage
5.0	VI–VII	Some structural damage, such as cracks in walls and chimneys
6.0	VII–VIII	Moderate damage, such as fractures of weak walls and toppled chimneys
7.0	1X–X	Major damage, such as collapse of weak buildings and cracking of strong buildings
8.0 and over	XI–XII	Total or nearly total damage

Modified after Gere and Shah (1984); in Yeats et al., 1997, Table 4-1, p. 73.

longitudinal and shear waves (*P* and *S* waves) for at least four stations not lying on the same straight line.

For example, consider a station situated at point *A* as shown in Figure 8-5. Assume that *P* and *S* waves are propagated in the crustal medium from the hypocenter at the constant velocities V_P and V_s, respectively. Suppose that a hypocenter of an earthquake is located at point *H*, the time of arrival to *A* of *P* waves from point *H* is t_p, and the time of arrival of *S* waves to *A* from point *H* is t_s. These travel times are not known; however, it is possible to determine the moments of arrival of the *P* and *S* waves to the station *A* and, consequently, it is possible to calculate the difference $(t_p - t_s)$. It is then possible to evaluate the distance d_A from point *A* to point *H*. This method is based on the following series of simple relations:

$$d_A = V_p t_p = V_s t_s = \left[V_s t_p + V_s\left(t_s - t_p\right)\right]$$

$$(V_p t_p - V_s t_p) = t_p\left(V_p - V_s\right) = V_s\left(t_s - t_p\right)$$

$$\frac{t_p}{\left(t_s - t_p\right)} = \frac{V_s}{\left(V_p - V_s\right)}$$

$$t_p = \frac{V_s\left(t_s - t_p\right)}{\left(V_p - V_s\right)} \tag{8-1}$$

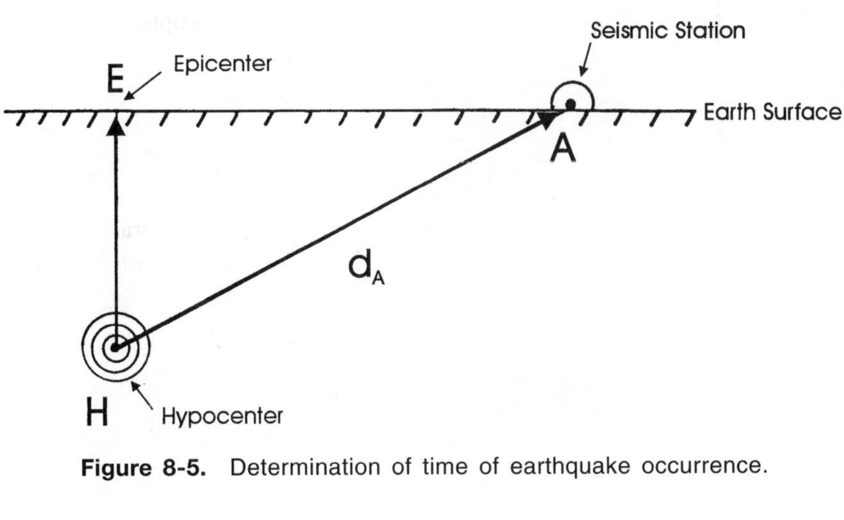

Figure 8-5. Determination of time of earthquake occurrence.

Introducing the notation $K_v = \dfrac{V_s}{\left(V_p - V_s\right)}$, a simple expression can be obtained:

$$t_p = K_v(t_s - t_p) \tag{8-2}$$

Therefore:

$$d_A = K_v V_p(t_s - t_p) \tag{8-3}$$

The graphs of Eq. 8-2 (so-called Wadati's graphs, Figure 8-6) are plotted for all seismic stations. They can be used to determine the time

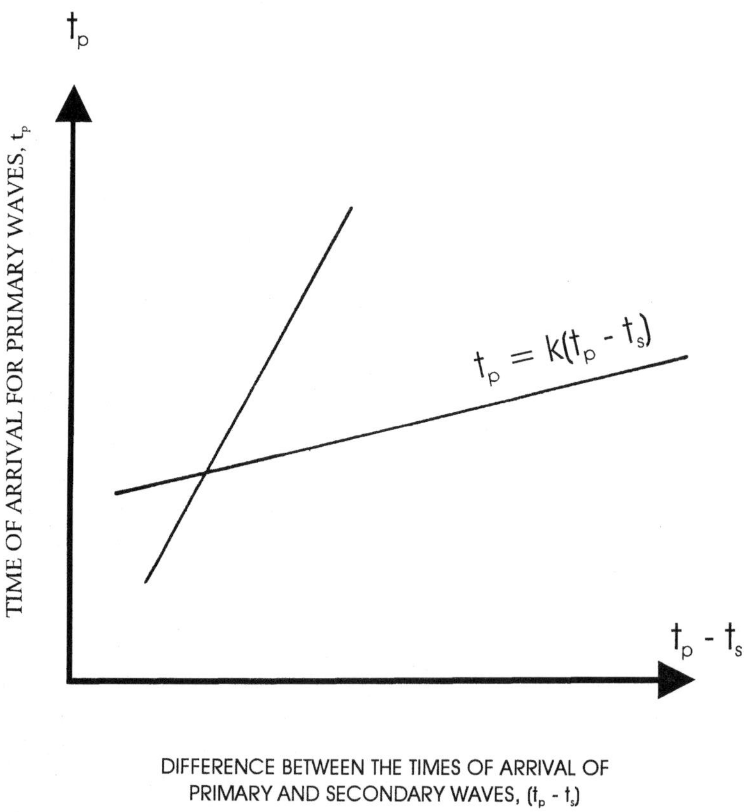

Figure 8-6. Wadati's graphs for two different seismic stations.

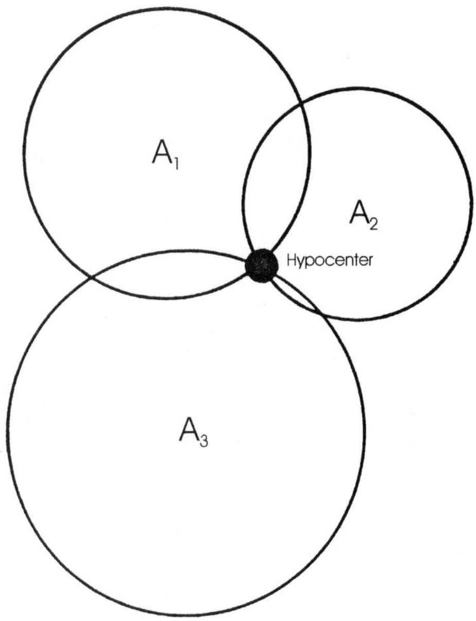

Figure 8-7. Determination of the hypocenter location.

of arrival of *P* waves based on the difference $(t_s - t_p)$ of the arrival times for *P* and *S* waves.

These calculations for the four stations allow construction of the four spheres (with centers at the geometric points representing the stations) to the surfaces of which point *H* must belong. The intersection of all four spheres locates point *H* and, consequently, allows determination of the hypocenter location. A simplified illustration of the method in the plane view is shown in Figure 8-7.

The time of occurrence of an earthquake can be calculated as the difference of local time of arrival of *P* waves and t_p. There are many other simplified methods to determine the location of a hypocenter based on simple geometric considerations. The velocities of propagation of *P* and *S* waves vary for various crustal media; however, their average ratio varies insignificantly and is approximately equal to $\sqrt{3}$ (see, for example, Karapetyan and Karapetyan, 1981). This means that the average ratio $\frac{V_s}{(V_p - V_s)} = 1.37$. This number can be used as a rough estimate to evaluate the distance from hypocenter.

CHAPTER 9

Forecasting Large Earthquakes

INTRODUCTION

Before developing a forecasting system, it is necessary to describe a suitable format for convenient presentation of the prediction results for public use. The format and components of a forecasting template are generally determined by the type of prediction and are different for a long- and short-term prediction.

TYPES OF FORECASTS

Long-Term Forcasts

A long-term forecast, which spans a number of years, must define seismic environmental conditions in terms of earthquake frequencies of various magnitudes for a particular area and maximum earthquake potential. It describes the seismic climate over a large geographic territory (e.g., southern California). These conditions are translated into requirements for housing and industrial construction for the area of interest along with other measures, which are necessary to prevent or mitigate consequence of possible earthquakes.

In the case of seismic science, the final goal of a long-term earthquake prediction is development of a system of seismic zonation that specifies past and future (potential) seismic events, their frequency, and possible consequences over certain areas with similar seismic conditions. Thus, long-term prediction of seismic conditions for a particular area requires preparing certain documents: seismic zoning maps, construction codes, emergency plans (instructions how to behave in the case of an earth-quake), etc.

Short-Term Forecasts

In short-term earthquake forecasting, it is necessary to predict large earthquakes within days or even minutes. An ideal forecast should enable the prediction of an oncoming earthquake with a magnitude greater than $M_R = 6$. Smaller earthquakes ($M_R < 6$) are recorded for general characterization of seismic activity in the region of interest, because predicting these earthquakes is usually not important from a damage viewpoint.

An earthquake hazard warning should be given if one specifies the location of the event (with the precision of a kilometer), time of occurrence (with the accuracy of a day), magnitude of the quake (greater than $M_R > 6$) and probability of occurrence (greater than 0.5).

To develop a short-term forecast, good prognostic technique must include all of the following components: (1) place, (2) time, (3) magnitude, and (4) probability of earthquake occurrence with acceptable accuracy. To date, short-term forecasting has not been developed by scientists.

A practical prognostication technique requires the following:

1. An adequate theoretical analysis to identify input and output variables
2. Detailed mathematical models specifying relationships among variables
3. An information system providing regular monitoring of input and output variables at the location of interest

Presently, scientists have only one suitable underlying physical concept for developing an earthquake prediction technique: the concept of accumulated stresses. The identification of input and output variables of the technique must be made based on their links to the accumulated stresses at the point of interest. For the purposes of earthquake prediction, these links can be described quantitatively by corresponding mathematical models. To date, there have been only a few attempts to develop such models and these failed because it was not possible to structure any attempt without specifying conveniently measured variables reflecting the crustal stress state. Without specified variables and structured mathematical models, a reliable forecasting system cannot be developed. Input variables must characterize premonitory phenomena that supposedly indicate oncoming large earthquakes.

A forecasting template should be presented to the public in the form of a forecasting bulletin. This bulletin should represent an output of the forecasting system in a form convenient for decision making and, in addition, impose requirements on prognostic technique. There are patterns that can be used for this purpose, such as the California short-term Earthquake Prediction Response Plan (1990), which contains the criteria for and format of long- and short-term earthquake forecast. This work started in 1974, when an advisory group on earthquake prediction for the state of California was formed. In 1976, this advisory group was named the California Earthquake Prediction Evaluation Council (CEPEC). This Council analyzes and assesses earthquake predictions for the Director of the State Office of Emergency Services (OES).

The Council has adopted the forecasting format of the National Academy of Science. Its Committee on Seismology (1975) determined the main components of a "good" prognosis:

> An earthquake prediction must specify the expected magnitude range, the geographical area within which it will occur, and the time interval within which it will happen with sufficient precision so that the ultimate success or failure of the prediction can be readily judged. Moreover, scientists should also assign a confidence level to each prediction.

As previously stated, for short-term prediction, the location resolution on the order of several kilometers is sufficient. A one-day advance notice timespan is considered excellent. The confidence level of prediction is usually expressed in terms of the probability of earthquake occurrence. The probability of earthquake occurrence is considered to be moderate if it ranges from 0.1 to 0.5 and high if it varies from 0.5 to 1.

Maximum Size Earthquakes

One of the major problems of seismic zoning is the ability to determine the maximum possible magnitude of an earthquake for a specific area of interest. Because of its nature, a great deal of extrapolation is inevitable. It is very difficult to apply a classic functional geologic approach to determine the maximum possible magnitude because of the unclear relationships among the variables involved.

Statistical methods linking the energy released by the earthquake and the parameters of seismicity over an area of interest seem to be more appropriate for this purpose.

To evaluate the maximum magnitude of an earthquake, consider the fact that large earthquakes occur, as a rule, over areas with intense seismic activity. Based on this fact, Riznichenko (1976) suggested considering the correlation between the long-term seismic activity *(A)* and energy *(E)* released by a maximum earthquake in order to estimate the magnitude of a potential maximum earthquake. A regression line can be constructed based on available statistical data on input *A* and output *E* variables. Then, this line can be extrapolated to the area of possible maximum seismicity in order to estimate the energy of a future event. An example of such a regression line, constructed on the historical statistical data over the eastern Tien Shan and Altai-Sayan region (Kazakhstan), is given in Figure 9-1. If there are reasons to

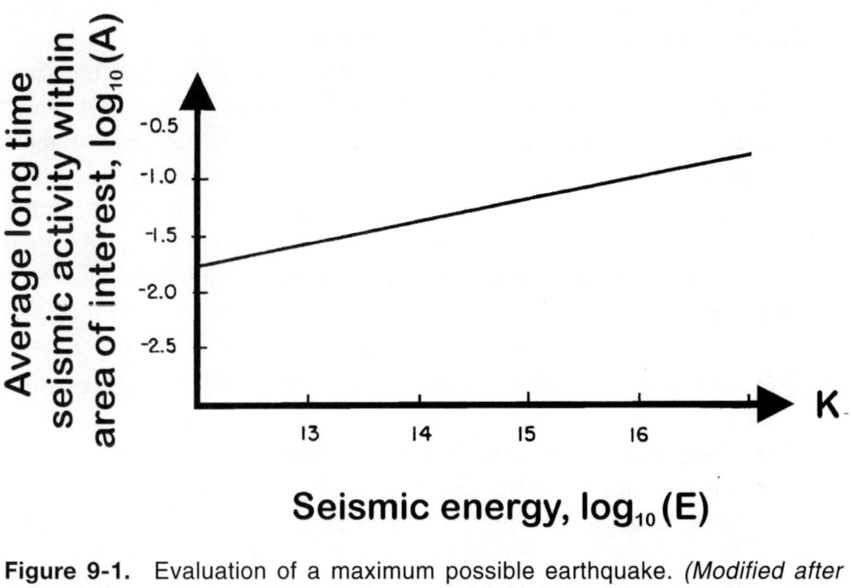

Figure 9-1. Evaluation of a maximum possible earthquake. *(Modified after Riznichenko, 1976.)* Seismic energy = energy released during a seismic event.

suspect an increase in seismic activity in the future, then this line can be readily extrapolated above $K = 17$ and used for approximating the potential maximum energy. It must be realized that there is a great deal of uncertainty in such an extrapolation. However, it does provide for quantitatively evaluating a future maximum event in order to develop construction requirements and emergency plans.

Forecasting Point

Unfortunately, the term "probability" when applied to express uncertainty in earthquake occurrence confuses both the public and researchers alike. When employing the probabilistic approach to measure the uncertainty of prediction, it is necessary to specify a random event of interest and include this event in the so-called probabilistic space. Then it is necessary to evaluate probability of this event.

A direct mathematical earthquake description can be accomplished by introducing a 5-dimensional vector (which can be represented by the point in the 5-dimensional space) of real numbers with the following components: time, three spatial coordinates, and magnitude. The working terms "forecasting space" (F-space) and "forecasting point" (F-point) formally describe the earthquake forecast. Using these terms, making a particular earthquake prediction means to chose a certain F-point from F-space. If a particular F-point is selected, then the time, place, and magnitude of the forthcoming quake are completely determined as its components. Currently the only reasonable method for specifying of this point is a statistical estimation of the components of the 5-D vector. As a result of this estimation, a 5-D point random estimate of the supposedly oncoming "real" F-point can be obtained.

To express a quality of the statistical point estimate, it is customary to use a confidence area (confidence interval in the one-dimensional case): an area to which a chosen estimate belongs with the preassigned sufficiently high probability (which actually specifies a desirable confidence level).

For example, such an area may be designated as a sphere S_R with the center at F of a certain radius R (Figure 9-2). Therefore, if the probabilistic notions are applied to introduce the confidence level of prediction, then a procedure of finding the radius of a sphere that corresponds to *a priory* assigned probability P must be developed. This means that the point F belongs to the area S_R with the given probability

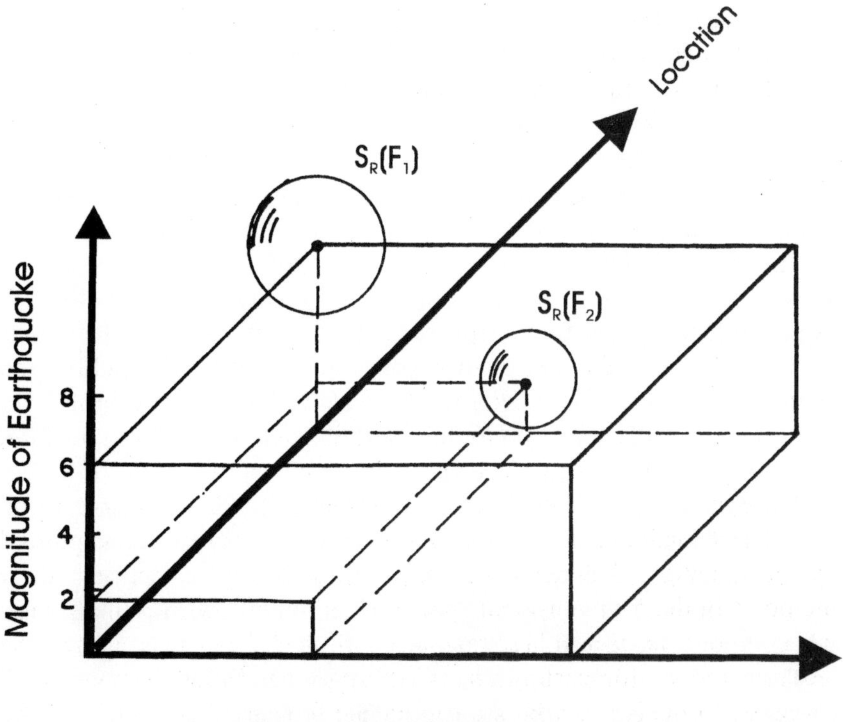

Figure 9-2. Forecasting space and points

P. Then, the value *P* can be taken as the confidence level for the earthquake prediction.

The problem of finding the radius cannot be solved at the present level of scientific development. In statistics, 90% or 95% probability threshold usually defines the confidence level. The California short-term Earthquake Prediction Response plan classifies the interval of high probabilities of earthquake occurrence as 50%–100%. This means that for earthquake prediction, confidence levels greater than 50% are considered high enough to issue an immediate public warning.

To construct the confidence interval for a point estimate corresponding to a given confidence level, the theoretical probability distribution for a random point estimate as a function of statistical data must be determined. Searching for the probability distribution is a daunting task even in the simplest, specially selected, one-dimensional theoretical cases. Those familiar with statistical procedures of the confidence interval construction, realize that it is impossible to construct a confidence 5-D area for the *F*-point based on the theoretical considerations. Under these circumstances, the development of a good comprehensive forecast of major seismic events is hardly possible. In order to overcome these obstacles, the description of a "good" forecast must be simplified based on some reasonable hypotheses.

The first hypothesis relates to the place of occurrence. From the previous discussion, it follows that the large faults (such as the San Andreas Fault in California, for example) are the locations of greatest concern, because they represent weakened zones with continual movement at the boundaries of the tectonic plates. This allows specifying the most probable places of future earthquakes in a given geographic area (e.g., Parkfield, California).

The second specification relates to the magnitude of a large earthquake. Taking into consideration that earthquakes with a magnitude greater than $M_R \geq 6$ can cause devastating damage and threaten the lives of thousands of people, only earthquakes with a magnitude $M_R \geq 6$ are considered.

The third simplification relates to the probability of occurrence of an oncoming event. If the estimate of this probability exceeds 0.5, this indicates a confident forecast and a hazard warning should be issued. Consequently, it is admissible to use just a qualitative indication of strong confidence in the prognostic format.

Summarizing these three proposed simplifications, one would accept as admissible a short-term earthquake forecast such as, "Tomorrow, in the Parkfield area, a large earthquake is expected with strong confidence of certain probability." In this forecasting format all of the important information on the forthcoming earthquake is presented in a simple and comprehensive manner.

In the 1995 bulletin prepared by the Southern California Earthquake Center, entitled "Commonly Asked Questions on Scientific Aspects of Earthquakes," a definite opinion regarding the possibility of scientific prediction of seismic events was presented. The question: "How close are the scientists to reliably predict earthquakes?," is answered in the bulletin (released to the general public) as follows:

> Scientists are far from being able to make reliable earthquake predictions in either the short, intermediate, or long term. Only with respect to the San Andreas Fault, for which there are enough

data, do we feel it is possible to make long-term probabilistic forecasts of the likelihood of future large earthquakes.

It must be remembered that for a long-term probabilistic forecast the probability of occurrence is usually evaluated on the basis of the statistical data as a relative frequency of the recurrence of events over a long time interval (100 years or longer).

The general public is very pessimistic in the assessment of the prognostic power of contemporary seismic studies. People believe that it is impossible to forecast the major seismic events. This perception has led to a skeptical approach to all proposed prognostic methodologies and to a considerable cut in funding for research on earthquake prediction. The allocation of funds for earthquake prediction studies in the 1990s was insufficient in comparison with those of the 1980s. Seismologists became very cautious in expressing opinions regarding the possibility of a short-term prediction. They also became cautious to release information on any new developments in earthquake prediction to the mass media, in order to avoid further criticism.

There is a considerable social need, however, for information on new developments in earthquake prediction. If scientists do not provide such information and do not formulate reasonable approaches to the solution of this problem, then, inevitably, enthusiastic amateurs will take over.

Since 1985, efforts of the scientific community in California have improved the state of earthquake preparedness dramatically. Knowledge of seismic history, seismic patterns, and accumulated evidence on premonitory events have considerably increased the prognostic potential due to the collaboration between scientists and state officials. The Northridge earthquake of 1994 provided a great deal of information leading to the reduction of the terrible consequences of seismic events. Earthquake prediction and its assessment is now conducted in California by the California Earthquake Prediction Evaluation Council (CEPEC), which must collect and validate the background information necessary to make a decision on implementing emergency plans.

Probability of Earthquake Occurrence

Evaluating the probability of occurrence is the most important and, at the same time, the most difficult problem of earthquake prediction. Before evaluating the probability of occurrence, the geographic area of interest must be specified as well as the prediction timespan. Then, past statistical data on earthquake occurrence with a certain range of magnitudes related to this geographic area can be used to estimate the parameters of a suitable probabilistic model.

Two probabilistic models (discrete and continuous) in the form of the Poisson's distribution and conditional probability model are commonly used to describe the probability of future earthquake occurrence. The stationary Poisson's model (Cornell and Winterstein, 1988) assumes that the probability of an earthquake occurrence, in a future time interval, Δt, does not depend on the time elapsed from previous events. If K is a random number of earthquakes, which occurred in the time interval Δt, then the probability of at least one earthquake in the time interval Δt is expressed by the following formula:

$$P(K \geq 1) = 1 - \exp(-N) \tag{9-1}$$

where N is the expected number of earthquakes in the time interval Δt.

Equation 9-1 is easy to apply. It is only necessary to estimate one parameter N, which can be accomplished by processing the statistical data for earthquake occurrence in the area of interest over a specified timespan.

Suppose, for example, that for a short-term forecast a one-day timespan is specified and the magnitude of seismic events is taken as greater than $M_R = 3$. Suppose also that 10 such events occurs during a 1,000-day interval. Then, the statistical estimate for N is 10/1,000 = .01. Thus, the probability of occurrence of an earthquake with a magnitude greater than $M_R \geq 3$ may be estimated by using Eq. 9-1:

$$P(K \geq 1) = 1 - \exp(-0.01) \approx 1 - .99 = .01$$

which means that the probability of occurrence of an earthquake tomorrow is about one hundredth.

The Working Group of California Earthquake Probabilities recommended applying a conditional probability model (1990) to introduce the dependence on the time elapsed since the last earthquake. If T represents a random time of recurrence, and if no earthquake occurred prior to the time t, then the dependence of the probability of earthquake recurrence in the time interval $[t, (t + \Delta t)]$ can be expressed in the form obtained from the general definition of conditional probability as follows:

$$P\big(t \le T \le t + \Delta t \,\big|\, T > t\big) = \frac{P\big(t \le T \le (t + \Delta t)\big)}{P(t \le T < \infty)} \tag{9-2}$$

For evaluating the probability of earthquake occurrence in any time interval *($t \le T \le t + \Delta t$)*, the assumption of the existence of a corresponding density function *f(u)* can be used. Then:

$$P\big\{t \le T \le (t + \Delta t)\big\} = \int_t^{(t+\Delta t)} f(u)\mathrm{d}u \tag{9-3}$$

where u is a variable of integration.

It is customary to use a log-normal probability density for *f(u)*:

$$f(t) = \left[MS\sqrt{2\pi}\right]^{-1} \exp\left\{\frac{\ln\left(\dfrac{t}{M}\right)^2}{2S}\right\} \tag{9-4}$$

which is defined by two parameters M and S. These parameters are determined, in turn, by a mathematical expectation, m, and standard deviation, s, of the recurrence time that are defined by the following formulas:

$$S^2 = \ln\left[\frac{s^2}{\left(m^2 + 1\right)}\right] \tag{9-5}$$

$$M = m\exp\left\{-0.5S^2\right\} \tag{9-6}$$

The statistical estimates of m and s are made based on the data on the time of occurrence of earthquake for the area of interest.

The denominator in Eq. 9-1 is "silently" assumed to be equal to 1, because sooner or later an earthquake event occurs over any territory.

The advantage of the conditional approach is that it allows for a convenient method to introduce the dependence of the probability of earthquake recurrence on the time elapsed since the last occurrence. In a similar manner, the influence of any time-dependent characteristic such as accumulated stress can be described (e.g., on the probability of future earthquakes).

Probabilistic Interpretation of the Precursor's Inclusion in the Decision-Making Process

Based on a belief in a particular precursor, it should then hold true that its appearance increases the chances of the forthcoming earthquake occurrence. From the Probability Theory viewpoint, counting on a precursory event means implementing conditional probability in the decision-making process in evaluating the probability of earthquake occurrence. This implementation only makes sense if a conditional probability of earthquake occurrence is greater than the unconditional probability for occurrence (provided that a chosen precursory event is realized). Otherwise, a chosen precursor is useless for prognostication purposes.

To analyze the situation, two schemes of decision making can be considered when evaluating the probability of earthquake occurrence in the form of decision-making trees: (1) without precursor, and (2) with precursor. These two schemes are shown in Figure 9-3. For the first scheme, only the probability $P(E)$ of the earthquake occurrence must be estimated. Introducing the precursory event Π into the process changes the one-stage scheme (two-branch tree) of the evaluation of the probability of earthquake occurrence (E) into a two-stage decision-making scheme (four-branch tree).

For the one-stage scheme, there are only two possible outcomes: E and $\sim E$. For the purpose of earthquake prediction, assign (evaluate, estimate) the probability of event E as $P(E)$. ("~" denotes an event complementary to E.) The only objective way to assign this probability is to use the corresponding relative frequencies.

For the two-stage scheme there are four possible outcomes: (Π and E), (Π and $\sim E$), ($\sim\Pi$ and E), and ($\sim\Pi$ and $\sim E$). There are two mutually exclusive outcomes favorable to the occurrence of the event E: (Π and E) and ($\sim\Pi$ and E). If the precursory event Π occurs, then, instead of probability $P(E)$, one is inclined to use a conditional probability $P(E/\Pi)$ (where $P(E/\Pi)$ is the conditional probability of earthquake occurrence provided that the precursory event Π has been already realized) as the corrected probability of occurrence of a future earthquake. Its definition may be inferred from simple relations:

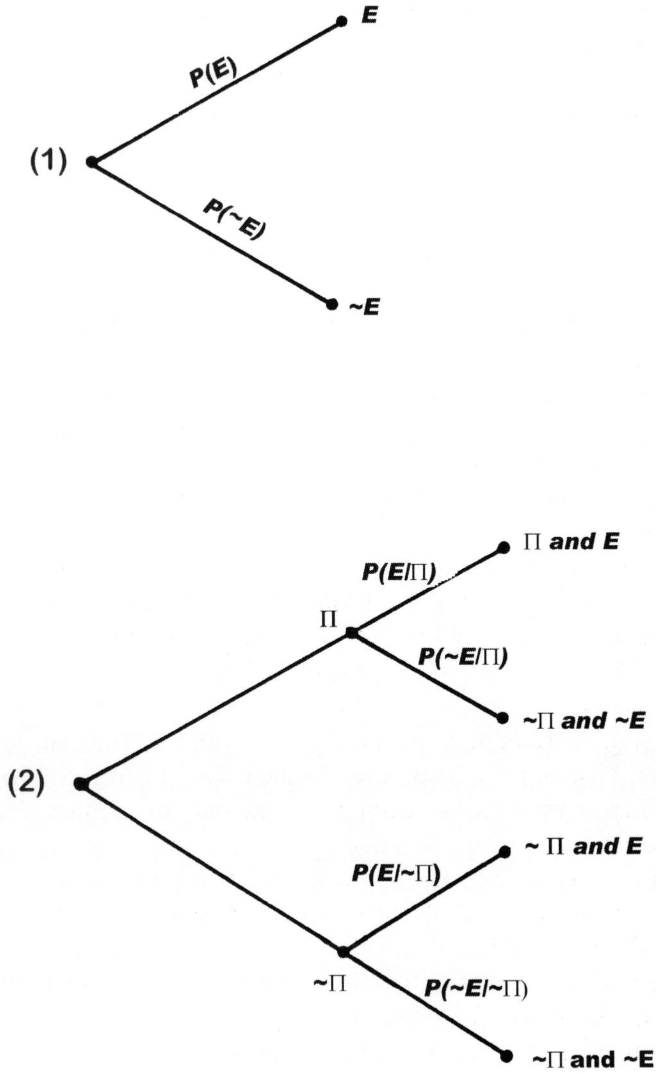

Figure 9-3. Decision making based on conditional probabilities where P = probability, E = earthquake occurrence, and Π = occurrence of precursor.

$$P(\Pi \text{ and } E) = P(\Pi) \times P(E/\Pi) \tag{9-7}$$

and

$$P(E/\Pi) = P(\Pi \text{ and } E)/P(\Pi) \tag{9-8}$$

When the idea of precursors is applied, the last conditional probability should be greater than the unconditional probability $P(E)$ to reinforce the confidence of an investigator. If, for example, a precursory event occurs every time an earthquake occurs, then $P(\Pi \text{ and } E) = P(\Pi)$ and $P(E/\Pi) = 1$.

$P(\Pi \text{ and } E)$ can also be written in the following form:

$$P(\Pi \text{ and } E) = P(E) \times P(\Pi/E) \tag{9-9}$$

Substituting Eq. 9-9 into Eq. 9-8 and assuming that $P(\Pi) > 0$, one obtains:

$$P(E/\Pi) = P(E) \times P(\Pi/E)/P(\Pi) \tag{9-10}$$

which is a version of the famous Bayes's formula for the considered specific case.

If a particular precursor Π is selected, then Eq. 9-10 enables the evaluation of conditional probability $P(E/\Pi)$ of earthquake occurrence. Equation 9-10 allows for comparison of informative power of various precursors: Obviously, preference should be given to precursor Π, which maximizes the conditional probability $P(E/\Pi)$.

CHAPTER 10

Ancient Foretellers

COSMIC EVENTS

In prehistoric times, earthquake prediction was the exclusive privilege of astrologers and soothsayers. Currently, the public gives little validity to earthquake prediction because of its many failures, such as the San Francisco earthquakes of 1906 and 1972. Analyzing earthquake prediction by astrologers, Gribbin and Plageman (1974) suggested the existence of links between cosmic phenomena, such as planet alignments, and major seismic events ("Jupiter Effect"). Their reasoning can be well characterized by the following quotation:

> "The gravity of the planets can affect the Sun, through tidal interactions, and disturbance on the Sun can influence the Earth through the changes in the magnetic field which links all the planets in the Solar System. Only on very rare occasions can these small effects add up to produce any dramatic results on Earth. But one of these occasions—an alignment of the planets that occurs only once every 179 years is due in 1982. We are convinced that this will trigger off . . . the earthquake activity on Earth, and by that time the California San Andreas Fault system will be under considerable accumulated strain. Geophysicists report that their measurements of that fault now show that it is overdue for a slip greater than that of 1906, and just needs a trigger. There can be little doubt, we feel, that the planetary and solar influence in the early 1980s, following the rare planetary alignment will provide that trigger. In particular, the Los Angeles region will, we believe, be subjected to the most massive earthquake experienced by a major center of population during this century."

The cited passage demonstrates that in the absence of sound theoretical approach there are an endless variety of possible pseudoscientific speculations regarding probable underlying processes and triggering causes for a large earthquake. Analyzing this quotation, it can be concluded, at first, that it predicted a large earthquake in 1982 (year of rare alignment of planets) in the vicinity of the San Andreas Fault. However, the last sentence in the previous quotation about the Los Angeles area casts doubt on the time of a future earthquake.

The question, which is inevitably asked by curious readers, is Why would the rare alignment of the planets trigger a large earthquake? With the same odds, one could postulate that this rare alignment of planets could organize inner processes in the earth's core that would mitigate the interference of plates' boundaries and considerably diminish the chances of earthquake occurrence.

Referring to the United States Earthquake Catalog for the year of 1982, we find out that no large earthquake occurred in California. In fact, only one significant earthquake, which had a magnitude of M_R = 5.1, occurred in February of 1982 in Northern California, and happened off the coast (U.S. Geological Survey Bulletin 1655, 1983). Therefore, it can be concluded that this particular prediction failed, and that predicting earthquakes cannot be made on the basis of remote celestial and planetary events.

Atmospheric Phenomena

Reports on unusual luminous phenomena in the lower layer of atmosphere on the eve of several major seismic events abound in ancient literature worldwide. There is a legend in Japan that claims a short rainbow of three colors (black, yellow, and indigo) is an ominous sign of an oncoming earthquake. Contemporary amateur foretellers also report seeing a short rainbow prior to earthquakes. At the time of the 1965–1967 Matsushira series of earthquakes in Japan, amateur photographers took many colored pictures of unusual lighting. The famous Japanese seismologist, T. Rikitake, and his associates conducted many electric and magnetic observations at the area of epicenters of these earthquakes. As reported by Rikitake (1976), however, they never observed any anomalous luminosity. Thus, it was concluded that the amateur pictures and reports on mysterious luminosity could not be attributed to earthquake occurrence and, therefore, luminosity cannot serve as an earthquake precursor.

A possible scientific explanation of these phenomena is linked to the hypothetical generation of strong electric fields along the paths of fluids migrating to the upper crustal layers (so-called streaming potential effect). The generation of streaming potential is closely related to the capillarity of rocks. In the capillaries of clayey rocks the positively charged ions (cations) form the so-called double layer (Figure 10-1). If a fluid is propelled through capillaries under the pressure drop, positively-charged particles of the mobile part of a charged layer move in the direction of flow generating the streaming potential. Inasmuch as the flow rate is proportional to the pressure drop, so is the generated streaming potential proportional to the pressure drop.

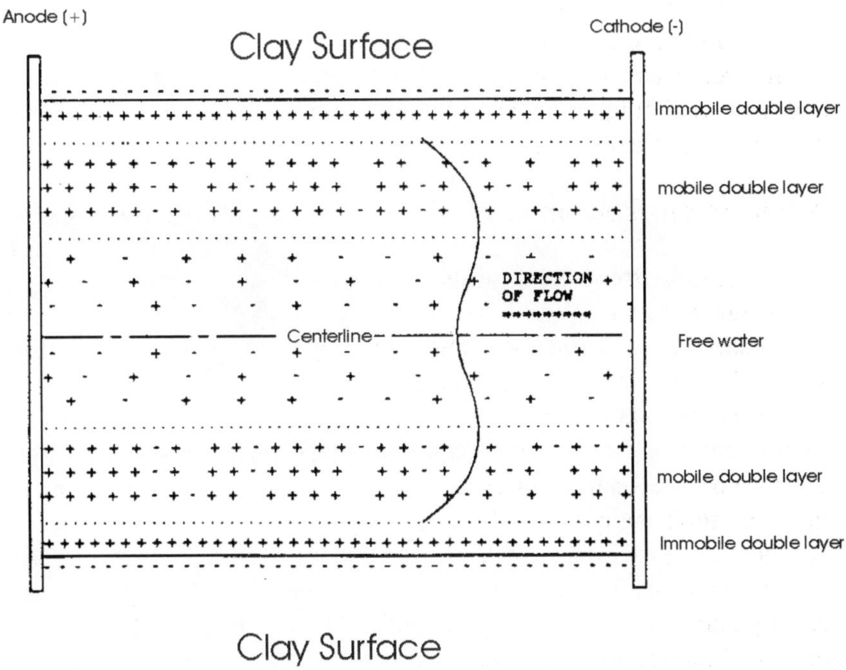

Figure 10-1. Schematic diagram of streaming potential, generated between two clay surfaces, caused by propelling fluid through capillaries due to the charged double layer.

If the fluid conductivity is S, dielectric constant is D, viscosity of fluid is μ, and the pressure drop is Δp, then the voltage V produced over the distance of pressure drop is equal to:

$$V = \frac{DZ\Delta p}{4\pi\mu S} \quad (10\text{-}1)$$

where the variable Z is a potential characteristic of the rock matrix, which is on the order of tens of millivolts for ordinary rock species.

Except for fluids of very low conductivity, it is difficult to expect a streaming potential of over tens of millivolts per meter. It seems improbable that these fields are able to generate luminosity in the atmosphere. Currently, they cannot be detected in the atmosphere by any instrumental means. Changes in the potential, however, probably can be recorded underground and, therefore, theoretically can be used as an earthquake precursor.

Electrical currents are also known to flow through sediments (see Serruya et al., 1967). These currents may result in the flow of positively charged mobile double layer of ions in interstitial fluids toward the cathodes (Figure 10-1). In addition, when fractures widen from seismic activity, the volumetric flow rate will increase. Thus, the rates of fluid flow in sediments and rocks may also serve as an earthquake precursor.

Abnormal Animal Behavior

Premonitory abnormal behavior of animals (wild and domestic) is of scientific interest. Anecdotal reports of such behavior abound. It is not surprising, therefore, that in the past, people kept animals as earthquake predictors. Thus, scientists worldwide have conducted a great deal of research on the subject of abnormal animal behavior on the eve of earthquakes.

Detailed studies on animal behavior were conducted after a series of nine great earthquakes in China with the magnitude higher than $M_R = 7.5$ (Ma et al., 1988). There were considerable anomalies in the behavior of animals before (and during) all nine events. Cattle, dogs, rats, fish, chicken, birds, pigs, and snakes changed their behavior before the land earthquakes, whereas many sea species behaved unusually on the eve of sea earthquakes (together with land animals along the seashore). Animals panicked (for example, cows jumped out of their stables in terror) and calmed rapidly without visible cause. In many cases, an earthquake then occurred.

The time lapse between the anomalous behavior of animals and the earthquake occurrence ranged from a few seconds to several months. Despite the variations in species and in timing, only three possible models of behavior change were considered in all of the reports: (1) changes in the activity level, (2) changes in vocalization, and (3) migration in space. Many reports have not been verified and seem to be speculative. However, the consistency in description and the considerable number of these reports cannot be ignored.

Among anecdotes on abnormal animal behavior, the story of a dog named Menelic is the most impressive. The story was presented in the book entitled *Earthquakes* (Russo, 1966). On the eve of the devastating Messina earthquake of 1908 in Italy, a troop of comedians was staying at the Mazotti Hotel in Messina. Two teenage acrobats, brothers Mickele and Alfredo Gambia, occupied the same hotel room with their dog Menelic. On the morning of December 28, 1908, the dog, which had been sleeping at the brothers' feet, suddenly began barking and howling. The brothers tried to silience her, but the dog continued barking and howling. Menelic then pulled the blankets away from the brothers, caught one of them by the arm, and pulled him out of the hotel. The dog continued barking, so that the whole troop of comedians woke up and gathered in the street. The dog leaped and ran around the people, pushing them out of town. When the procession of comedians left town, the earthquake occurred, turning the town of Messina into ruins.

Wordwide observations show that the main behavioral anomalies of domestic animals are concentrated around a one-day timespan before a seismic event and are linked to large active faults. Before the Songpan earthquake in China in 1976, for example, three time periods were observed, during which pronounced abnormalities occurred one day prior to the seismic event along the Longmenshan fracture zone. The Chinese investigators (Ma et al., 1988) reported that the time and location of the anomalous animal behavior was consistent with the anomalies in physical and chemical phenomena, such as electric resistivity of rocks, groundwater level, escape of ground gases, radon emission, and so on.

Another mysterious story of anomalous animal behavior concerns snakes, which began to crawl out of their holes 2 to 10 days before the Spitak earthquake in 1988. On the eve of the earthquake in

Figure 10-2. Snakes that ceased hibernating and crawled out of their holes prior to the earthquake. *Vipera lebetina* (left) and *Colober ravergieri* (right) are shown above. Other snakes in the area were *Natrix teyellata, Colober caspius,* and *Typhlops vermicularist.*

Armenia, which occurred in wintertime, the hibernating snakes (Figure 10-2) crawled out of the holes near the fault zone and died on the snow-covered ground. One of the possible hypotheses explaining the behavior of these snakes is the supposition that they are sensitive to changes in the concentration of various gases near the fault zone due to upward gas migration on the eve of the earthquake.

The numerous examples of abnormal animal behavior on the eve of seismic events suggests the possibility of applying these anomalies for forecasting purposes. Scientists, however, are skeptical because these stories are difficult to verify. Additionally, the time lapse between the animal behavior modification and the occurrence of seismic events varies considerably in these reports (from minutes to months). For this reason, it is not possible to use abnormal animal behavior as a reliable earthquake precursor. It is believed that most behavioral abnormalities can be attributed to the influx of gases (e.g., CO_2, radon, etc.) to the surface along faults and fractured zones.

CHAPTER 11

Messages from the Earth's Crust

INTRODUCTION

The main concept behind short-term prediction lies in detecting and evaluating anomalous changes in continuous observation data on earthquake preparation processes. Searching for appropriate precursory phenomena and revealing the patterns of premonitory events related to these phenomena are the primary considerations in developing earthquake prediction techniques. There is a strong consensus in the scientific community today that excessive accumulation of stresses is the primary cause of all seismic events. Consequently, measurements of strains and the resulting earth-surface tilts can provide important information on prequake conditions. Continuous surveillance of strains and tilts is to detect anomalous changes in the systematic records of strains and crustal movement, which may uncover the patterns that precede earthquakes

PHYSICAL MODELS OF RUPTURE

Success in short-term earthquake prediction depends on whether or not a chosen precursory phenomenon can indicate the time and place of a major rupture. The problems that accompany the development of an earthquake prediction technique can be illustrated by the simple physical experiments with homogeneous and heterogeneous fragile media (Mogi, 1985). For example, consider a piece of glass as a physical model of a homogeneous medium. If bent, the glass plate can break suddenly, without any premonitory signs that might indicate the time and place of a rupture.

Response of the medium changes completely when one uses granite instead of glass. Granite is also a fragile material and, if a considerable force is applied bending the granite plate, it breaks into small pieces. Before breaking, however, the place of the future fracture is clearly displayed by a web of microfractures, and the time of occurrence is indicated by the pronounced increase in the acoustic emission (increasing volume of the rattling sound). Theoretical curves in Figure 11-1

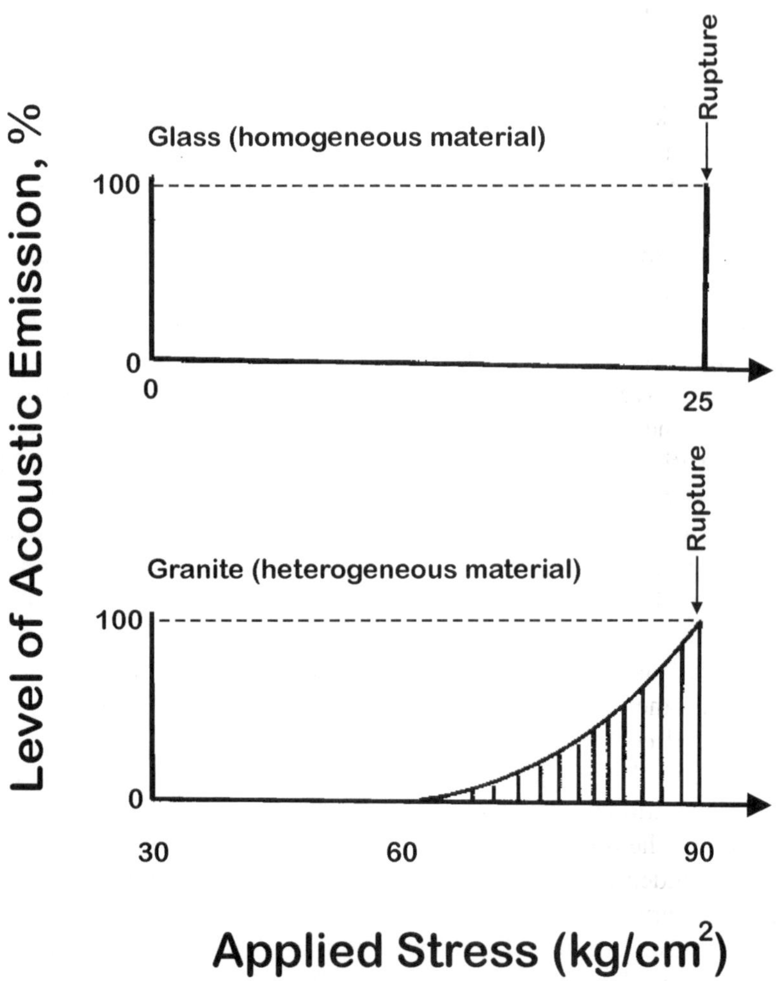

Figure 11-1. Developing acousting emission due to medium fracturing *(Modified after Mogi, 1985, fig. 3.1, p. 17.)*

show the difference in the behavior of two different materials: homogeneous (glass) and heterogeneous (granite) before rupture, when stress increases at a constant rate with time. Comparing these two cases, it can be concluded that there is a possibility of predicting the moment of breakdown if the tested sample contains inclusions of different materials making the rock heterogeneous.

Inasmuch as the earth's crust is essentially heterogeneous, some indications of an oncoming earthquake, such as foreshocks and crustal movement, prior to the principal rupture can be expected. If some portions of the crust are heterogeneous, whereas the others are rather homogeneous, then in some places it is possible to discover anomalies in the rock petrophysical characteristics prior to the earthquakes, whereas in other places these anomalies are not manifested.

Statistical data have clearly shown that large earthquakes occur repeatedly in the vicinity of existing faults. The faults represent weakened zones that are prone to breakdown and repeatedly produce earthquakes. In addition, big faults, such as the San Andreas in California, allow the crustal movement of tectonic blocks on the opposite sides of the rupture along faults. The "slip-stick" character of their movement produces collisions of blocks that continually generate the seismic activity along the faults. In this situation, the irregularities of fault boundaries are essential to evoke some premonitory phenomena.

To illustrate the fact that even slight fault boundary irregularities create some precursors that can be monitored in situ, Japanese seismologist Mogi (1985) carried out a simple physical experiment with a system modeling an artificial fault. The modeling system was developed as a rectangular parallelepiped granite sample cut at an angle and then composed again. Parts of the system were kept together by applied compression forces as shown in Figure 11-2.

In Figure 11-3, the experimental sample is shown. The numbers in this illustration indicate the location of strain gauges. If the lateral pressure is held constant and the vertical force is gradually increased, then a sudden slip followed by a sudden stop (sticking) occurs along the boundary plane. This experiment imitates the relationship between the changing strains and seismic motion prior to a "slip-stick" type earthquake. Figure 11-4 shows the changes in parallel strains with time in the homogeneous fault plane when the compression force is gradually increasing in the vertical direction. It can be seen that for

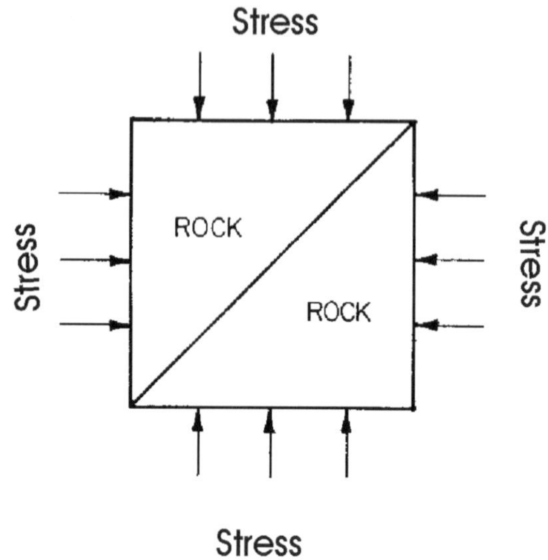

Figure 11-2. Modeling plane slipping. *(Modified after Mogi, 1985, fig. 3.2, p. 18.)*

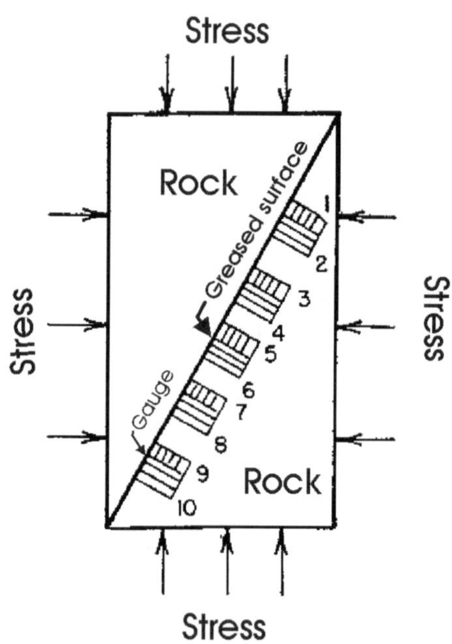

Figure 11-3. Slipping along boundaries with irregularities. *(Modified after Mogi, 1985, fig. 3.3, p. 18.)*

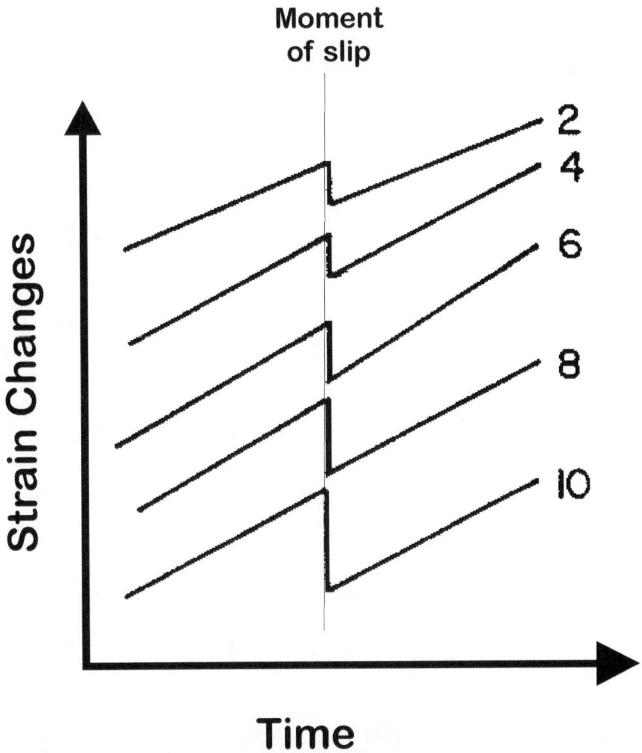

Figure 11-4. Strain changes at the location of gauges. *(Modified after Mogi, 1985, fig. 3.4, p. 19.)*

all the observation points there is a sudden change in the strain of the same kind because of the release of stress by slip. At the same time, there are no visible anomalies in the changes of stress for all the points of observation.

If some heterogeneity on the surface of a section is created, then different patterns of changes in the strain graphs can be observed. Figure 11-5 illustrates the processes of strain changes if friction of a certain fault section is locally decreased by applying a small amount of grease (in nature, an analogous situation can be created by excessive rainfall) in the vicinity of gauge No. 6. The patterns of changes in strain are considerably different for this gauge, and those close to it, in comparison with other gauges. This experiment illustrates the role of heterogeneity in developing predictive techniques.

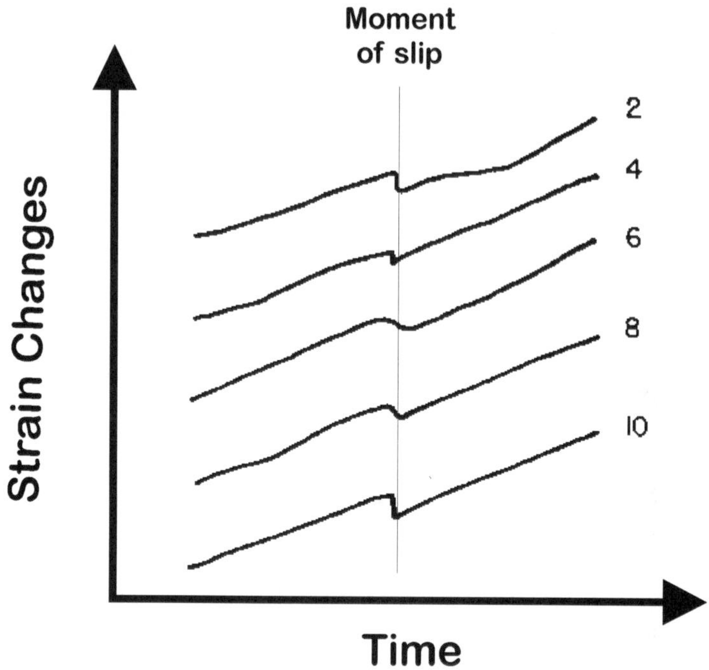

Figure 11-5. Strain changes before and immediately after the slip of the granite blocks. *(Modified after Mogi, 1985, fig. 3.5, p. 20.)*

In Figure 11-6, typical distributions of strains are shown for which there were no precursory anomalies prior to a sudden slip. The primary slip occurred in the vicinity of gauge 4 after the strain increase became sluggish and then began to decline. In the vicinity of gauge 6, on the other hand, the strain increased just prior to the slip, suddenly decreasing afterwards. There were no indications prior to the slip at positions 1 and 7. Thus, if only the strains at these positions were monitored, then a sudden slip along the fault in this particular case could not be predicted.

This spatial variation of stresses has been observed many times in the case of real earthquakes. It was important, however, to obtain a typical spatial distribution of strains in a specially designed experiment as a result of a local change in friction. This pattern of the spatial distribution of strains can be explained as follows (Mogi, 1985): When a local slip starts at the vicinity of gauge 4, it releases some local tension

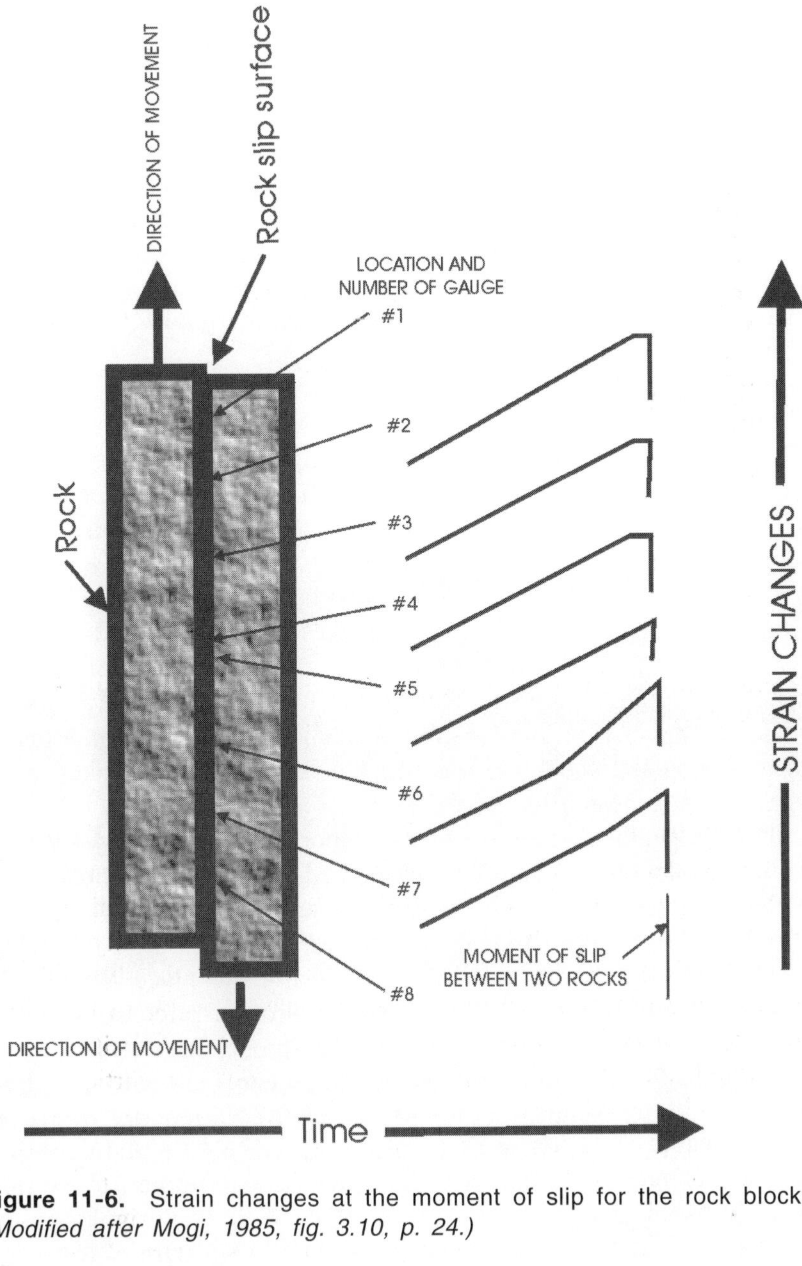

Figure 11-6. Strain changes at the moment of slip for the rock blocks. *(Modified after Mogi, 1985, fig. 3.10, p. 24.)*

at position 4. Because the whole piece of rock is immobilized, however, the stress must increase in some other area, which happens in the vicinity of gauge 6. Thus, a local slip provokes the increase of stress at position 6, which, in turn, leads to the total slip of the whole piece of rock. This example demonstrates that even for faults that are considered to be homogeneous, closer inspection may reveal some points of heterogeneities that may serve as the triggers for a slip. As a result of this slip, a sudden shock can occur that can be felt as a major seismic event. In addition, this experiment shows that the real cause of a quake is the rapid dynamic redistribution of stresses rather than their gradual static increase.

The Parkfield region in central California is a typical area where earthquakes can occur repeatedly in the vicinity of an existing fault (San Andreas Fault) as a result of this "slip-stick" sequence of events. In this area, the earthquakes of $M_R \geq 6$ magnitude have occurred in 1857, 1881, 1901, 1922, 1934, and 1966. An earthquake of this magnitude, therefore, occurred, on average, approximately once every 22 years. Thus, it can be concluded that in the vicinity of large faults, frequent recurrence of strong earthquakes can be expected.

Giusti (1995) found that the hypocentral distributions of the 1934 and 1966 earthquakes were related to the same portion of the Parkfield region. Judging from the location of main shocks and the distribution of following aftershocks, it was concluded that the earthquakes originated at the same portion of the fault.

The above facts lead to several important conclusions. First, the faults represent the so-called weakened zones. This increases the probability of rupture in their vicinity and, consequently, increases the probability of earthquakes. Secondly, if the faults of interest are heterogeneous, it is possible to predict the place and time of rupture by monitoring the strains at several observation points within the area. All anomalies are recorded in the process of changing stress. For homogeneous faults, these anomalies can be weak and highly localized. This requires more detailed observation of strains and deformations. Thirdly, mechanistic (e.g., tilting, changing strains, dislocation of tectonic blocks) precursory phenomena regularly reappear at the same faults.

These conclusions can be considered as the basis for developing a mechanistic earthquake prediction theory. If the fault of interest is heterogeneous, then the mechanistic characteristics of rocks are very useful as precursory phenomena. A fault with pronounced irregularities at the boundaries of tectonic blocks was observed at the Izu Peninsula in Japan. This can be explained by the fact that the tectonic area of interest is highly heterogeneous. It is a large tectonic site situated between Sagami and Suruga troughs.

For the San Andreas Fault, only a few precursory events were observed (stress, strain, and changing tilts) mostly because of homogeneity of the fault structure. It is hoped, however, that carefully planned observations with large number of strain-meters can reveal beforehand peculiarities of processes accompanying earthquake preparation. The method and strategy of monitoring is schematically shown in Figure 11-7. The left part of the illustration shows hypothetical positions of the strain-meters in an area with considerable heterogeneity, as in the case of multiple faulting. In this case they need

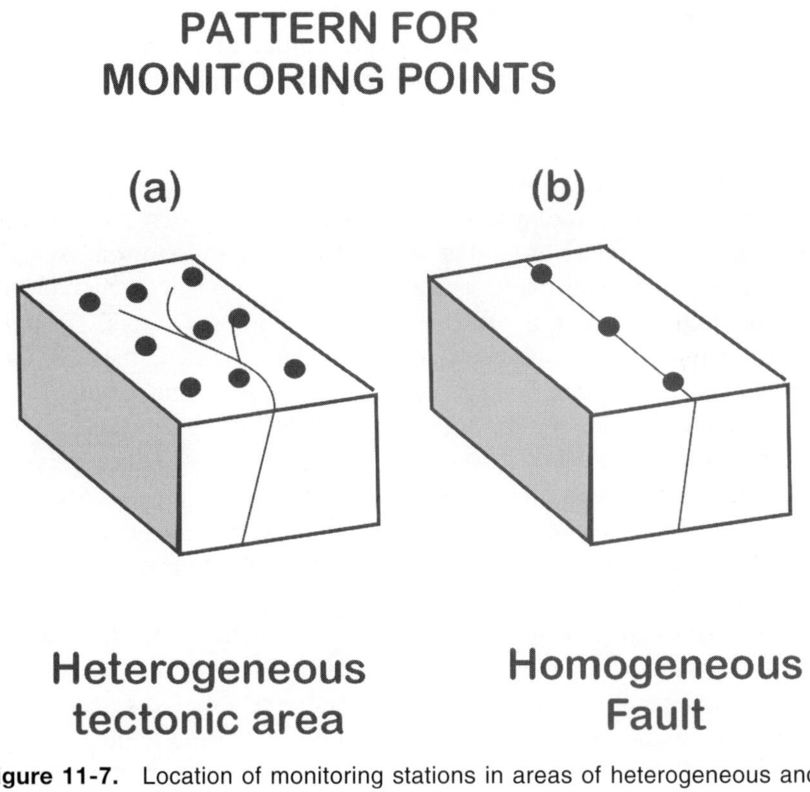

Figure 11-7. Location of monitoring stations in areas of heterogeneous and homogeneous faults.

to be uniformly distributed over the area of interest. The right part of the figure represents a monitoring scheme in the vicinity of faults with a prevailing main fault. In this case, the sensors should be positioned uniformly along the extension of principal rupture. If a sufficient number of strain-meters are placed along the fault's extension, it will facilitate recording even small variations in the patterns of the strain changes (Mogi, 1988).

If microfractures precede a major seismic event, various precursory phenomena can be expected to appear, such as foreshocks and crustal deformation. As previously stated, in this case considerable crustal heterogeneity must be expected. The more complex the tectonic structure, the more probable the appearance of the precursory anomalies in strains and tilts over the area of interest.

When the prevailing fault has a simple structure (like the central part of the San Andreas Fault), a seismic slip along the fault's plane should be expected. In this case, it is difficult to expect the appearance of foreshocks; however, there is a possibility to observe other precursory phenomena if the monitoring system is comprehensive enough to locate the point of slip. For example, the frictional forces along the fault could be monitored.

When attempting to develop a successful strategy to identify precursory phenomena, remember that different areas vary considerably in their petrophysical properties. For this reason a monitoring strategy should be developed based on understanding geological peculiarities of a particular region.

As a result of accumulated stresses being released, there is local crustal movement, which often precedes a major seismic event. Therefore, monitoring the movement of crustal blocks can provide valuable information on the seismic state of an area of interest. This monitoring is achieved by observing the position of tectonic blocks in space. The most advanced method of data collection related to positioning of tectonic blocks is based on the measurement of tilts for a set of characteristic points of the crustal surface.

Regular measurements of the crustal deformation prior to and after the San Francisco earthquake of 1906 showed that accumulated stresses were the cause of this earthquake. Special benchmarks were installed in the northern part of the San Andreas Fault to monitor the changes in strain. Direct measurements of deformations on opposite sides of the fault showed that, as a result of earthquake, the sides slipped along each other for a distance of 6 m. Regular measurements of the displacement of opposite sides of the fault plane showed that they move continuously along each other with an average velocity of 2.5 cm/year. If accumulated stresses and their release have an uniform character, then approximately 100 years are needed for the displacement of this magnitude. This process, however, is nonlinear and accelerates considerably on the eve of seismic events. Thus, the information on the rate of deformation (rate of displacement) can be used to develop a precursory pattern for a large earthquake.

TILT AND STRAIN OBSERVATIONS

The increased unbalanced crustal stresses can move crustal blocks in arbitrary directions that can be manifested in the changes of tilts. The Japanese scientists pioneered the study of the relationship between the patterns of changes in tilts and seismic activity in 1910. In 1930, they discovered a strong correlation between the frequent changes in tilts and seismic activity (Rikitake, 1979). Since then, experiments with tiltmeters were conducted for various locations, which produced useful information about the patterns of crustal movements during the periods of breakdown.

By the time the Tottori earthquake in Japan occurred in 1943, a special tiltmeter was installed in a vault of the Ikuno Mine, which was located about 60 km from the epicenter of the event. An anomalous tilting at this position was recorded 6 hours prior to the principal shock. Seven hours before the $M_R = 5$ Matsushiro earthquake of 1965, the tiltmeter at the Matsushiro observatory recorded frequent anomalous tilting in the observation mine. In these cases, anomalies were observed several hours prior to the principal shock. For many other observations, the anomalies occurred days and even months before the major seismic events. Anomalies in the tilt data are well correlated with the anomalies of crustal strains for the locale of interest, which make them an informative precursor of major seismic events.

In New Zealand, since 1972 regular monitoring of tilts in mines has revealed some anomalous patterns in the changes of tilts. During the same period, two strong shallow earthquakes (with $M_R = 5.5$ and $M_R = 6.2$) occurred at a distance of 65 km from the place of observation. The anomalies in tilts were observed one month prior to these earthquakes. However, the third deep earthquake (M_R = 5.9) occurred without any visible anomalies in the tilts. These facts lead to a hypothesis that tilt anomalies can be used mostly for prediction of shallow earthquakes, preparation of which can be manifested at the earth's surface.

For most data, the timespan of premonitory phenomena is longer than expected. Why does the earth's crust start to react as much as a month prior to the event? The results of rock mechanics experiments can be considered as a plausible explanation. Within a heterogeneous medium, local stress concentration occurs as the axial stress increases and, when a minor rupture occurs, the concentrated stress is instantly redistributed. If the stress increases again, local stress concentration reoccurs until the principal rupture occurs initiating the quake. The duration of precursory phenomenon may reflect the duration of this repeated process, which is important for predicting the time of earthquake occurrence.

Tilt anomalies occur frequently prior to large earthquakes. The time of their occurrence varies from several minutes to several months prior to the earthquakes. Many considerable anomalies, however, were not followed by major seismic events. This makes the problem of evaluating related anomalies very important. The question arises as to what pattern of anomalies can indicate the redistribution of stresses leading to the occurrence of large earthquakes. It is important to note that the results of monitoring and related conclusions strongly depend on the positions of tiltmeters. The selection of monitoring sites requires a great deal of preliminary investigation on the local structural peculiarities for an area of interest. Prediction is more reliable if the described tilt measurements are combined with the basic displacement observations that can be performed with GPS (Global Positioning System).

GPS provides a convenient means for measuring land surface subsidence. Chilingarian and coworkers (Katz et al., 1994) showed that land subsidence is strongly correlated with seismic activity, especially for areas with intensive oil and gas production. Thus, the measurements of land subsidence (or the surface profile) for these areas complement the tilt data creating a general picture of stress distribution and redistribution in rocks.

GPS consists of a set of orbiting satellites that continuously transmit radio signals. A set of receivers on the earth's surface process the radio signals, and the corresponding software determines the signal's time of travel from each satellite visible to a particular receiver. Then, the time of travel is translated into an instant distance between the receiver and satellite. For determining the coordinates of the absolute position of a receiver it must be visible from four satellites. The computed distances from the four satellites can be taken as radii of the spheres with the centers located at the satellites. Mutual intersection of these four spheres determines the point of the receiver, thereby providing its coordinates on the earth's surface. This method of locating the receiver is called "absolute positioning." The absolute positioning method is very similar to the technique of locating the earthquake hypocenter based on the measured distances from four seismic stations.

The accuracy of absolute positioning using GPS is not sufficient for surface subsidence (surface profile) monitoring. To improve the accuracy of positioning, the differential navigation technique can be used. This technique requires the simultaneous use of two closely positioned receivers. The location of one receiver is used as a reference benchmark. The position of the second receiver is always determined by adding the differences of its coordinates with respect to the first receiver. These differences are determined by the differences in traveling times from the two receivers to a particular satellite. The differential method of positioning is illustrated by Figure 11-8.

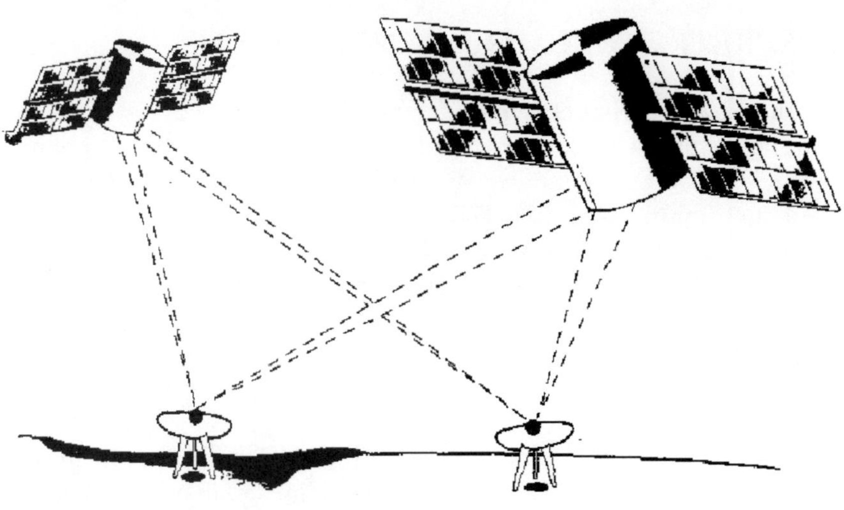

Figure 11-8. Differential method of positioning based on GPS observations. *(Modified after Chilingarian et al., 1995, fig. C-1, p. 455.)*

New monitoring systems, such as radar satellites, can considerably improve the quality of the earth's surface images. Radar images of the surface include bands that indicate elevation shifts of the surface as small as an inch. These satellites use radio waves to scan the earth's surface, which allows gathering information about the earth's "skin wrinkles" even at night and/or in bad weather.

Unfortunately, the long-length radio waves provide only a low resolution of the objects. To overcome this obstacle, scientists developed the technique of portraying based on the continuous scanning of the same place by the moving satellite. The radio echo gives the exact distance from the scanned point to the satellite. If the same point is scanned from a slightly different satellite position, under a different angle, it affords the possibility of creating a 3D image of a particular location (see Chilingarian et al., 1995).

The first radar satellites of this class were used by U.S government in 1988 for long-term orbital espionage. These satellites were also used for global topographic mapping using interferograms. These topographic maps were used for accurate missile launching during the 1994 Gulf War.

Over time, the surface shifts detected by the interferograms can be interpreted as changes in rock stresses. An example of this interpretation for the Landers fault in California is presented in Figure 11-9.

DISCRETE STRUCTURE OF THE CRUST

The observed patterns of the wrinkles on the earth's surface is similar to the "typical" wave patterns on the surface of a granular layer, discovered recently by Behringer and Pak at Duke University (Pak and Behringer, 1993). The basic experiments were fairly simple: A tray with an inch-thick layer of ordinary sand was placed on the table of a vibromachine generating vertical vibrations (*Discover,* Nov. 1996, pp. 105–116).

> Suddenly the sand appeared to come to life, rising for an instant off the tray, then collapsing again, rising and collapsing, like a formless rag doll. Gradually the tray picked up speed. In those instants when the sand was suspended in air, it began to take shape. Soon it formed the perfectly rounded hills and valleys of a sine wave.

Figure 11-9. Wrinkles on the earth's surface. *(Modified after New York Times, September 3, 1996.)*

> When the tray reached a maximum speed, the sine wave became square, which was an extraordinarily unexpected event.

There is no satisfactory theoretical explanation for this observation and no suitable model to describe the process of ripple formation. Even though this phenomenon cannot be explained, the fact remains that the earth's surface demonstrates shape patterns, which are similar to the granular material under a periodic disturbance. This provides a plausible approach to explain the "wrinkles on the earth's surface."

It is important to note that the presentation of the earth's crust as a solid continuum is a convenient abstraction rather than a scientific fact. As was shown by Sadovskiy (1979), the earth's crust consists of blocks of several predominant sizes. These crustal blocks are kept together by gravity force, frictional forces, and lateral stresses. Their mutual positions may provide some information about the distribution of stresses. These mutual positions of blocks can be monitored by a set of tiltmeters. Thus, the tilt measurements can indirectely provide information on the distribution of stresses in the earth's crust. Based on these considerations, anomalous patterns of tilts may be used as precursors of seismic events.

Uniform granular material, consisting of grains of the same size and shape (e.g., spherical), which are organized in a large self-supporting system, exhibits extremely heterogeneous patterns in distribution (and redistribution) of stresses in space and time. An example of such a self-supporting system containing uniform granules is a grain silo (storage facility for grain). A silo is usually designed and constructed under an erroneous assumption that the total weight of grains is uniformly distributed in space, with uniform pressure on all confining surfaces.

> In fact, when the grains come to rest against one another, they form intricate, quasi-self-supporting structures. That is why adding more grains to the top of a silo often does not increase the pressure delivered to the bottom, but rather increases pressure outward against the sides of the silo (*Discover,* Nov. 1996, pp. 105–116).

These grain structures can deliver a huge force to practically all points in the silo, in an unpredictable manner. Pressure may build up on one of the metal rings that encircle the silo and break it, thus rupturing the corrugated steel shell. According to Behringer (1996, p. 108), this is exactly what happened in Whitby, Canada, in 1990: 25 cows were buried under 500 tons of feed grain as a grain silo suddenly burst.

The grain assemblage in the silo is prone to forming a self-supporting static structure. Normally, the grains in the silo are motionless even if the bottom chute is open. The flow of grains tends to stop after a small portion is released from the silo. A conical self-supporting structure is formed, which supports the total weight of the grains in the silo when the chute is open. Figure 11-10 illustrates a pseudo-stable, static state of grains in a silo.

The above description of granular material behavior is similar to the readjustment of crustal blocks in the earth. Pseudo-static assemblage of the crustal blocks suddenly ruptures when an earthquake occurs, after which individual blocks abruptly move to find a new position of equilibrium for the entire assemblage. The crustal material involved

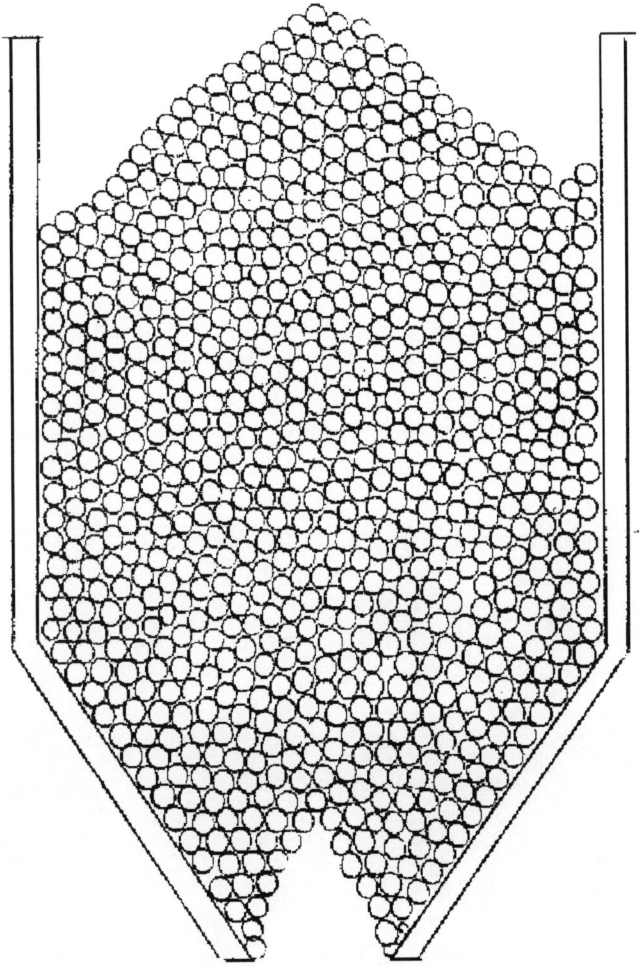

Figure 11-10. A pseudo-stable assemblage of grains in a silo.

in the preparation of and participation in the seismic event is not continuous but rather discrete. The crust consists of the rock blocks (chunks) of various sizes that are held together by gravity forces, friction, and lateral pressure. This circumstance sheds light on the chaotic nonuniformity of stress distribution in the earth's crust. The assemblage of rock blocks can create an intricate distribution of stresses in the crust that can deliver enormous forces far beyond the area of origin.

Uniform granular material exhibits extremely nonlinear, nonuniform transfer of applied stresses (in the space occupied by the granules). This was confirmed by Behringer's (1997) experiments on oscillating strains in the granular material caused by a periodic disturbance.

In his experiments, Behringer used an assemblage of small, flat polyurethane discs as a model of uniform granular material. These discs have the ability to change the manner in which they polarize light under the changing stress. The light penetrates freely along the line of maximum stress. Behringer placed the discs on a platter with a rotating hub at its center. As the hub rotated, it squeezed the discs that contacted it. In turn, these discs pushed other discs on the platter. The polarized light passed through the discs exhibiting the lines of greatest stresses. Watching the picture from above, Behringer could observe these stress chains flickering like flashes of lightening (Figure 11-11).

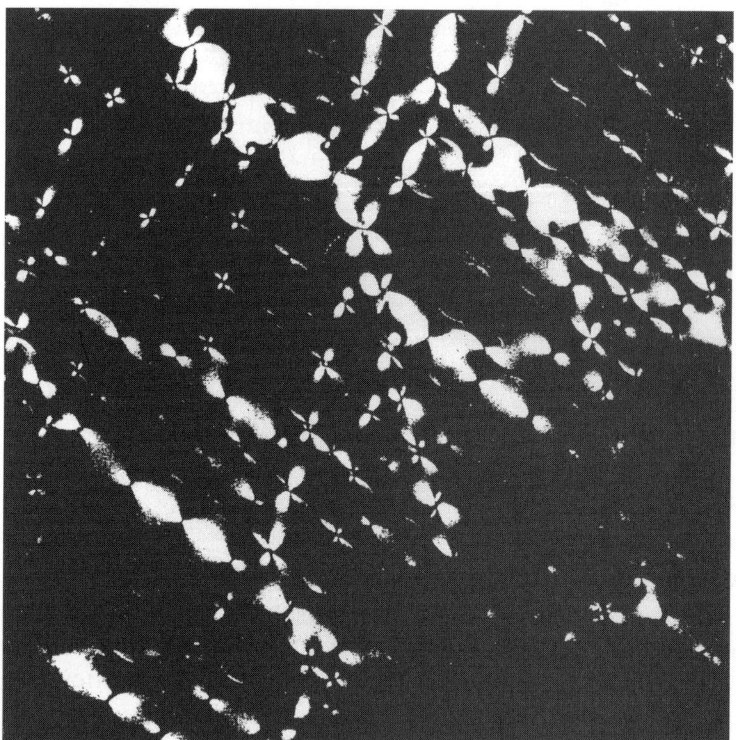

Figure 11-11. Flickering chains of stresses in the layer of polyurethane disks. *(Modified after Behringer, 1997, fig. 3, p. 996.)*

In the molten hot mantle of the earth, one can boldly hypothesize the existence of gigantic circular fluxes of matter and energy [this hypothesis is supported by the experiments with seismic waves (O'Nions et al., 1980)]. One of the well-known consequences of this continuous rotation is the continental drift. The circular fluxes can occur in various plane sections of the earth's sphere simultaneously causing large fragments of the crust to squeeze and relax periodically.

Because of the irregular shape and size of crustal blocks, the maximum stress paths are considerably more random than those for granular material. In addition, because of multiple sharp points and edges, the earth's material (the assemblage of blocks) can pack in many different ways. This makes any particular packing configuration very improbable and extremely unstable.

If changing (flickering) maximal stresses in a chain of blocks cross the point of contact of sharp edges of two large crustal blocks, these sharp edges may be broken in a small area around the point of contact. Consequently, the entire assemblage of crustal blocks loses the state of rest and moves slightly to form a new pseudo-stable state. Taking into account the huge masses of participating blocks, one can easily envision the power of the resulting blow.

These logical considerations supported by the experimental results on the behavior of granular material suggests a dynamic origin of the earthquake rather than the static concept of accumulated stresses. Alternating maximum stresses cross the earth's crust in various directions. Due to extreme nonuniformity of the mixture of blocks forming the crustal assemblage, these flickering chains form bizarre paths in the crustal medium. If this dynamic stress is applied to a weakened zone (such as a fault or fracture), with already accumulated static stresses, it is highly probable that the coherent rock blocks in this zone will rupture. The resulting shift of the blocks accompanied by a huge release of energy in various forms (mostly in the form of seismic waves) is felt as an earthquake.

If this dynamic picture of stress fluctuations in the earth's crust is correct, then monitoring the strains for the purpose of earthquake prediction is clearly unproductive. First, it is very difficult (if not impossible) to monitor a three-dimensional field of strains over a large territory. Secondly, because of the random dynamic changes in stresses, it is not possible to make any reasonable extrapolation of the value of stresses even in close proximity to the point of observation.

Thus, it is necessary to select another precursor that reflects the changes in local stresses and, at the same time, possesses some integral properties. An example of such a precursor is the rate of upward gas migration and its composition (concentration of various gases) at the point of interest.

Including more potential precursors in an existing monitoring system does not necessarily increase the chances for successful earthquake prediction. First of all, in many cases the signals generated by different precursors are contradictory and can be confusing. Secondly, to obtain primary information sufficient for a reliable short-term forecast, the area of interest must be covered with a dense network of gauges. If the intent is to measure many precursory parameters, then it is necessary to limit the extent and density of the monitoring network because of obvious budget limitations. Regular monitoring of a few informative parameters, which reflect the process of preparation and occurrence of earthquake, is sufficient for a short-term earthquake prediction technique.

All potential precursors are linked to each other because the cause is the same. Most parameters are difficult to measure. Therefore, before development and implementation of a monitoring network, it is first necessary to identify the most informative precursor (with the most distinct pattern of anomalies on the eve of major seismic events), taking into consideration the cost and convenience of measurements at many locations over large territories. The most appropriate indicator of the pronounced changes in the geodynamic conditions of a site is the "rate" of upward migration of gases (such as methane, carbon dioxide and radon), especially in the vicinity of faults and fractured zones. At the same time, chemical analysis of well waters and the content and composition of various gases should also be made.

Electrical Phenomena Associated with Rock Strain

There is some evidence that electrical rock properties change when the stresses applied to the rock are altered. There are two phenomena of interest:

1. Changes in the electrical resistivity as a result of the rock stress alteration.
2. Generation of electrical fields by piezoelectric effect.

Rocks and sediments conduct electric current in varying degrees, and appreciable changes in conductivity from one bed to another are quite common. For example, resistivity (reciprocal of conductivity) for shales varies from 1 to 100 ohm-m, whereas that of sandstones varies from 10 to 10^5 ohm-m. The conductivity is primarily a function of the electrolyte content of interstitial water and texture of the sediment (for example, tortuosity). The electrical properties of various minerals and rocks are given by Guyod (1944, p. 5).

There are many sources of potential difference in the ground and the principal ones include: (1) electrochemical action, (2) contact of dissimilar substances, and (3) electromagnetic induction (Chilingar et al., 1966). Any dissimilarity existing between two substances, which are in contact with each other, results in a potential difference between them. It is not necessary, however, for a current to flow between substances having potential differences. If there is a potential difference in a homogeneous substance, there must be a current flowing through it somewhere. Current will only flow in a closed circuit; otherwise a battery or a capacitor situation is present.

In addition to telluric currents, there appears to be a steady vertical electric current in the earth's crust having a density of 1–100 amp/km^2 (R. Ambronn, 1926; in Guyod, 1944, p. 97).

Streaming potential (electromotive force) is generated across a porous membrane when an electrolyte is forced (by pressure) through the porous medium. On the other hand, application of direct current to a porous medium filled with solutions results in steady movement of positively charged ions toward the negative electrode (Chilingar et al., 1965, 1970).

The change in resistivity is a function of changes in the porosity, fluid saturation, and chemistry of interstitial fluids. Therefore, measurements of electrical resistivity can provide information on porosity that can be altered as a result of variations in stresses. The changes in permeability, in turn, change the migration rates of gases to the surface. This presents one possible explanation for the radon gas migration to, and changes of its concentration in, the near-surface layers (radon anomalies) on the eve of major seismic events.

It is difficult to explain the generation of strong electric fields (usually DC fields) in the ground. The existence of these fields may explain the lights in the atmosphere on the eve of strong earthquakes. These fields may be generated by the piezoelectric effect, which manifests itself in pronounced redistribution of electrical charges in some minerals when strong pressure is applied. They can be developed and maintained *if* the resistivity of rocks is of an order of 10^9 ohm/meter (the resistivity of quartz) and the mineral crystals are statistically oriented over several kilometers (which seems to be highly improbable). Because the realistic values of resistivity are in the range of 10 to 10^4 ohm/meter, Morrison (1976) stated that it is difficult to expect the development of significant DC fields in the ground.

The piezoelectric fields associated with high-frequency components of seismic waves may form AC fields. This does not apply to shallow quakes, for which seismic waves and associated changes in stresses would decay very quickly and could not be registered at the surface. It is necessary to note that these fields can be generated by seismic waves at the time an earthquake occurs or as a result of foreshocks and, for this reason, cannot serve as premonitory events.

VAN Method of Earthquake Prediction

The VAN method, named after Greek Professors Varotsos, Alexopoulos, and Nomicos, uses Seismic Electric Signals (SES) as short-term precursors of earthquakes (Uyeda, 1993). A solid medium containing dipoles (linear particles with oppositely charged ends) generates electric currents if the changing applied stress exceeds some critical threshold (Varotsos et al., 1996). In general, this critical threshold differs from the rupture stress level. In many cases, however, it is less than the rupture level and SES can be used as a precursor of the rupture event. In the crustal medium, Professors Varotsos, Alexopoulos, and Nomicos discovered sensitive zones (sometimes located tens of kilometers away from the site of interest) for which SES can be easily registered. Several sensitive zones can be attributed to the same epicentral area. Mapping these zones for a particular suspected epicentral area, so-called selectivity maps can be produced.

The selectivity maps are a means of locating a future focal zone of SES occurrence. During the experimental stage of the studies, the VAN group discovered and tested an important relationship among focal distance, R, magnitude of future earthquake, M, and the observed intensity of SES ($\Delta V/L$), where ΔV is the voltage amplitude of SES and L is the dipole length. The relationship is given by the following formula:

$$\log(R \times \Delta V/L) = aM + b \tag{11-1}$$

where a and b are constants that will be determined from the statistical data for the sites of interest.

Selectivity maps allow for determining the distance R; whereas the intensity of SES ($\Delta V/L$) is measured instrumentally. Once these variables are defined, Eq. 11-1 can be applied to estimate the magnitude M of an oncoming event.

This method has been tested in Greece since 1982 and has demonstrated a considerable rate of success. Between January 1, 1987, and June 15, 1995, 14 large earthquakes occurred over the territory where this method was tested. Ten of the quakes were successfully predicted. Because of the high rate of success, this method has been scrutinized by the world's scientific community for the last decade. Special discussions were dedicated to this method, and the recommendation to extend it to other seismic territories was made (Uyeda, 1996). Unfortunately, this did not happen because utilization of the method requires meticulous preliminary work in searching for the sensitivity zones, development of the instrumental base, and establishment of the monitoring network for detection and analysis of SES.

Wave Velocity Ratio as a Precursor

Under normal conditions, the ratio of wave velocities (V_p/V_s) does not vary considerably. For a particular area of the earth's crust, this ratio is almost constant and its mean value for various directions of propagation is approximately equal to $\sqrt{3}$ (Karapetyan and Karapetyan, 1981). Regular measurements of this ratio on the eve and at the time of major seismic events have been made in several different geographic areas. The graphs of $[V_p/V_s(t)]$ as time functions demonstrate a distinct pattern of considerable fluctuations of this ratio on the eve of major seismic events. These experiments were carried out independently in the United States, Russia, and Japan.

Large amount of information on the velocity ratio was collected by the prominent Armenian seismologist, N. Karapetyan, in connection with the catastrophic Spitak earthquake of 1988 (Karapetyan, 1991). Karapetyan tried to use the (V_p/V_s) ratio to develop a prediction technique for 20 significant earthquakes that occurred in the hypocentral area of the Spitak earthquake from January 1967 through

December 1988. During the two months preceding the events, this ratio gradually decreased and then increased to the primary (or a slightly higher) value immediately prior to the earthquake (Karapetyan and Karapetyan, 1981). This pattern of change is illustrated in Figure 11-12. Selected average values of this ratio for several significant seismic events in the vicinity of the principal Spitak earthquake are presented in Table 11-1. It seems that this ratio can provide prognostic information not only on the time and place of occurrence, but also on the energy released in the hypocentral area of an earthquake.

Table 11-1
The (V_p/V_s) Ratio in the Hypocentral Zone of the Spitak Earthquake Showing the Time of Earthquake Occurrence, Energy Class[1], and V_p/V_s Ratio (Mean Value)

Year	Date	Hour	Minute	Class, K[1]	V_p/V_s Ratio
1967	1/30	01	20	13	1.762
1973	6/27	23	24	10–11	1.712
1977	1/10	12	26	10	1.706
1978	1/17	19	12	11	1.714
1979	7/03	05	48	9	1.697
1979	12/18	23	20	9	1.692
1980	5/07	03	41	9	1.694
1980	5/09	03	37	9	1.703
1982	1/17	10	27	10–11	1.732
1986	3/12	17	53	10–11	1.686
1986	3/26	22	25	10	1.664
1986	3/29	19	11	9–10	1.669
1986	7/27	05	59	9–10	1.669
1987	4/15	15	48	9–10	1.665
1988	6/22	01	18	11	1.671
1988	9/03	15	55	9	1.669
1988	11/11	17	55	8–9	1.702
1988	12/03	15	54	7–8	1.718
1988	12/06	15	27	9	1.748
1988	12/06	18	34	7–8	1.750
1988	12/07	07	41	16	1.817

[1] K = *lg* E, *where* E *is the released energy expressed in joules.* $lg\ E = 4 + 1.8\ M$ *for* $M < 6$ *and* $log\ E = 8 + 1.1\ M$ *for* $M \geq 6$.
After Karapetyan, 1986, 1991.

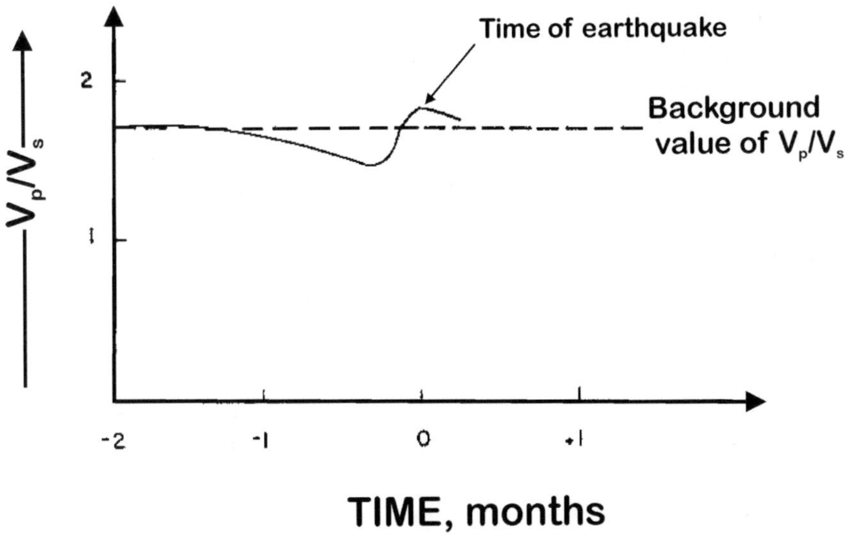

Figure 11-12. Pattern of changes in V_p/V_s ratio on the eve of an earthquake.

The data presented in Table 11-1 explicitly show a gradual decrease in the V_p/V_s ratio on the eve of the major Spitak earthquake (1988) with an abrupt jump-like increase at the time of the main shock. Two hypotheses were proposed to explain this phenomenon (Karapetyan and Karapetyan, 1981). The first hypothesis suggests that the change in the acoustic properties of crustal medium occurs as a result of microfracturing, which appears at the stage of earthquake preparation. The second assumes that the acoustic properties change due to variations in the composition and pressure of pore fluids in the hypocentral zone of future earthquakes.

CHAPTER 12

Fluids Talk

HYDRODYNAMIC ANOMALIES

There are numerous examples where major seismic events are accompanied by pronounced anomalies in the dynamics of ground fluids. The Lisbon (Portugal) earthquake of September 1, 1755, for example, has led to dramatic changes in the hydraulic regime of wells over the territory of Western Europe. In Chekhia and France, many wells dried out, whereas in other areas new groundwater sources emerged with huge debit of water.

Rikitake (1976) collected information on groundwater anomalies from Japanese legends, which have accumulated over the centuries. Some examples include:

1. When the water flow of a fountain decreases, an earthquake is approaching.
2. When the well water level suddenly decreases, an earthquake may be approaching.
3. A sudden lowering or rising in the level of the groundwater may be a precursor of an earthquake.
4. When well water suddenly becomes muddy, an earthquake may be approaching.

The contemporary stress accumulation theory for predicting earthquakes provides a satisfactory explanation for these anomalies in groundwater levels and rates of discharge on the eve of major seismic events. The intense stress load variations in the tectonic blocks participating in the earthquake preparation cause considerable changes in the formation fluid pressures. As a result, well water levels in the vicinity of the area of active preparation of an earthquake are changing, which indicates an alteration in the underground fluid pressure. Thus, water wells can be considered sensitive gauges of the fluid pressure at the points of monitoring. There are many well-documented cases of water level changes from Russia, Japan, and China, which provide data on the temporary changes in water levels for selected wells before and after earthquakes.

The ability to recognize a pattern of change in groundwater levels preceding a principal shock is the most important problem of premonitory observation. To discover this pattern, it is necessary to establish a time series of systematic observations in groundwater levels at the same location at several (fixed) control points. Relevant information can be found, however, for only a few seismic areas due to the lack of water wells.

The Russian Academy of Sciences initiated regular monitoring of groundwater levels on the Island of Kunashir in the extremely seismic area of the Kuryl Islands (Sadovskiy et al., 1977). Three wells were designated especially for the purpose of earthquake prediction. The water table levels in these wells have been measured continuously and automatically recorded since November of 1976. In order to establish the relationship between the levels of groundwater and earthquake occurrence, it was important to evaluate the influence of weather factors such as atmospheric pressure and precipitation. The specially conducted investigation revealed that the water level fluctuations attributed to these factors did not exceed 3–4 cm, with the absolute accuracy of water level measurements of 1–2 cm.

Several graphs of the water level changes over time for three experimental wells are presented in Figure 12-1. According to these graphs, the water levels gradually decreased several days prior to the significant earthquakes. The rate of decrease slowed down considerably right before the shock. Usually, the earthquakes occurred when the groundwater level was at a minimum or starting to ascend.

Figure 12-2 shows a characteristic pattern of the groundwater level changes on the eve of earthquakes. Experiments on earthquake prediction based on recognition of this pattern were conducted on the Island of Kunashir in 1976 and 1977. Sadovsky et al. (1977) reported that this pattern of water movement allowed successful forecasting of several large earthquakes. Earthquake warnings, however, were terminated after several false alarms, which spread panic among the people of the far eastern provinces of Russia.

Systematic data on the dynamics of groundwater were collected in the case of nine major earthquakes in China, which occurred between 1966 and 1976 (Tangshan, 1976, with two strong (M_R = 7) aftershocks;

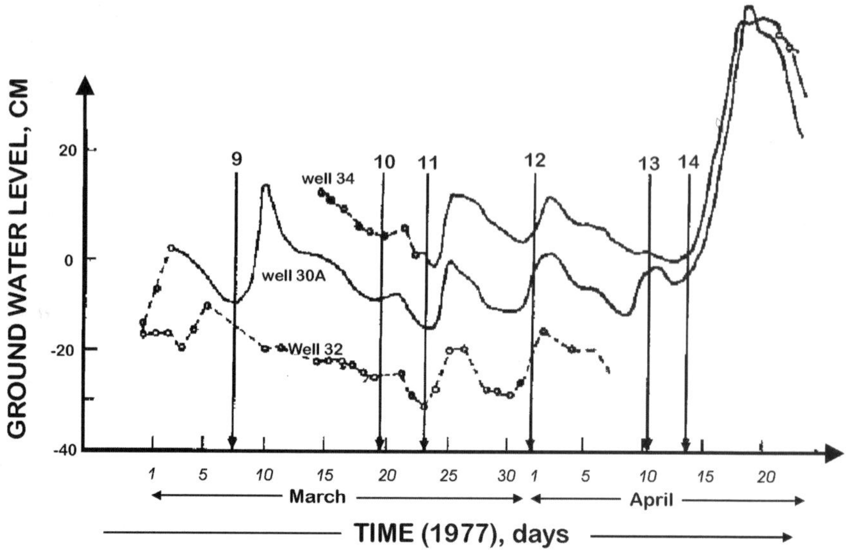

Figure 12-1. Changes in groundwater levels (with respect to normal) preceding earthquakes in the three experimental wells on the Island of Kunashir, Kuryl Islands. *(Modified after Sadovskiy et al., 1977.)* Numbers 9–14: original numbers of earthquakes.

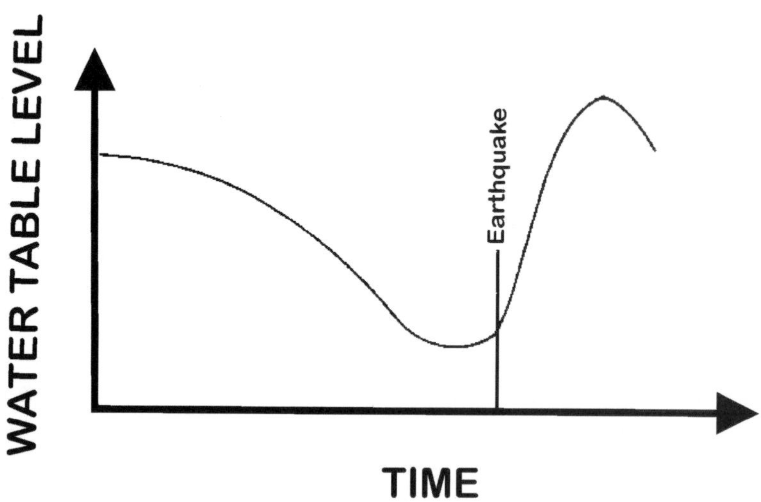

Figure 12-2. Pattern of changes in groundwater level on the eve of an earthquake.

Xingtai, 1966; Bohai, 1969; Haicheng, 1975; Longling, 1976; Songpan, 1977; Dulan, 1971). These earthquakes occurred within a confined area with a well-developed network of groundwater wells for which a special observation network was developed and implemented long before the earthquakes ocurred (Ma et al., 1988).

The strongest earthquake (Tangshan, 1976), followed by two devastating aftershocks, occurred within the Beijing—Tianjin—Tangshan observation network (Figure 12-3). A large amount of data on groundwater levels was collected in this area. The pattern of water-level fluctuations was the same for all wells in the vicinity of the epicenter (Ma et al., 1988). The water level decreased slowly during the four-year period prior to the main shock. For the last three preceding months, this decline accelerated in several wells and, in general, the groundwater level became the lowest during the entire history of observations. Three to four days prior to the principal shock the water level in the epicentral area drastically increased—some wells even became artesian. The principal shock occurred at the time of maximal acceleration of the rise in water level.

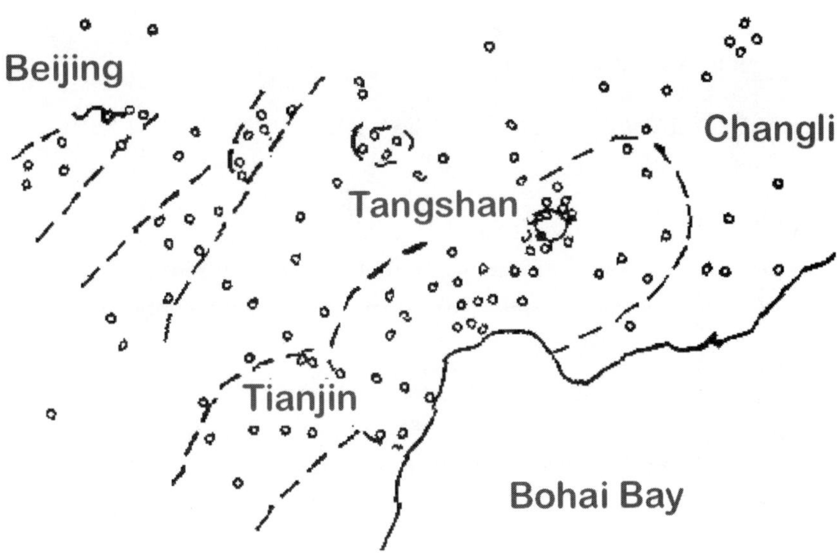

Figure 12-3. Observation network of groundwater wells in China *(Modified after Ma et al., 1988.)*

Many pronounced hydrodynamic anomalies accompanied the Tangshan earthquake (Ma et al., 1988). Three hours before the first aftershock with M_R = 7.1, the water level in the Biokou well increased rapidly by 8 cm on the background of long-lasting decline trend. The same pattern was observed prior to the second aftershock with M_R = 6.9 on December 15, 1976. Distinct anomalies were also observed in the adjacent areas. For instance, the water level at the Lunnan meteorological station drastically increased, surpassing regular variations of this parameter. At the Panshan well, located at the north end of the Tancheng—Yingkou fracture zone, the water level declined at a steady rate for several months prior to the main shock, moved upward several hours prior to the earthquake, and continued to rise after its occurrence.

The above anomalies are illustrated in Figure 12-4, where variations in groundwater debits in the Tangshan mine are shown. Analyzing these anomalies, two conclusions can be made: (1) main seismic events are often accompanied by pronounced changes in the water level (debit) in wells situated in the vicinity of the epicenter, and (2) there is no regular pattern of changes in the groundwater level (or debit) on the eve of a major seismic event. Of course, the patterns vary considerably for various wells and geologic scenarios. In order to establish a characteristic pattern, each particular well must be individually investigated.

CHEMISTRY OF GROUNDWATER

Several Japanese investigators reported changes in the chemistry of groundwater before and after the Kobe earthquake of 1995. It was previously suggested that changes in the pH of underground water should be monitored continuously in the selected water wells because of changes in the migration rate of CO_2 to the earth's surface prior to earthquakes and the resulting chemical reactions:

$$CO_2 + H_2O \leftrightharpoons H_2CO_3$$

$$H_2O_3 + H_2O \leftrightharpoons H_3O^{+1} + HCO_3^{-1}$$

With increasing CO_2 content, there is an increase in both $[H_3O^{+1}]$ and $[HCO_3^{-1}]$. The net result is a decrease in the pH and an increase in the

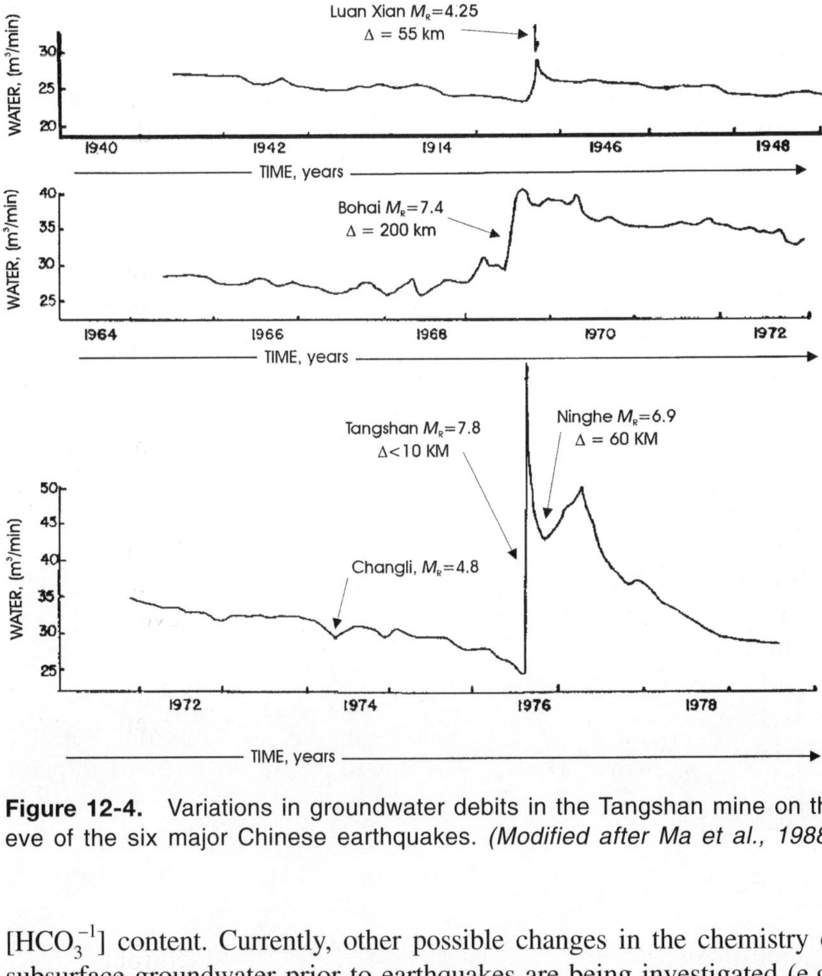

Figure 12-4. Variations in groundwater debits in the Tangshan mine on the eve of the six major Chinese earthquakes. *(Modified after Ma et al., 1988.)*

$[HCO_3^{-1}]$ content. Currently, other possible changes in the chemistry of subsurface groundwater prior to earthquakes are being investigated (e.g., the effect of decomposition of migrating CH_4 in the presence of bacteria). This may give rise to additional CO_2, which will further decrease the pH of groundwater (rise in the acidity of groundwater).

CHAPTER 13

Whispering Gases

INTRODUCTION

For many reasons, we do not believe that the greater the number of potential precursors examined, the better the developed forecasting system. First, different premonitory phenomena frequently send contradictory messages, which can confuse decision makers. Secondly, this considerably delays the final determination of the imminence of earthquake that can be crucial in short-term forecasting. Finally, because of budget limitations that are faced by those starting to develop and implement a monitoring network system, it is necessary to choose between expanding the list of monitored parameters and the density of the monitoring network. It is recommended that only a few informative parameters be monitored, which reflect the process of preparation of an earthquake (at the nodes of a regular detailed monitoring network) in order to develop a short-term earthquake prediction technique.

Representative precursors are supposedly related to each other because they are symptoms of the same cause. Some precursors are actually functions of others. Specific resistivity of ground, for example, acts as a highly sensitive strain gauge just before an earthquake. The rates of gas migration to the earth's surface, however, can be substituted for the resistivity measurements because resistivity reflects changes in ground porosity and, consequently, permeability. The latter controls the rates of upward gas migration. As previously noted, these migration rates reflect the crustal conditions in the vicinity of the point of monitoring and can be easily measured in contrast with the resistivity. The latter is very sensitive to the measurement location and demands specially-designed instruments and experiments.

Many potential premonitory parameters are difficult to measure. Thus, before developing and implementing a monitoring network, it is necessary to first identify the most informative precursor (with the most distinct pattern of anomalies on the eve of major seismic events), taking into consideration the cost and convenience of measurements at many locations over large territories. We believe that the most appropriate indicators of the pronounced changes in the geodynamic conditions on the eve of earthquakes are the rates of upward migration of gases (such as methane, carbon dioxide, and radon), especially in the vicinity of preexisting faults and fractured zones. Obviously, the gas concentrations in well water must be continuously monitored in conjunction with the chemical analysis of these waters (especially pH).

RADON GAS CONCENTRATION AS A PRECURSOR

In the early 1980s, an intensive discussion on the value of radon gas as a precursor of earthquake occurrence appeared in scientific literature. Anomalies in radon concentration in underground water prior to earthquakes in various parts of the world have been observed on numerous occasions. A thorough investigation on the radon gas precursor was conducted in the early 1960s by seismologists in connection with the disastrous Tashkent (Uzbekistan) earthquake (Abduvaliev et al., 1986). It was discovered that the radon concentration in the groundwater increased considerably on the eve of the earthquake.

Similar radon anomalies in the soil and in the lower layer of atmosphere were observed repeatedly in various geographic regions (i.e., the San Andreas Fault in California, the Western Fergana in central Asia, China, Japan, etc.). In many cases, however, there was a decrease in the radon concentration. This created doubt as to whether the radon gas concentration could be used as a potential precursor. This pessimism was not justified (Chilingar et al., 1996), however. The patterns of radon gas variations were observed repeatedly on the eve of strong earthquakes as shown in Figure 13-1.

One explanation for the considerable variation in the radioactive radon gas concentration on the eve of a seismic event involves changes in the permeability of faults and fractured zones due to stress redistribution in the process of preparation of earthquakes. In the late 1970s, Sadovskiy (1979) showed that predominant sizes of crustal blocks form during the process of crustal disintegration. The vertical dimension of some blocks may be on the order of the depth of the

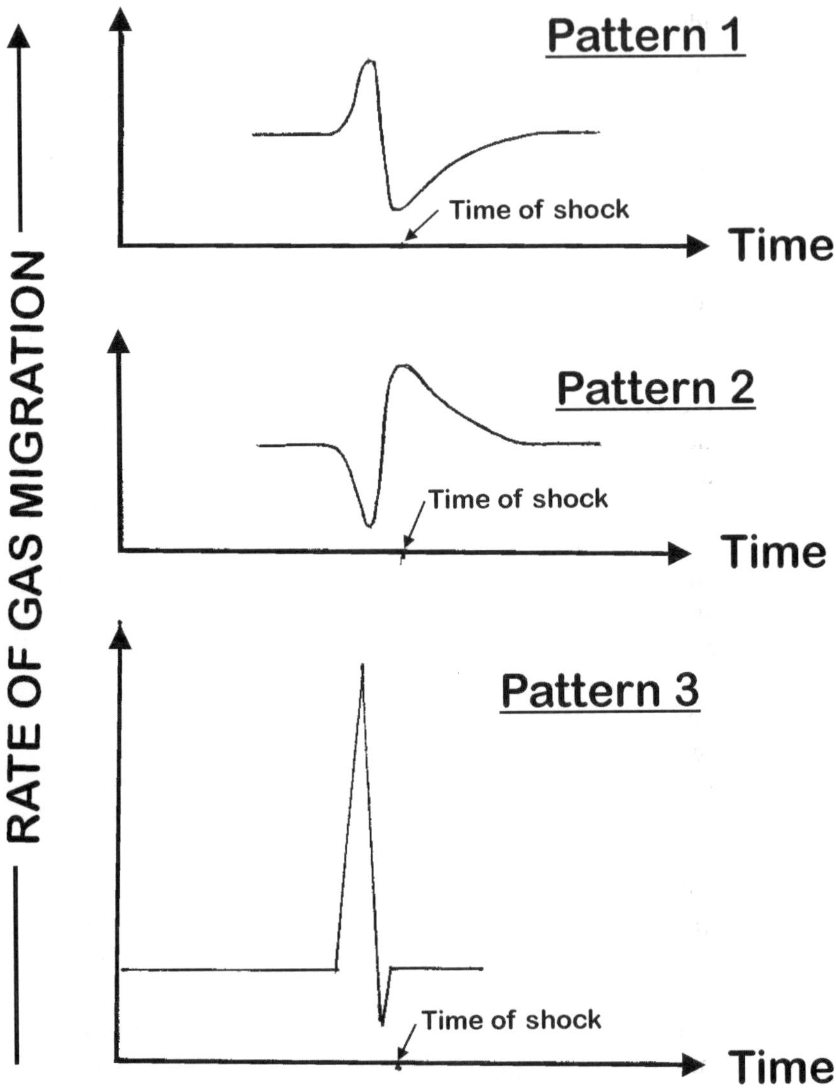

Figure 13-1. Patterns of change in the radon concentration in the soil gases and groundwater before and after an earthquake.

crust. The boundaries of these blocks represent weakened zones, where faults and fractured zones can be formed with resulting high permeability. These permeable zones are avenues for the migration of the earth's fluids (especially gases) from the hot layers of the crust to the discharge space (i.e., to the earth's atmosphere). For example, the rates of upward helium flow for these zones can exceed 100 times the usual rate of flow. The volumetric rates of the migration of gases, however, can also decrease as a result of changing stresses, which can reduce the widths of the faults and fractures with subsequent reduction in permeability. There are considerable fluctuations in the migration rates of gases in seismically active areas, especially on the eve of major seismic events.

Moving upward from the lower (hot) layers of the crust, the various gases (filling the pore space inside the rocks) can act as carrier gases, dragging the entrapped heavy, radioactive radon gas molecules and carrying them in the direction of the earth's surface. At the earth's surface, gases accumulate either in soil gases or in groundwater. This possibly explains the formation mechanism of the radon anomalies on the eve of seismic events. The vehicle for this process is a vertical flow of lighter gases (e.g., CH_4 and CO_2) from the lower layers of the earth's crust or from the oil and gas fields.

This mechanism explains not only the radon gas anomalies on the eve of seismic events, but also the nonuniformity of radon gas distribution in the soil gases, underground waters, and the lower layer of atmosphere around the earth's surface. For example, considerable differences in seismicity for different geographic areas can explain why radon concentrations in the groundwater and atmosphere in southern Kazakhstan and central Asia are 15–20 times greater than those of western Siberia and Yakutia in Russia (Abduvaliev et al., 1986).

In Figure 13-2, the concentrations of radon gas are given for the time period from October 18, 1980, to December 11, 1980, corresponding to the two western Fergana (central Asia) strong earthquakes (with $M_R \approx 5$). In these graphs, strong peaks (up and down) of the radon gas concentration correspond to the moments of seismic events. These variations were recorded at observation stations located more than 150 km from the epicenter of events. The anomalies in the radon gas concentration can be explained by the development of weakened zones in the vicinity of boundaries of tectonic blocks.

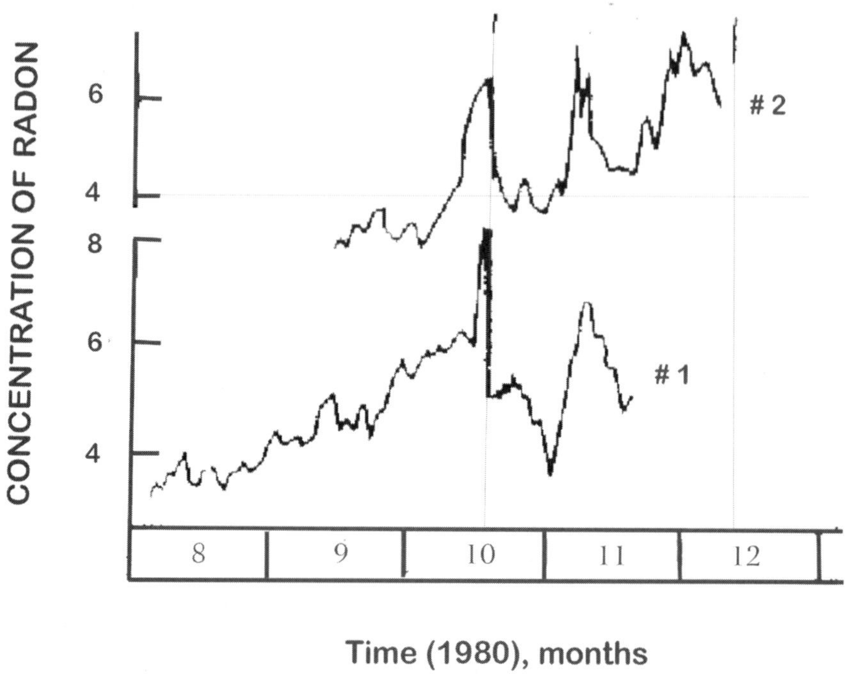

Figure 13-2. Radon concentration anomalies accompanying earthquakes in the experimental well in the western Fergana area (central Asia). *(Modified after Abduvaliev et al., 1983.)* 1 = count of alpha particles; 2 = count of β particles.

Initially, there is a development of faults and fractured zones as a result of increased fluctuation in the rock stresses. This leads to the alteration of rock permeability and the development of highly permeable zones in weakened sections (or, vice versa, to reduction of permeability as a result of narrowing of faults and fractures). The upward flow of gases is accelerated through the faults and fractured zones (fractures have a very high permeability; see Chilingarian et al., 1992). Gases are then discharged into the surface soil layer, underground and surface water bodies, and the lower layer of the earth's atmosphere. Under normal conditions, the variations in the radon gas concentration are not considerable and mostly lie within the range of natural long-term fluctuations. There is also some insignificant migration of heavy radon gas by diffusion.

As a result of additional stresses accumulated prior to earthquakes, some fractures become oriented forming coherent conduits. Immediately prior to, and at the moment of a seismic shock, the stresses are instantly released. As the formation pressure drops considerably, the adsorbed gases are released into the pore space. Also, some dissolved gases separate from liquid. Along the conduits, the flow rates of gases (CO_2, CH_4, etc.) increase considerably and, correspondingly, the rate of the entrapped radon gas migration exceeds the usual limits. In other areas, the permeability can be considerably reduced as a result of tectonic shifts and destruction of conduits. This can be manifested by a decrease in the migration rates of gases to the earth's surface.

At the moment of principal rupture that originates an earthquake, the adsorbed gases are released and migrate to the earth's surface. The resulting changes in the radon gas concentration can be easily recorded by radioactivity measuring devices.

The most pronounced anomalies are observed on the eve of, and immediately after, a seismic event as shown in Figure 13-3. In most cases, the seismic event occurs after the concentration of radon gas reaches its extreme (maximum or minimum depending on the monitoring point). The patterns of changes of the radon gas concentration over time are similar to the patterns of changes in concentrations of other gases (for the same points of observations). Apparently, light gases (e.g., CO_2 and CH_4) serve as the carriers for the heavy radon radioactive gas. The pronounced anomalies in the radon gas concentrations coincide with the anomalies in the flow rates of other gases (see Figure 13-4). Changes in the concentrations of light gases prior to afterschocks of the Dagestan earthquake of May 14, 1970, are shown in Figure 13-5.

CARBON DIOXIDE GAS MIGRATION AS A PRECURSOR

Sources of underground CO_2 gas (besides the plant root respiration) are: (1) decomposition of organic material near the surface, (2) degradation of organic material, (3) metamorphism of carbonate rocks (limestones and dolomites), and (4) the mantle.

The $\delta^{13}C$ of CO_2 derived from the mantle ranges from –4.7 to –8.0‰ (relatively depleted) (see Irwin and Barnes, 1980, p. 3115). Organic matter is more depleted in ^{13}C than mantle CO_2 (–12‰ or lower for CO_2 derived from the breakdown of land plants or from petroleum).

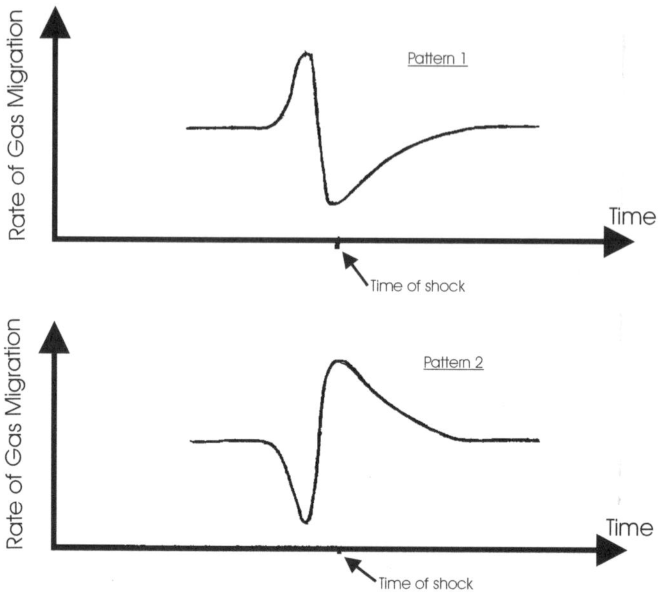

Figure 13-3. Possible patterns of changes in the rate of gas migration to the earth's surface on the eve of a seismic event. *(After Chilingar et al., 1996, fig. 2, p. 135.)*

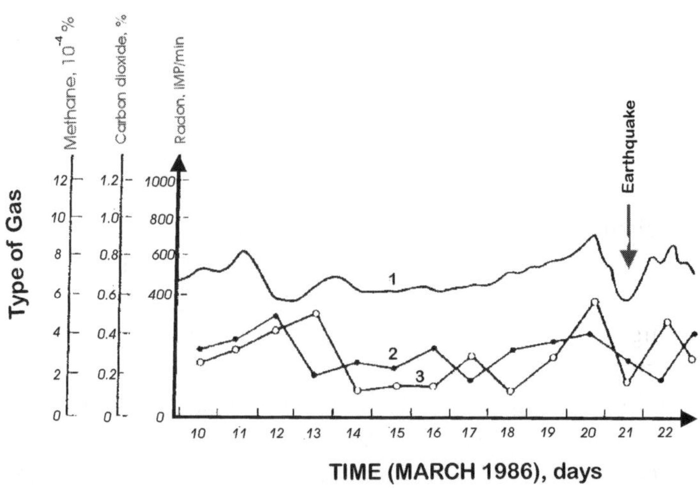

Figure 13-4. Changes in concentrations of (1) radon, (2) carbon dioxide, and (3) methane in the subsurface soil gases prior to the Ura-Tyubinsk earthquake, Central Asia, on March 21, 1986; $M = 4.6$. *(Modified after Abduvaliev et al., 1990, fig. 3, p. 1336; courtesy of the Russian Academy of Sciences.)*

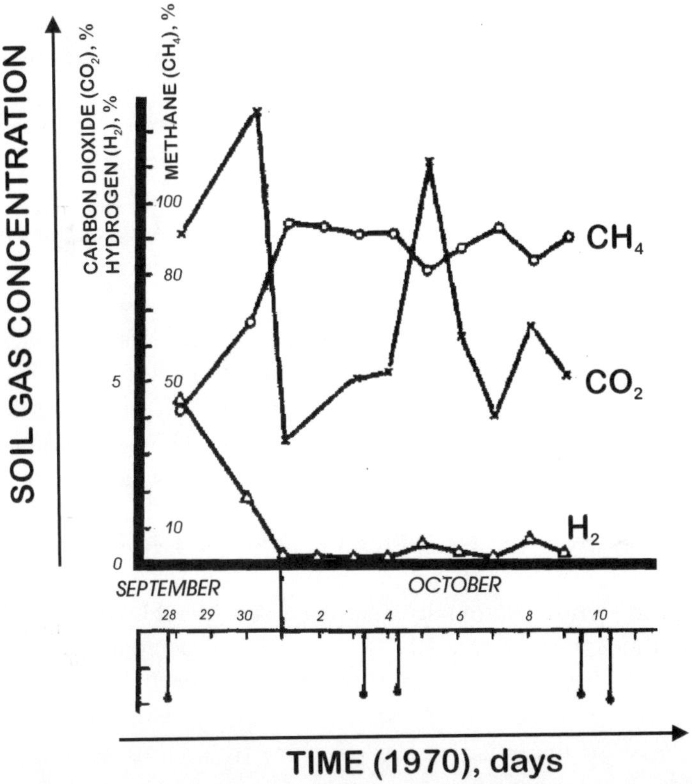

Figure 13-5. Changes in the concentrations of light gases (CH_4, CO_2, H_2) in the groundwater prior to aftershocks (vertical segments) of the Dagestan earthquake of May 14, 1970. *(Modified after Khitarov et al., 1974, fig. 1, p. 167.)*

Major seismicity zones worldwide are characterized by CO_2-rich springs. Thus, the monitoring of CO_2 content in water could be useful in earthquake prediction. Obviously, in the absence of springs, CO_2 gas also migrates along faults and fractured zones, and variations in its rate of migration should be monitored in the future.

Influx of CO_2 gas into spring waters gives rise to high HCO_3^- (bicarbonate anion) content (known as "soda water"). Springs that contain greater than 1,000 ppm HCO_3^- are associated with areas of seismic activity. In Russia, CO_2-rich springs occur in zones of intense seismic activity in the Caucasus mountains, Kamchatka, Sakhalin, and the Kurile Islands. Tectonically stable east European and Siberian platforms are devoid of CO_2-rich springs.

CO_2 gas can also act as a carrier for other gases. According to Irwin and Barnes (1980), radioactive components with short half-lives could be practical earthquake harbingers.

As a result of regional or contact metamorphism, carbonates (limestones and dolomites) are converted to calc-silicate minerals with resulting release of CO_2. According to Irwin and Barnes (1980), extensive terranes of pumpellyite-bearing metagraywacke indicate previous production of CO_2 resulting from metamorphism. The $\delta^{13}C$ content of CO_2 of this origin is close to 0‰.

The importance of monitoring the CO_2 content lies in its ability to produce high pore pressures, which reduce the effective frictional strength of rocks (Hubbert and Rubey, 1959). This may cause slip on faults. Thus, the presence of CO_2 in regions, where earthquakes are practically unknown, may indicate a potential seismic hazard.

CHANGES IN CONCENTRATIONS OF LIGHT GASES

The earth's crust continually releases a considerable amount of gases into the atmosphere. Vernadskiy (1934, 1940) metaphorically used to call this process the "gaseous breath of the earth." It reflects the chemical reactions occurring inside the deep layers of the earth's crust and the state of the existing conduits for discharging these gases into the earth's atmosphere.

The distribution of surface emanation of gases is not uniform over the lithosphere and varies in time. Fluctuations in the composition of gases and migration rates to the earth's surface provide important information about the tectonic processes preceding seismic events. The increase (or decrease) in the migration rates of gases and pronounced changes in the composition of gases can indicate ruptures in the various layers of the crust as a result of tectonic activity. Variations in the gas migration to the earth's surface can manifest themselves as local weather changes.

Among the numerous stories of unusual weather phenomena preceding earthquakes, stories can be found about mysterious mist and fog. In the book entitled *Earthquake Prediction* (1976), Japanese seismologist Tsuneji Rikitake tells of the mysterious heavy air, known as "Chiki," rising from the ground on the eve of earthquakes. According to the legend, several days before the 1802 earthquake ($M_R = 6.6$) that struck Sado Island, Niigato Prefecture, Japan, a businessman wanted to go sailing. A boatman told him that the weather was extraordinary:

misty in a strange manner with mountain slopes covered by mist, whereas the upper parts of the mountains could be clearly seen. The businessman recalled what his father used to tell him: "Prior to an earthquake, a strange air called "Chiki" (which literally means air from the earth) comes up from the ground." They walked away from the town for a distance of about 15 km, when a strong ground shock occurred.

Miners in the area foretold the earthquake three days prior to its occurrence by the appearance of "Chiki." When "Chiki" rose in the mine, the workers could hardly see each other. They left the mine immediately and did not come back until the earthquake occurred.

There are many well-documented occurrences of the "Chiki phenomenon" in Japanese scientific literature. The documents were revised by Musha and published in a three-volume edition under the title *Dai Nihon Jishin Shirio* (Imperial Earthquake Investigation Committee) during World War II.

Numerous observations confirm that on the eve of major seismic events, pronounced changes in the rate of gas migration to the earth's surface do occur (Chilingar et al., 1996). This process is manifested mainly in the vicinity of faults and fractured zones because they represent avenues for the migration of gases (due to high permeability). Faults and fractured zones also often represent the places of contact of the boundaries of tectonic blocks involved in an oncoming seismic event. These weakened zones are the most sensitive to stress changes on the eve of earthquakes. Changes in the stress lead either to the formation of new channels (faults or fractured zones) for the migration of fluids to the earth's surface, or to the destruction of old ones. Both processes contribute to the changes in the rates of upward gas migration (increase or decrease) prior to earthquakes.

The integral characteristics of a medium that immediately reflect changes in rock stresses are porosity, formation fluid pressure, and permeability of the rocks. These characteristics are most sensitive to the stress changes along the faults between the crustal blocks participating in a seismic event. Consequently, on the eve of major seismic events the changes in the natural gas migration rate along the faults occur as a result of rapidly altered permeability and pressure fields, especially in the presence of oil and gas reservoirs. Possible patterns of these changes are given in the Figure 13-3. Supposedly, in the vicinity of an epicenter, examples of both patterns for the same event can be found in approximately equal numbers.

It is necessary to monitor the concentration of various gases in the subsurface soil gases or groundwater near the faults or fractured zones (or, more accurately, its time derivative, which varies directly with the rate of migration). The representative normal value of the migration rate for some specific area can be obtained by averaging this parameter over the area of interest.

Unfortunately, in the areas of oil fields there have been no systematic observations on the rate of natural gas migration along the faults and fractures in seismically active areas. Many researchers, however, have observed pronounced changes in the rate of migration of radon gas from the deep layers of the crust. They reported an increase in the concentration of radioactive radon in the groundwater and subsurface soil gases near the fault zone, or in the adjacent atmosphere, prior to earthquakes. Therefore, it is surmised that the primary cause of the increased concentration of radon is the pronounced increase in the rate of migration of other gases (see illustrations in Figure 13-3), which shows the simultaneous changes in concentrations of radon, CO_2, and CH_4 gases in the subsurface soil layer near the fault zone). On the other hand, if the crustal movements result in the lower permeability of the fault or fractured zones, then the rate of migration of radon and other gases (carrier gases) decreases.

The correlation between the concentrations of radon and methane gases in the subsurface soil layer is very high for seismically active areas. Based on these results, one can erroneously conclude that these parameters are interchangeable as potential precursors of earthquakes. However, changes in the concentration of lighter gases as an earthquake precursor are preferable because the intensity of emanation of radon depends on the presence of its carrier gases, especially methane and CO_2. The primary reason in choosing the radon gas concentration as an earthquake precursor is the convenience of measuring its radioactivity. Intensity of radioactivity can be easily translated into the radon gas concentration in soil air and water.

VARIATIONS IN CONCENTRATIONS OF GASES (SOIL AIR AND GROUNDWATER) NEAR FAULTS AND FRACTURED ZONES

Altered permeability of the faults and fractured zones, especially near the hypocentral zone, is one of the possible causes for the changing patterns in the upward migration of various gases to the earth's surface. The pressure and temperature changes within the hypocentral zone are other possible causes. Accumulation and flickering of stresses in the seismically-active areas result in the rupture of the rocks and the appearance of multiple fractures and faults. One of the indicators of this process at the stage of stress accumulation is foreshocks or tremors preceding a principal quake. As a result of the altered thermodynamic conditions, the most volatile products are released in the hypocentral area. These additional discharges influence the composition of gases migrating to the earth's surface. Most often, data on the composition of gases is determined after an earthquake has occurred. There are, however, several examples in seismic literature of regular observations of the variations in volumetric composition of various gases (and carbon isotopes).

Following the Dagestan earthquake of 1970 in the Caucasus mountains, for example, researchers systematically analyzed the volumetric composition of various gases emanated to the atmosphere and variations in their carbon isotopes. The variations in the isotopic composition of carbon in CO_2 and CH_4 during the aftershock period reached 2%. The most considerable variations were observed at the time of strong aftershocks. It seems that these variations may be due to the formation of new conduits in the crustal rocks.

During the period of observations, several strong aftershocks occurred. The observation results are presented in Figure 13-5, which shows strong variations in the composition of gases (Khitarov et al., 1974; Kravtsov and Fridman, 1974). These gases consisted mostly of methane during the period between aftershocks. At the moment of shocks, the composition of gases changed; in particular, the share of CO_2 and H_2 increased considerably.

CHANGES IN THE PERFORMANCE OF OIL FIELDS

There is a relationship between the rate of oil production from sandstone reservoirs and earthquake activity (Kouznetsov et al., 1998). Minor seismic activity (seismic vibrations) prior to large earthquakes can decrease the water cut (water/oil production ratio) and increase the relative permeability to oil due to a reduction in oil-water interfacial tension (Kouznetsov et al., 1998). The gas/oil ratio also increases as a result of seismic vibrations facilitating formation of free gas (degassing of oil effect). This freed gas also becomes available for migration. Consequently, any changes in the performance of oil fields (e.g., rates of oil, water, and gas production) must be carefully monitored and evaluated by petroleum engineering experts.

HYDROCARBON SEEPAGE IN THE MARINE ENVIRONMENT

There are many areas in offshore California where natural gas migrates to the bottom of the ocean. Time-spatial changes in the rates of hydrocarbon seepage in the vicinity of offshore oil and gas wells can be used for several purposes. The primary goal is to determine the influence of oil and gas production on the marine environment. Research conducted by the UCSB showed that intense oil and gas production in the Santa Barbara Channel (California, USA) considerably reduced the seepage of tar and organic gases (Luyendyk et al., 1996). This improved the environmental conditions of this world-renowned recreational area. The hydrocarbon gases, such as methane, are precursors to smog-forming ozone, which is a health hazard. The information on hydrocarbon seepage is collected on a regular basis in the Santa Barbara Channel, which is an area of intense seismic activity. These data can be used to develop earthquake precursors.

The Santa Barbara Channel has the greatest intense marine hydrocarbon seepage in the world, after the Caspian Sea. Most of the tar on the Santa Barbara beaches and more than half of the tar on the Los Angeles beaches originate from the offshore natural seeps in the South Ellwood Anticline (South Ellwood and La Goleta seeps) in the Santa Barbara Channel. The total gas seepage from the South Ellwood Field was around 5 million cubic feet per day in 1994.

According to Luyendyk et al. (1996), about 8% of this seepage is being captured by seep tents (seep containment devices) (Figure 2-9). Prior to the Northridge earthquake of 1994, there was a significant surge of hydrocarbon seepage that was attributed to a "malfunction" of the measuring devices. This leads to the belief that the continuous volumetric measurement of gas rates in the seep tents and detection of possible anomalies can serve as another method of earthquake prediction. Some avenues of gas migration to the surface in this area are presented in Figure 13-6 (also see Gurevich et al., 1993).

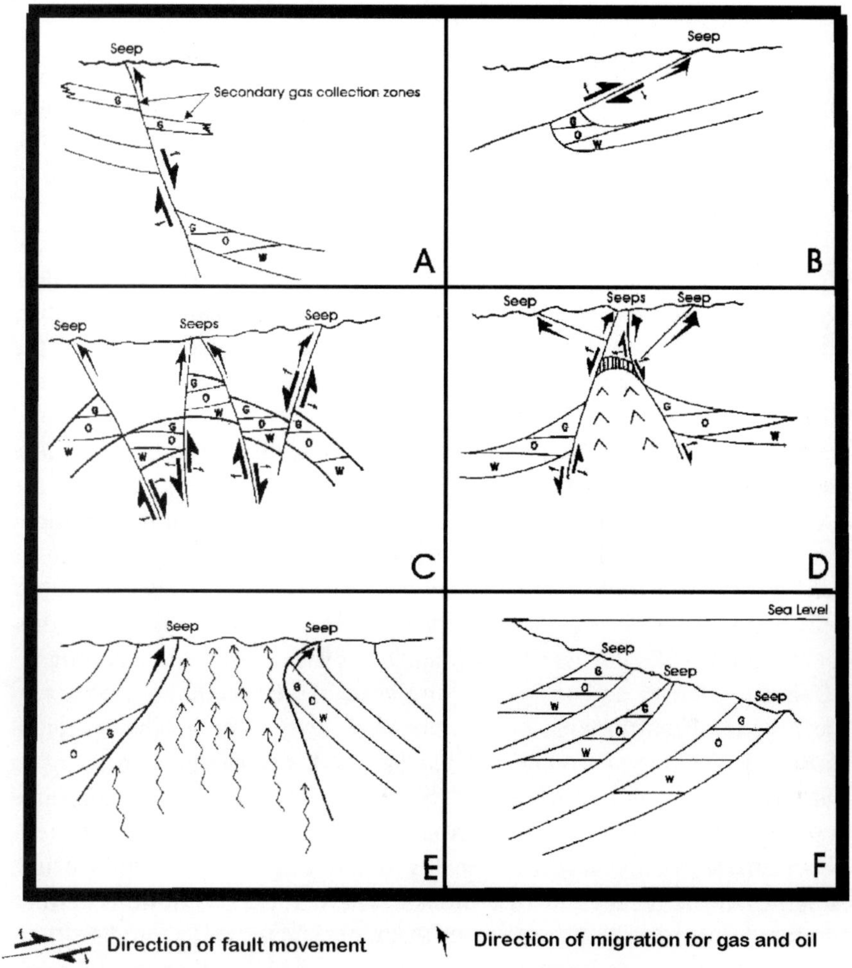

Figure 13-6. Locations of typical surface seepages in relation to oil fields. A = gas migration along a normal fault. B = Gas migration along a thrust fault. C = a faulted anticline with several gas seepages along faults. D = seepage along faults created by a salt dome or plug. E = seepage along a diapir. F = seepage along an angular nonconformity. The presence of fractured zones are not shown in these figures. Curved arrows show directions of gas migration.

PRIMARY CAUSES OF GEOCHEMICAL ANOMALIES

The following is a summary of the causes of the previously described geochemical anomalies on the eve of earthquakes. The first cause is the possible generation of gases at the hypocenters. These gases are discharged into the soil gases, groundwater, and the earth's atmosphere through preexisting and newly formed conduits (faults and fractured zones). It is very difficult, if not impossible, to evaluate the volume of gases generated at the epicentral areas of earthquakes.

Secondly, as a result of rapid changes in stresses, the physical properties of rocks in the vicinity of a hypocenter can be altered. There is a sharp reduction in stresses in the hypocentral zone immediately after the shocks, as a result of instant rupture and dislocations of the crustal blocks. As a result of a decrease in pressure, volatile gases adsorbed on rocks can be rapidly desorbed. Newly formed faults and fractured zones provide good conduits for the fast migration of these gases to the surface. The inert gases, such as He, Ar, and N_2, are especially sensitive to crustal perturbations. The observed data demonstrate pronounced changes in the rates of migration of these gases before and after major seismic events.

The third possible cause for the increased amount of gas migration to the earth's surface is an intense acoustic emission accompanying accumulation and, especially, release of accumulated stresses in the process of earthquake preparation and development. The principal rupture is accompanied by a dense network of fractures forming in the proximity of a hypocenter. Each of these fractures generates acoustic vibrations of various frequencies (from tenths of hertz to tens of kilohertz). This acoustic energy dissipates in the earth's crust, shaking and shifting the rock blocks. The acoustic vibrations reduce the bonds of adsorbed gases with the rock matrix and stimulate their release. This process seems to contribute considerably to the increase in the migration rate of gases to the earth's surface.

CHAPTER 14

Progress in Developing a Forecasting System, Preferential Precursors, and Monitoring Network

DECISION MAKING IN EARTHQUAKE PREDICTION

Earthquake prediction is a challenging problem requiring a great deal of interdisciplinary research. In principle, there are two major approaches to developing a methodology for earthquake forecasting. The first approach is to search for intricate cause-effect links between earthquakes and the events accompanying them, trying to establish some functional relation among quantitative characteristics describing the events. The second approach is to study the statistical data pertaining to earthquake occurrence in order to find out how preceding and accompanying events influence the frequency of earthquakes. These two approaches do not contradict each other and can be combined in a joint decision-making forecasting system. A corresponding flowchart for the decision-making process is presented in Figure 14-1. This decision-making process was applied in the Tokai District of eastern Japan and at Parkfield, California.

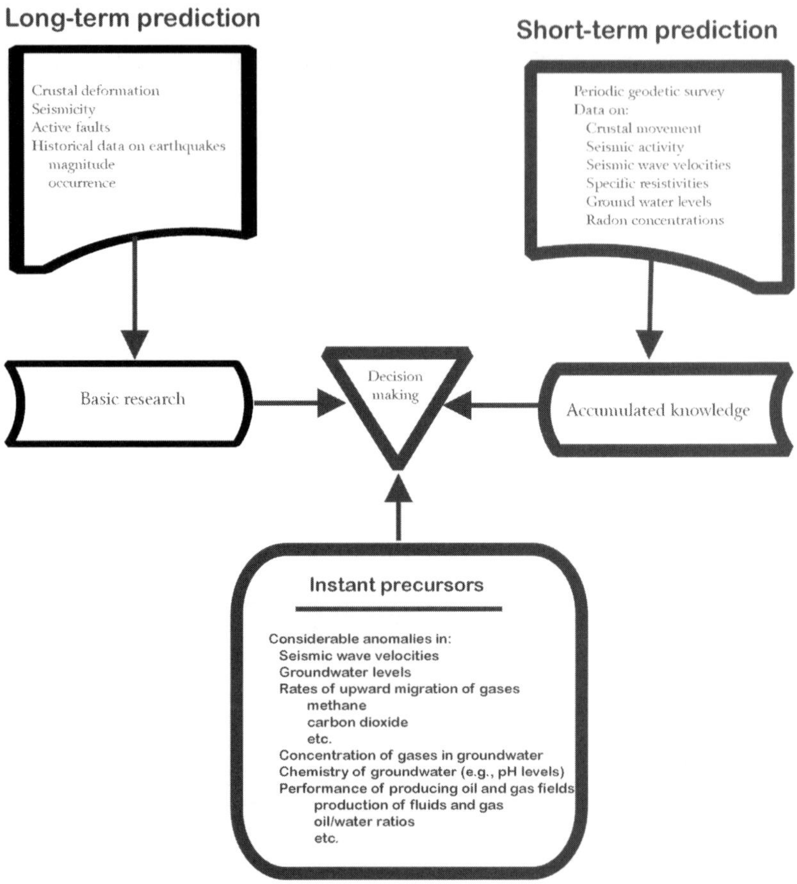

Figure 14-1. Decision-making process on earthquake prediction.

In 1969, the results of regular surveys gave evidence on the accumulation of stresses in the coastal region of Shizuoke Prefecture, eastern Japan. Additional evidence of a possible earthquake came from the analysis of historical and geological data, which showed that the focal zone of the disastrous Ansei-Tokai earthquake of 1854 extended into the coastal region. These circumstances led the authorities of Prefecture to develop a monitoring network and establish a Coordinating Committee on Earthquake Prediction. This committee regularly applies the procedure of the earthquake prediction based on the decision-making system similar to the one shown in Figure 14-1.

The city of Parkfield, California, is located at the central part of the San Andreas Fault in an area with high seismic activity. This area was selected as an experimental site by American seismologists and is considered the world's largest earthquake prediction experiment conducted on a continual basis. The following describes one experimental attempt to predict a large quake in this area (Giusti, 1995). On October 19, 1992, an earthquake with $M_R = 4.7$ occurred in Parkfield. After analyzing the available information (stresses, tilts, etc.) the California State Office of the Emergency Services released a warning about the possible occurrence of an earthquake with $M_R = 6$ or greater in the Parkfield area during the next 72 hours, with the estimated chances of occurrence being one out of three. This alert was also issued because of the high level of background seismic activity during the month of October. The predicted earthquake, however, did not occur. The alert was cancelled on the evening of October 21, 1992. Further study revealed that the October 19th event was a separate earthquake, which did not signal the occurrence of a larger quake.

Another similar situation emerged on November 14, 1993. After a swarm of small earthquakes on November 12, 1993, warning of an imminent earthquake of $M_R = 6$ or greater was issued, and seismologists gathered at Parkfield, California anticipating a possible event. Even though a high level of consensus was reached with respect to a large quake occurring, again, the predicted earthquake did not occur (Giusti, 1995).

These examples of unsuccessful prediction demonstrate that accumulated knowledge from traditional earthquake precursors, such as stresses, strains, and tilts, is insufficient to predict an earthquake. The weakest feature of these and other similar forecasts is, of course, the evaluation of the confidence level (the probability of occurrence of the future earthquake). Cautious forecasters insured themselves, however, by setting the chances of a presumable Parkfield earthquake at a 33% level. It should be noted, however, that before a warning is issued, a confidence level of at least 50% is recommended and the ability to show how the estimate was obtained.

As discussed in Chapter 9, the only possible way to increase the probability of an earthquake occurrence prediction is to reevaluate it as a conditional probability $P(E / \Pi)$ for earthquake occurrence under the condition of a precursory event E appearing. Therefore, major improvement of the short-term forecast can be achieved if the monitoring network provides a decision-making system with updated information on the chosen precursor (precursors).

SELECTION OF SHORT-TERM PRECURSORS

In developing a short-term earthquake prediction, it is necessary to focus on the precursory events to raise the confidence level of forecasting. It is unlikely, however, that short-term forecasting will evolve in the near future based on precursors directly linked to the primary causes of earthquakes.

Accumulating and transitory (flickering) stresses, which can result in the rupture of the earth's crust, are difficult to monitor. In addition, the random intricate paths of flickering stresses drastically reduce the value of point observations. Under these conditions, even the densest monitoring network can fail to detect a possible locale of rupture because of the nonlinear random transfer and distribution of stresses. It is possible for the rupture to occur only a few meters away from a monitoring gauge, which indicates normal crustal conditions.

The same sequence of events can result in very different consequences. For example, earthquakes occuring in the vicinity of an active existing fault can, with the same probability, trigger a new major event or release accumulated stresses, thus ensuring no seismic activity. On June 27, 1997, referring to *Science* magazine, the *Los Angeles Times* informed readers that the Loma Prieta earthquake of 1989 relaxed the stresses on the Hayward fault in northern California, which virtually eliminated creeping movement along the fault for the next seven years. The advent of another large earthquake due to this fault (which had not ruptured since 1868) probably has been delayed because of this reduction of movement.

On the other hand, there is the example of the North Anatolian fault in Turkey, where the first earthquake occurrence of 1939 triggered a sequence of 10 severe ($M_R \geq 6.7$) earthquakes along the fault between 1939–1992. This example provides an unique opportunity to study how one large shock sets up a sequence of others (Stein et al., 1997).

The same seismic event can reduce the risk of rupture in one area and considerably increase the chances of rupture at another. The outcome crucially depends on the local geologic conditions. For example, the Northridge earthquake of 1994, which possibly reduced the chances of a very large quake in the San Fernando area, may have increased the probability of earthquake occurrence in the Ventura basin (California).

These examples, together with the theory of flickering stresses, lead to the conclusion that sufficient information cannot be ontained for short-term earthquake prediction based on monitoring stresses and seismicity alone. Accumulated stresses, tilts, and seismicity may only outline a suspicious area and may contribute to a long-term forecast specifying the seismic environment for a particular area such as the San Andreas Fault, for example.

For short-term forecasting, it is necesary to identify precursors that (1) are directly linked to the primary causes of earthquakes (their origin) such as changing stresses, (2) reflect the condition of medium (e.g., configuration of crustal assemblages, rock strength, permeability, and degree of fracturing), and (3) display crustal conditions not only at the point of monitoring but also at close proximity to it.

It appears that the only precursor that satisfies most of these requirements is the rate of upward migration of gases (radon, CH_4, CO_2, He, etc.). Their concentration and composition can be conveniently measured in water wells, together with the chemistry of well waters (e.g., pH). The patterns of changes in the rates of migration of different gases can be completely different for different locales as shown in Chapter 13. For this reason, it is extremely important to monitor these changes on a regular basis in the seismically active areas, such as the areas surrounding the San Andreas Fault. Only regular observations of these parameters, along with the measurements of strains and tilts, can reveal characteristic patterns for a particular location.

DEVELOPMENT OF A MONITORING NETWORK

The most detailed and goal-oriented network was developed (and is being continuously improved) in Parkfield, California. It consists of many monitoring stations that continuously measure the seismicity, stresses in the rocks along the fault, position of tectonic blocks with respect to each other and in space, and tilting of the earth's surfaces. Additionally, the water table level is continuously monitored in 18 water wells along the fault in the Parkfield area. The measurement and transmission of the information is performed with the latest technological tools, such as laser strain and tilt meters and the Global Positioning System (GPS). Several prominent institutions (CalTech, US Geologic Survey, and UC Berkeley) process all the available information and prepare periodic reports, which are available to both the public and professionals. It is recommended that these measuring devices be complemented with gas content analyzers (in soil, water, and air), and pH meters and chemical analyzers in the well water. As a result, viable predictive techniques may be developed.

In effect, current monitoring techniques can only be deemed "basic research." Without the additional monitoring techniques previously listed, the gathered information probably would not allow a reliable forecast of a large earthquake at the present time. The collected information, however, may help formulate plausible hypotheses on earthquake development and occurrence. Continuous monitoring of the seismicity by the present network provides both the scientists and general public with important information. Analyzing historical information on major seismic events and their economic consequences, the system can estimate the damage caused by a particular event. This is necessary for determining the amount of emergency response necessary in the case of an earthquake and the rapid response of federal and state institutions, such as FEMA and CEPEC, responsible for financial and other aid.

Currently a real-time seismological network have been developed as a tool for rapid access to seismic information in California (Gee et al., 1996). In southern California, Cal Tech in cooperation with the USGS developed the Broadcast of Earthquakes system. This system uses information from the southern California seismographic network for rapid notification of seismic events and general periodic information on the parameters of seismicity (magnitude of the seismic events, time and location of occurrence, and the amount of energy released).

In northern and central California, UC Berkeley and the USGS developed and implemented the Rapid Earthquake Data Integration System. This system is designed for rapid determination of the various parameters of seismicity for these areas. It provides automatic information on the location, time, and magnitude of major seismic events via the Internet.

IDENTIFYING DISTINCT PATTERNS

In the previous chapters, a variety of phenomena that supposedly provide prognostic information on an upcoming earthquake was discussed. For scientists searching for distinct precursors of a large earthquake, the main question is not the nature of the phenomenon, but the stability of a pattern of irregularity that signals the final stage of a developing earthquake. From theoretical research and laboratory studies, it is conventionally assumed that the main rupture is preceded by a series of discrete stages of development of small fractures that in turn create larger fractures, either by merging together or increasing in width, length, and/or depth. In many cases, these branching fractures demonstrate distinct repeating geometric patterns that signal the eve of shock.

Professor C. G. Sammis and associates (1995) at the University of Southern California analyzed the behavior of heterogeneous materials, such as granite (modeling the earth's crust material), under stresses changing in time. When the stress increased in this physical model, isolated micro-fractures first appeared. If loading increased, these micro-fractures deepened and created new ones. Some of them merged, creating bigger cracks. Then, the final rupture occurred producing the principal fault with a surrounding network of smaller cracks (Figure 14-2).

Figure 14-2B (circled) repeats the complete pattern of Figure 14-2A. For a mathematical description of this situation, Professor Sammis and associates referred to fractal geometry. Fractals are geometric structures whose parts repeat the whole. Figure 14-2 represents an example of such a fractal structure. They found the fractal structure that describes spatial distributions of earthquakes and their associated faults in many geographic areas such as California.

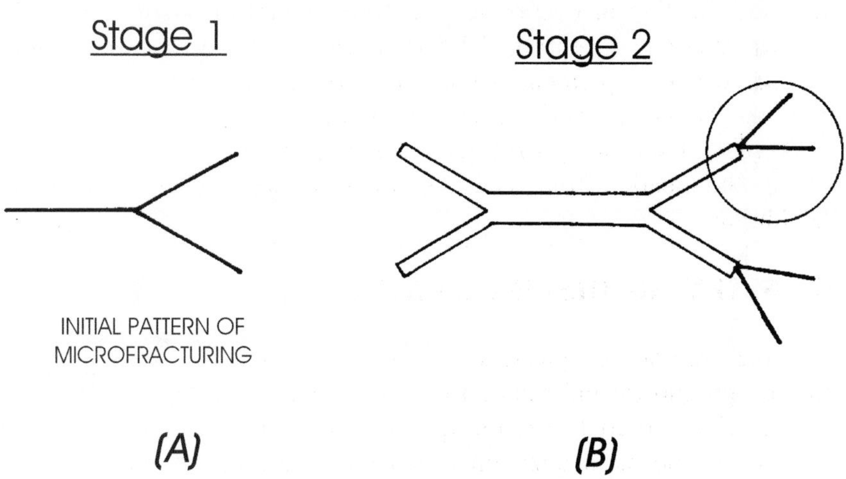

Figure 14-2. Fractal patterns in the development of cracks.

Another area of research in the search for earthquake precursors is the variations in patterns of seismic activity. Seismic activity is roughly characterized by the amount of energy released by the seismic waves within a certain geographic area. One of the distinctive patterns observed during earthquake development is a cyclic (stage by stage) seismic activity (i.e., a period of activity of small magnitude is followed by a period of quiescence). At the next stage, magnitude of the activity increases but its duration is shorter. The period of quiescence becomes shorter and represents the preparation for a stronger seismic activity with shorter duration, until the final burst of seismicity immediately precedes the occurrence of a large earthquake. This periodic pattern is presented in Figure 14-3. This pattern of seismic activity occurred on the eve of the Loma Prieta quake (M_R = 7.1) in northern California on October 18, 1989.

In 1995, Professor C. G. Sammis and associates applied a fractal description together with phase-transition mathematics to predict a major quake in Parkfield, California, which is a field research center for American seismologists. They specified the timespan of 160 days (from February 1 to July 9) for the future earthquake occurrence. In spite of the fact that this prediction failed, the scientists were not discouraged. Sammis stated that: "Earthquake prediction has a long

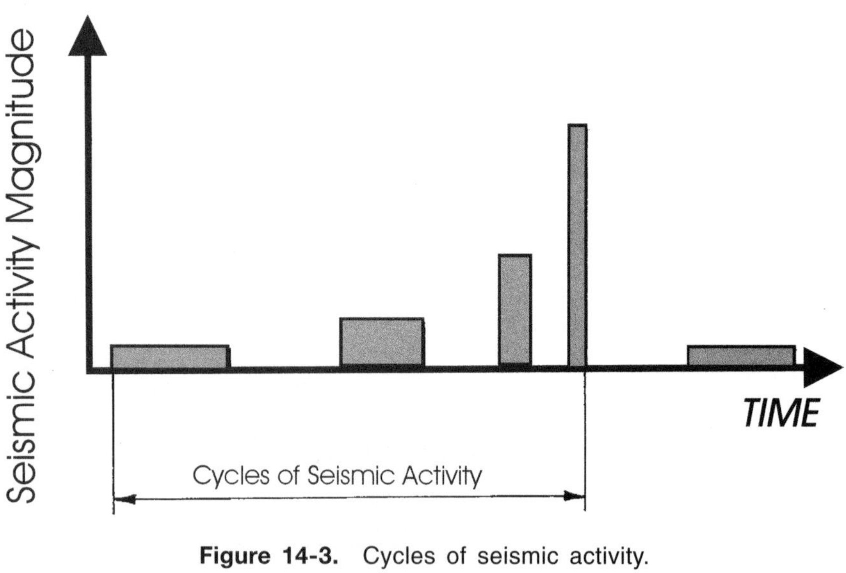

Figure 14-3. Cycles of seismic activity.

history of low success, and it would certainly be remarkable if this were reversed in a single stroke."

Patterns in seismic activity are not self-evident and are difficult to establish. These and other physical characteristics (e.g., degree of seismicity) can be investigated automatically by using neural networks. Katz and Aki (1989) pioneered the application of neural networks for earthquake prediction at the University of Southern California. Neural networks are especially efficient in working with complex data of a poorly defined nature.

Neural networks mimic, in a simplified manner, the algorithms of information processing in the brain. As in the brain, neural nets are commonly organized into several layers of nodes, with the output of the lower layer of nodes feeding the higher level of nodes. In the brain, this processing structure results in a progressively more abstract representation of the input stimuli as the information is being processed through the successive layers. In Figure 14-4, an example of

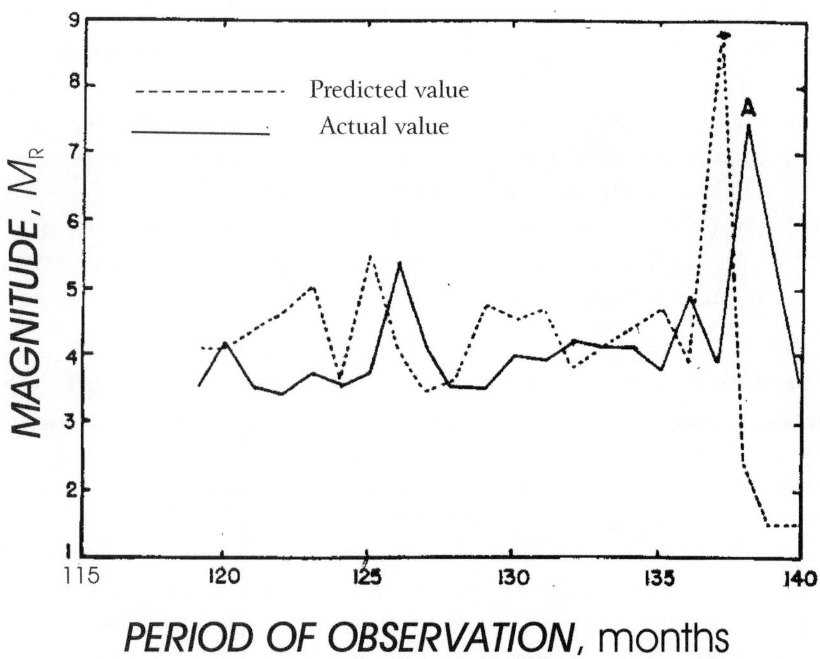

Figure 14-4. Example of neural network earthquake prediction based on integral characteristics of seismicity. *(Modified after Katz et al., 1994.)* A = actual, * = prognostic.

prediction on the basis of integral characteristics of seismicity is given. An improvement in prediction may be possible if the premonitory process is used as a component of a more complex process together with other correlatable components, such as the rate of subsidence and gas migration to the earth's surface.

CONCLUSIONS

The development level of contemporary seismic science, in general, and the present-day advances in earthquake prediction, in particular, are not sufficient, according to many scientists, to develop and implement a practical short-term forecasting system. We have presented several possible approaches to solving the forecasting problem. The following lists just a few of the revolutionary discoveries of the last decade and how they have influenced our vision of the problem:

1. An understanding of the dual nature (discrete and continuous) of the earth's crust evolved from discovering the predominant sizes of tectonic blocks and the laws of seismic wave propagation.
2. Flickering stresses complement the concept of accumulated static stresses with a dynamic random component.
3. An understanding of the mechanism of occurrence of deep earthquakes based on indirect inferences from statistics and physicochemical laboratory experiments.
4. An understanding of the migration rates of various gases (CO_2, CH_4, radon, etc.) along faults (and fractured zones) and their relative compositions.

Study of the sizes of tectonic blocks has attracted attention to the "weakened" zones (i.e., faults and fractured zones), which are formed at the boundaries of these blocks. These zones probably constitute the most probable areas of earthquake occurrence.

Dynamics of random flickering of stresses in the crustal assemblages considerably reduced the value of point stress and tilt measurements as primary information for developing predictive techniques. Thus, it is necessary to measure the parameters that reflect the properties of crustal medium in the proximity of point of measurement, such as the rates of upward migration of various gases.

The theory that deep-focused (below a depth of 600 km) earthquakes are caused by the transformation of the mantle matter has left little hope for their possible prediction (at least in the near future). This conclusion must be reached because no experiment can be designed to determine the specific place, scope, and time of transformation of olivine to spinel.

Instrumental basis for earthquake prediction has been considerably improved during the last decade. Potentially dangerous areas, such as Parkfield, California, were covered with seismographic networks providing information not only on current seismic events but also on potential precursors. The operation of these systems is being continually improved by the implementation of novel measurement techniques and telecommunication means, such as the Global Positioning System.

Given the above disappointments, there are positive signs indicating progress in developing earthquake prediction methods. The concept of precursors, for example, is a good working idea for the continuous search of patterns of irregularity in various parameters signaling an oncoming earthquake. Multiple synchronized point measurements of parameters, that reflect crustal stress variations in close proximity to the monitoring point (such as the rate of upward migration of various gases), must be implemented for this purpose.

Even if some short-term forecasting techniques fail, the attempts to develop new ones will inevitably be continued. As previously discussed, the pronounced changes in the rates of upward migration of gases (CO_2, CH_4, radon, etc.) in the vicinity of faults and fractured zones and their concentrations can be used in conjunction with other earthquakes prediction techniques. In addition, the chemistry and level of water in water wells, which are abundant in many seismically active areas, should be continuously measured. The pH (negative logarithm of H^+ ion concentration) of well water, for example, may change depending on the degree of influx of CO_2 gas.

Systematic measurements of these parameters, together with other integral characteristics of the seismic state of a tectonic system of interest, can be used as a database for the earthquake forecasting system. In order to evaluate the spatial scope, the number of control points, and the necessary frequency of observations, it is necessary to develop and implement an experimental gas migration rate monitoring system for already established seismic research stations.

PART III

Principles of Gas Migration

Gas bubbles at La Brea Tar Pits, Los Angeles, California.

CHAPTER 15

Gas Migration

INTRODUCTION

Gas migration can occur in the earth's crust as either a continuous or discontinuous phase through porous, water-filled media to the surface. The direction of gas migration in water-saturated formations is generally vertical (upwards or toward the surface) as the primary force moving the gas is the difference between the specific weights of gas and water (buoyancy). There are many examples around the world where gas migration to the surface, from oil- and gas-bearing formations, has created serious environmental hazards (i.e., explosions, fires, noxious odors, and potential emission of carcinogenic chemicals). In highly populated areas such as Southern California, previously undeveloped surface areas overlying oil fields have become more valuable and are currently being developed. As a result, environmental problems often occur today where there were few in the past. This migrating gas may be entrapped under paved streets, parking lots, and structures leading to severe environmental problems. An example of this type of problem occurred on March 24, 1985, when the Ross Dress for Less Department Store in the Fairfax area of Los Angeles, California, exploded, causing injury to over 23 people and the destruction of the store. The migrating gas accumulated in the ground under the paved areas and commercial structures. The entrapped gas was able to migrate into a confined space within the structure and was ignited by a spark generated by a time clock when an employee was checking out. The fire burned for days through cracks in the sidewalks and around the foundations of structures. It was difficult to extinguish the fire because the migrating gas continuously reignited the flames.

A serious environmental hazard can also exist because natural gas migration (primarily methane) can serve as a carrier for carcinogenic chemicals such as benzene. These problems also relate to the development of sites overlying gas storage reservoirs. The underground storage of natural gas in depleted oil fields (reservoirs) has now become a common practice in many urban areas. The reason for increased use of gas storage reservoirs is that peak demand deliverability (e.g., during cold weather) can be satisfied without dramatically increasing the size of interstate transmission lines. Underground gas storage facilities in abandoned oil fields, however, have demonstrated a long history of gas migration problems. There are many examples of environmental hazards associated with the escape of gas from these reservoirs. Wellbores, faults and fractures, as well as leaking (fractured) reservoir caprocks can contribute to the migration of gas. The risks of gas leakage must be continuously examined over the entire lifetime of a storage project. Wellbores can develop leaks as a result of corrosion of the casing, deterioration of the cement seals, and mechanical damage resulting from seismic activity.

In summary, it is imperative to understand the mechanisms and processes involved in gas migration from underground storage to the earth's surface in order to reduce the risk of injury and property damage in urban areas. Furthermore, changes in the rate of gas migration, as discussed previously, give important clues as to local tectonic activity.

SOURCES OF MIGRATING GASES

Natural gases, a mixture of light-hydrocarbon and non-hydrocarbon gases migrating from petroleum reservoirs, vary in chemical and isotopic composition depending on their origin and migration history. Here, natural gas refers to gas present in petroleum reservoirs. Crustal degassing and thermogenic or biogenic decomposition of organic matter can generate methane, the major constituent of gases. Common sources of gases include low-temperature bacterial fermentation (Rice and Claypool, 1981), thermogenic breakdown of deeply buried sedimentary organic matter (Schoell, 1983), and the mantle (Gold and Soter, 1980, 1982). The thermogenic decomposition of organic material is a large source of methane. Methane gas is also the major component of natural gases associated with petroleum deposits. The non-hydrocarbon gases often present in migrating gases are the "acid" and "inert gases." Carbon dioxide and hydrogen sulfide are referred to as "acid gases" as they can form acids on contact with water. Nitrogen and helium are referred to as "inert gases." Normally, gases are partially or completely saturated with water vapor. The source of sulfur found in hydrogen sulfide and mercaptans is from the organic material from which they were derived. Unless sulfur is present, gases are colorless and odorless.

Biogenic Methane Gases

Methane is the primary component of gases resulting from the biogenic decomposition of organic material under low temperature and pressure. For biogenic and thermogenic decomposition of organic materials, almost all of the hydrocarbon gases are composed of methane with very few heavier hydrocarbons (i.e., ethane, propane, etc.) (Figure 15-1). Figure 15-2 is a schematic diagram for the conversion

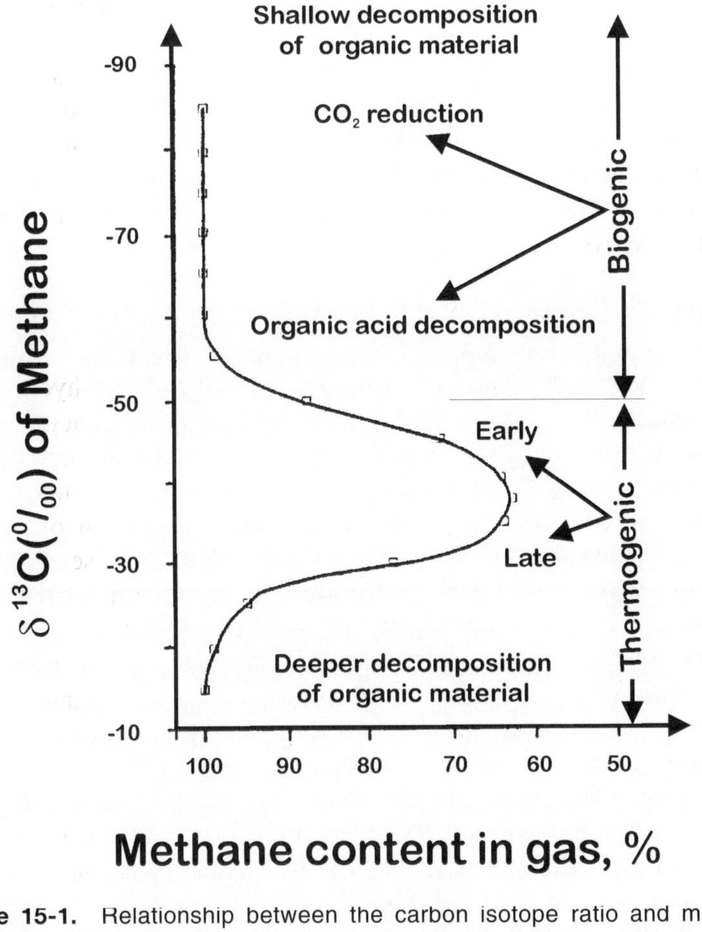

Figure 15-1. Relationship between the carbon isotope ratio and methane content $[C_1/(C_1–C_2)]$ in gas. *(Modified after Kaplan, 1994, p. 305, fig. 13.)*

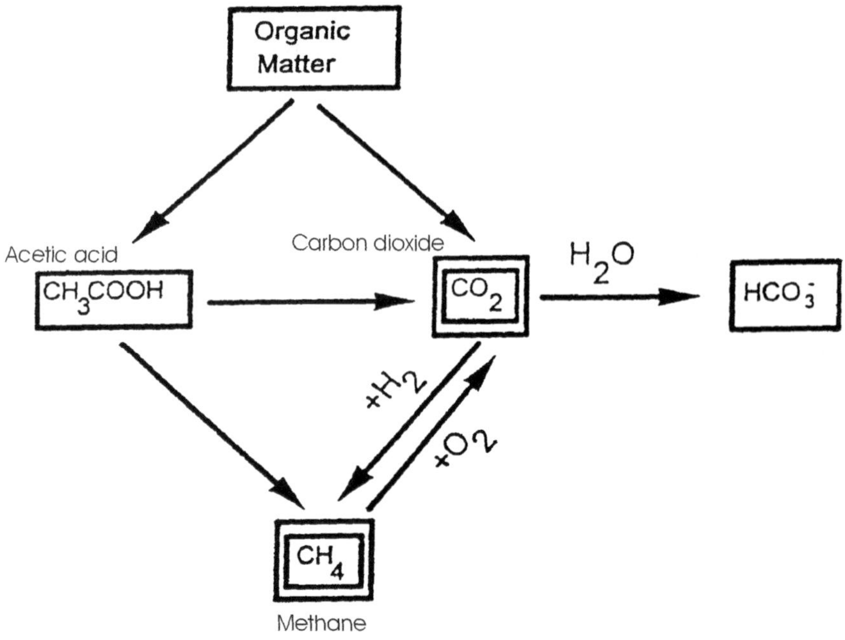

Figure 15-2. Schematic diagram for the conversion of organic matter to methane and carbon dioxide. *(Modified after Kaplan, 1994, p. 298, fig. 2.)*

of organic matter to methane and carbon dioxide. Biogenic decomposition only occurs in shallow sediments where organic matter is exposed to low pressure and temperature without oxygen.

Kaplan (1994) noted that the landfill environment was ideal for the biogenic generation of methane and hydrogen sulfide. Methanogenic bacteria (*Archaeobacteria*) are responsible for producing methane (Woese and Wolfe, 1985). Ward et al. (1978) stated that biogenic methane involves a consortium of microorganisms. The first step in the process involves the enzymatic hydrolysis of complex biochemicals discharged into the environment. When some oxygen is present in the decompositional environment (i.e., in water) aerobic respiration may rapidly convert the organic material into carbon dioxide. If oxygen is excluded during rapid burial, metabolic degradation proceeds by fermentation. The methanogenic bacteria can disproportionate the acids into methane (see Table 15-1).

In biogenic systems, the type of environment influences the method of methane formation. Mah and Sussman (1967) noted that primarily

Table 15-1

Chemical Reactions Leading to Methane Formation

Acetic acid fermentation: $CH_3COOH \rightarrow CH_4 + CO_2$

CO_2 reduction: $CO_2 + 4H_2 \rightarrow CH_4 + 2H_2O$

Mixed reactions:

(a) $CH_3(CH_2)(n-1)COOH + nH_2O \rightarrow CH_3(n-1)COOH + CO_2 + (n+1)H_2$

(b) $nCO_2 + 4nH_2 \rightarrow nCH_4 + 2nH_2O$

After Kaplan, 1994, fig.1, p. 298.

methane forms through fermentation of sewage sludge. Claypool and Kaplan (1974) suggested that in marine environments (Figure 15-3), methane forms by reduction of CO_2. In fresh-water environments (i.e., swamps, lakes, or rice paddy fields) the process is that of fermentation. In saline environments and ruminating of herbivores, methane is produced by respiratory oxidation of hydrogen with CO_2. Mah et al. (1977) have suggested that methane forms by a combination of two metabolic processes (Figure 15-2.) Kaplan (1974) described the microbiological successions in an aqueous environment, such as a lake, swamp, or estuary (Figure 15-4), and suggested that the same microbiological succession can also apply to the rice paddy field or landfill site in an environment with high rainfall. In this case, the methane will form at a depth where the respiratory organisms, which can derive the maximum energy, are no longer able to compete. Methanogenesis by reduction of CO_2 is more efficient than methanogenesis by fermentation.

Thermogenic Methane Gases

Thermogenic decomposition of organic material may give rise to oil and gas under high temperature and pressure. As shown in Figure 15-1, this type of decomposition creates not only methane, but also other light hydrocarbons. The methane content in the hydrocarbon portion of the gas is directly related to the temperature and pressure histories to which the organic material has been subjected. In order to differentiate the biogenic gases from thermogenic, it is necessary to know

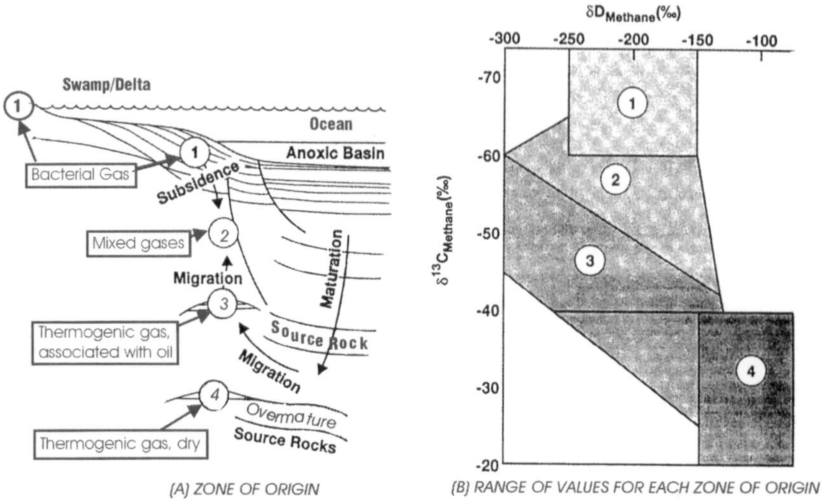

Figure 15-3. The isotopic composition of gas as related to its origin. The left panel depicts the geologic context of the gas origin (e.g., shallow bacterial, thermogenic, etc.). The right panel shows the range of anticipated values for carbon (*y*-axis) and hydrogen (*x*-axis) isotopic composition of methane in different genetic types of gases. Inasmuch as the PDB and SMOW are enriched in the ^{13}C and D in comparison to natural gases, the latter have negative $\delta^{13}C$ and δD values. *(Modified after Schoell et al., 1993, p. 338, fig. 1.)*

the ranges of isotopes of thermogenic and biogenic methane (Table 15-2). In fact, the isotopic values indicate the highest thermogenic pressure and temperature that the gas was exposed to. Examples of ranges for isotope ratios of carbon and hydrogen for various living and biologically derived materials, biological and nonbiological minerals, and gases are presented in Figures 15-5 and 15-6.

Isotopic Values of Gases

The origins of biogenic and thermogenic gases can be determined by isotopic gas analysis (the isotopic ratios of carbon and hydrogen). As shown in Figures 15-7 and 15-8, the isotope ratio of carbon ($^{13}C/^{12}C$) can be plotted versus the isotope ratio of hydrogen ($^{2}H/^{1}H$ or D/H). Clustering of data indicates the type of process in which the methane was formed. The heavy isotopes of ^{13}C and D (deuterium) are the rare species and constitute 1% of all carbon and 150 ppm of

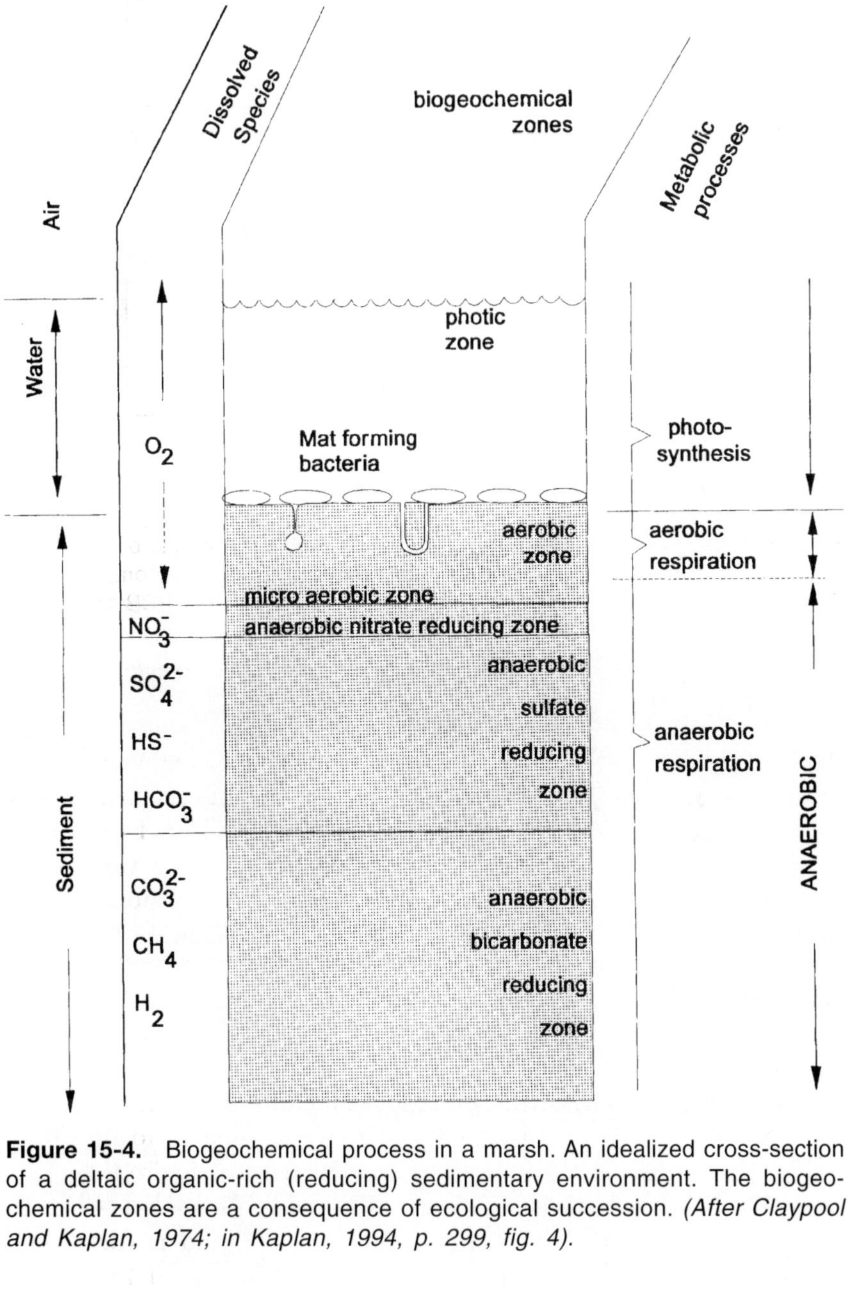

Figure 15-4. Biogeochemical process in a marsh. An idealized cross-section of a deltaic organic-rich (reducing) sedimentary environment. The biogeochemical zones are a consequence of ecological succession. *(After Claypool and Kaplan, 1974; in Kaplan, 1994, p. 299, fig. 4).*

Table 15-2
Range of Carbon and Deuterium Isotope Ratios for Methane Derived from Different Origins

		$\delta^{13}C$	$\delta^{2}D$
Biogenic	CO_2 reduction	–100 to –60	–250 to –150
	Organic acid decomposition	–60 to –45	–350 to –250
Thermogenic	Immature (low temperature)	–50 to –40	–250 to –220
	Optimum maturity	–40 to –30	–220 to –160
	Highly mature (high temperature)	–30 to –15	–160 to –90

After Kaplan, 1994, table 3, p. 301.

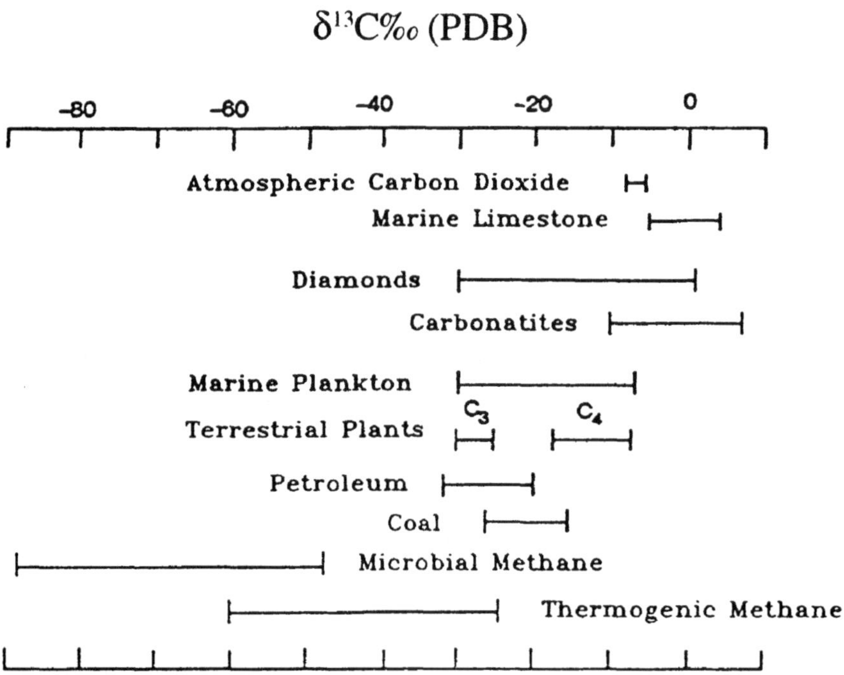

Figure 15-5. $\delta^{13}C$ values of selected materials. *(After Kaplan, 1994, p. 301, fig. 5.)*

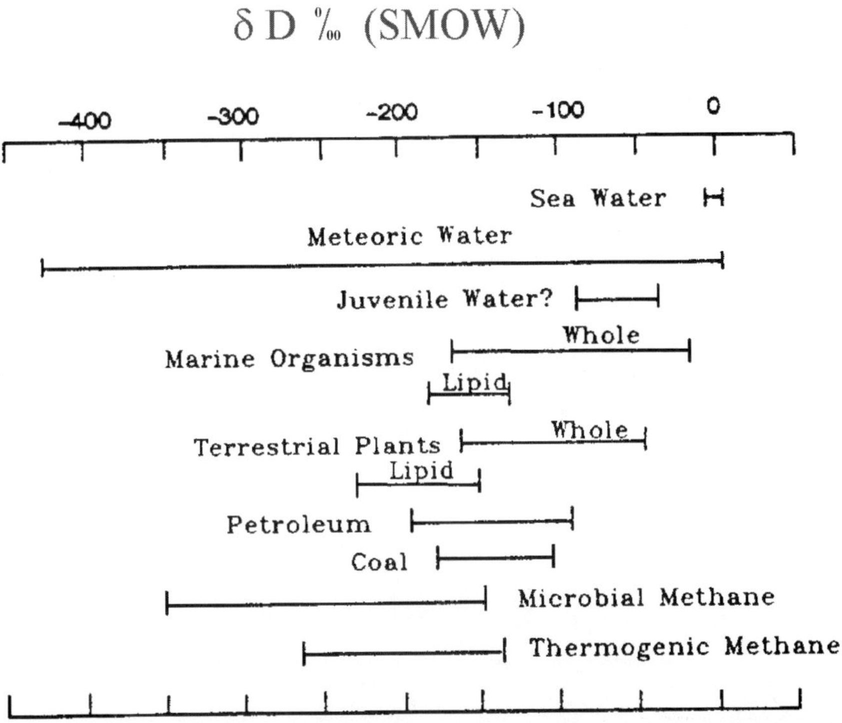

Figure 15-6. The δD values of selected materials. *(Modified after Kaplan, 1994, p. 301, fig. 5.)*

all hydrogen, respectively. Small variations in $^{13}C/^{12}C$ and $D/^{1}H$ ratios can be measured very accurately with modern mass spectrometers. For convenience, ratios are expressed in δ-notation as part-per-thousand ($^0/_{00}$) variations relative to the PDB (Peedee Belemnite; carbon) and SMOW (hydrogen) isotope standards as follows:

$$\delta^{13}C\left({}^{o}/oo\right) = \left[\frac{\left({}^{13}C/{}^{12}C\right)_{sample}}{\left({}^{13}C/{}^{12}C\right)_{PDB}} - 1\right] \times 1000 \tag{15-1}$$

and

$$\delta D\left({}^{o}/oo\right) = \left[\frac{\left(D/{}^{1}H\right)_{sample}}{\left(D/{}^{1}H\right)_{SMOW}} - 1\right] \times 1000 \tag{15-2}$$

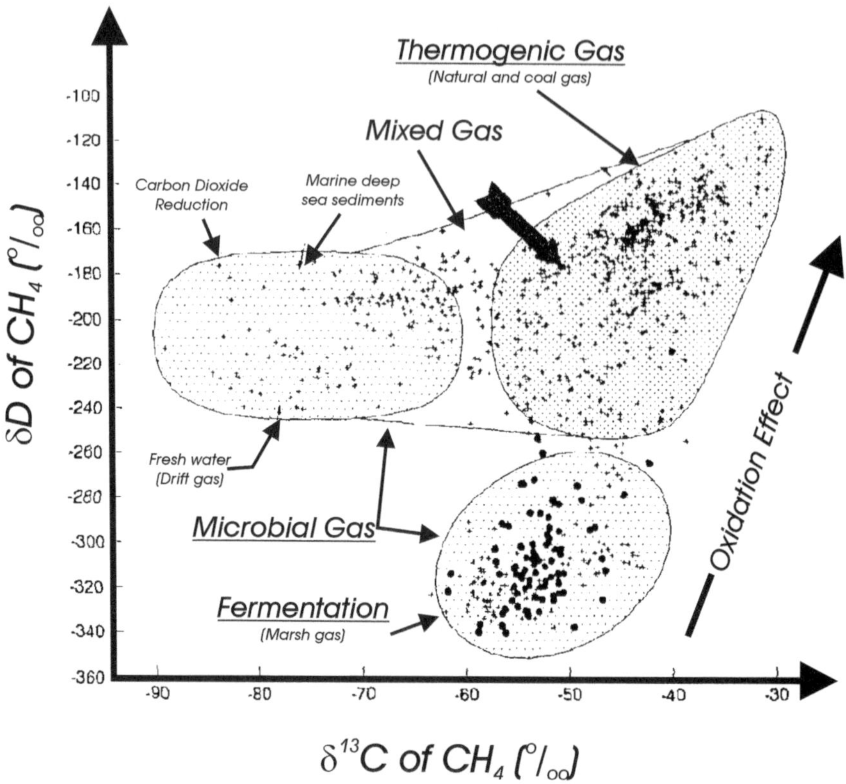

Figure 15-7. Carbon and hydrogen isotopic compositions of methane from various sources. The heavy arrow points to a specific thermogenic gas sample. *(Modified after Schoell, 1980, fig. 3.)*

$\delta^{13}C$ and δD values for a gas may vary widely depending on the origin of the gas (Figure 15-3). Biogenic gases are grouped by the isotope ratios in one portion of the graph, whereas the thermogenic gases are grouped in another area. These ratios do not change significantly with repressurization, solution, dissolution, or other physical processes that take place during the gas migration to the surface. These ratios, which reflect the origin of gases, are stable and independent of reservoir and sampling conditions. Consequently, the ratios are useful as tracers of origin and detection of gas movement from one part of a petroleum reservoir to another. As shown in Figure 15-1, the product of biogenic decomposition is almost all methane.

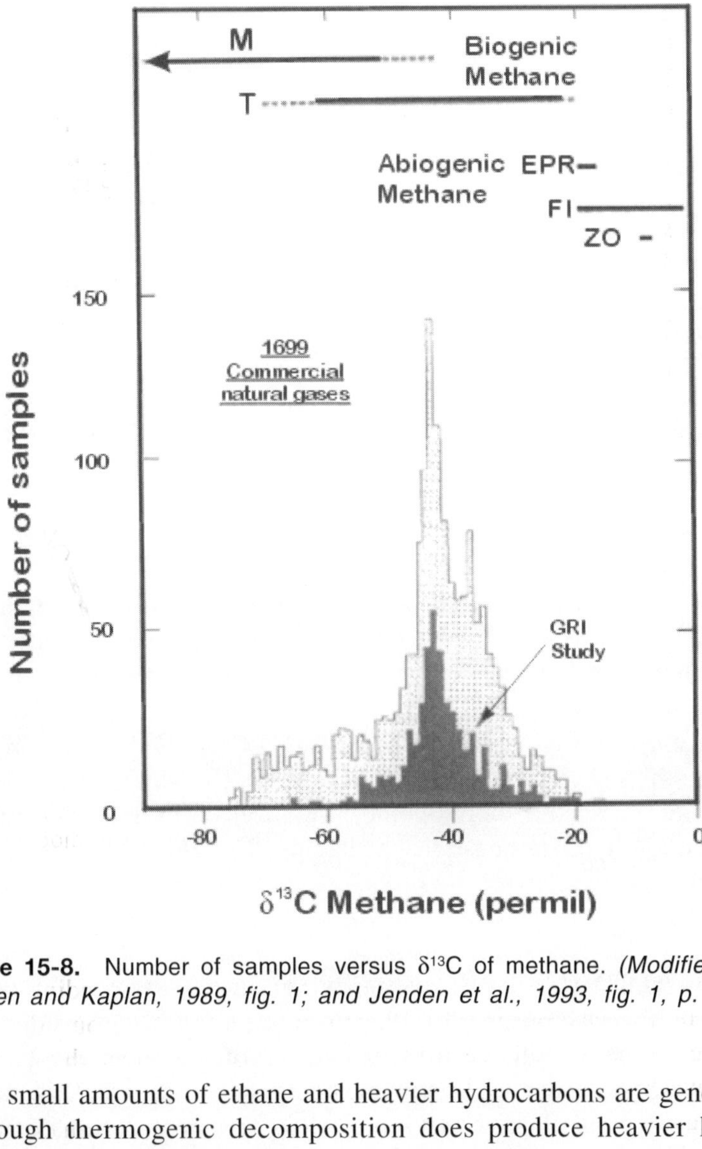

Figure 15-8. Number of samples versus $\delta^{13}C$ of methane. *(Modified after Jenden and Kaplan, 1989, fig. 1; and Jenden et al., 1993, fig. 1, p. 37.)*

Only small amounts of ethane and heavier hydrocarbons are generated. Although thermogenic decomposition does produce heavier hydrocarbons such as ethane in the earlier stages, in the later stages (higher pressures and temperatures) it also primarily produces methane.

According to Price and Schoell (1995), as natural gas migrates from its site of origin to the reservoir traps above, it may fractionate becoming much richer in methane. Table 15-2 lists typical ranges of isotopes for the biogenic and thermogenic decomposition of organic material. Isotopic ratio yields information on the past history. The composition and the isotope ratios of a gas may help determine the source of the gas.

Isotope compositions of light hydrocarbon gases can help identify the origin of surface gas seeps. Gases formed by bacterial alteration of shallow-buried sediments, peats, and organic-rich shales are composed almost entirely of methane (Figure 15-7). These gases (methane $\delta^{13}C < -60$ ‰) are isotopically light (Table 15-2).

Nonhydrocarbon Gases

Poreda et al. (1988) stated that at least two mechanisms are responsible for the origin of CO_2 in the subduction gases. Decarbonation or oxidation reactions produce a high-CO_2, low-$^3He/^4He$ gases, whereas high-$^3He/^4He$, CO_2-rich gas may originate from the mantle. He isotope ratios provide information on the input of magmatic gases into crustal systems. The magmatic He component has $^3He/^4He$ ratio (R) ranging from 0.5 to 3.8 times the atmospheric value (R_A); also it has high $^3He/^4He$ (R_A = 0.6–3.9) and low $CH_4/^3He$ ($<10^6$) ratios. The crustal He component has low $^3He/^4He$ (0.01 R_A) and high $CH_4/^3He$ ratios (for details see Poreda et al., 1988).

The nitrogen-rich gases in the California Great Valley may originate from mixing (1) hydrocarbon-rich gas with $^3He/^4He \leq 0.1\ R_A$ (derived from microbial and thermogenic alteration of sedimentary organic matter), (2) hydrocarbon-rich gas having $^3He/^4He \leq 2.75\ R_A$ (derived from the localized pyrolysis of sediments by magmatic volatiles), and (3) nitrogen-rich gas with $^3He/^4He \approx 1\ R_A$. The nitrogen-rich component is characterized by methane $\delta^{13}C < -25$ ‰ and N_2/Ar ratios as high as 22,000 (for details see Jenden et al., 1988).

In most soil-gas samples, the presence of nitrogen, along with oxygen, would indicate contamination of the sample with air. In general, nitrogen is an inert trace component and is likely derived from the thermal alteration of sedimentary organic matter. Dry thermogenic gases produced in the Sacramento and northern San Joaquin basins of California contain up to 87% nitrogen (Jenden et al., 1988). Investigation indicates that this nitrogen has a complex origin involving mixing from multiple sources. It appears that a portion of the nitrogen is derived from the igneous intrusives and subducted metamorphic rocks. The methane $\delta^{13}C$ values as heavy as -15 ‰ indicate that this gas high in nitrogen originates from the deep-crustal, metasedimentary rocks.

Helium isotope ratios may identify the gas as being a product of mantle degradation. Poreda et al. (1988) noted that magmatic He appears to be a common constituent of gases from tectonically active areas. The associations of mantle He and volcanism suggests that the heat supplied by the igneous intrusions could have played a role in producing the methane by thermal alteration of organic sediments. There is no evidence to invoke a mantle origin for methane.

Mixing of Gases

Schoell et al. (1993) prepared cross-plots of the isotopic ratios for methane versus percentage of nitrogen for mixtures of injected (B) and native (A) gases. The straight line is called the "mixing line" and represents a mixture of gases A and B (Figure 15-9). The triangles represent the composition of the gas that is moving into the domain of the native gas, whereas the circles represent the native gas that is present in the pores. The squares represent mixtures of the two gases. Isotopic analysis provides an excellent method of identifying and quantifying the degree of mixing. The precision to which one can quantify mixtures of two different gases is dependent on the precision of the analysis and on the consistency of composition of native and migrating gases. Composition of the sample from observation well (d), which plots directly on the mixing line, contains approximately 40% injection gas and 60% native gas.

In the case of gas storage, the native or reservoir gas has a relatively constant composition, whereas that of the injected gas can vary with time, depending on the source. Figure 15-10 is an example presented by Schoell et al. (1993) showing how the isotopic ratios of carbon and hydrogen can be plotted to determine the composition or mixing ratio of the native and injected gases. In this case, the isotopic ratios show that there are two different native gases. This is not unusual in reservoirs that are divided into fault blocks. The data in Figure 15-10 would indicate that there is no communication within the reservoir between the well (a) and the other three wells (d, f, and g) as shown by isotopic ratios.

The typical gas sampled at the surface is a mixture of biogenic gases (native) generated at the surface and thermogenic gases (migrating)

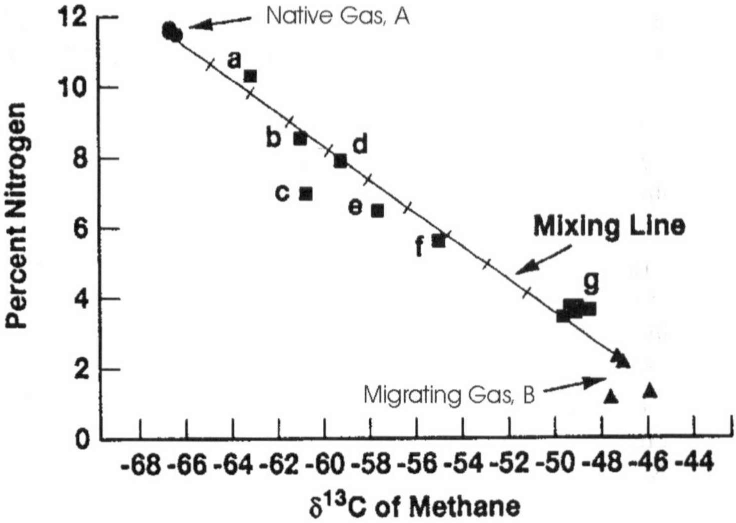

Figure 15-9. Nitrogen concentration plotted versus carbon isotope ratio for samples from a gas storage field. *(Modified after Schoell et al., 1993, p. 342, fig. 6.)*

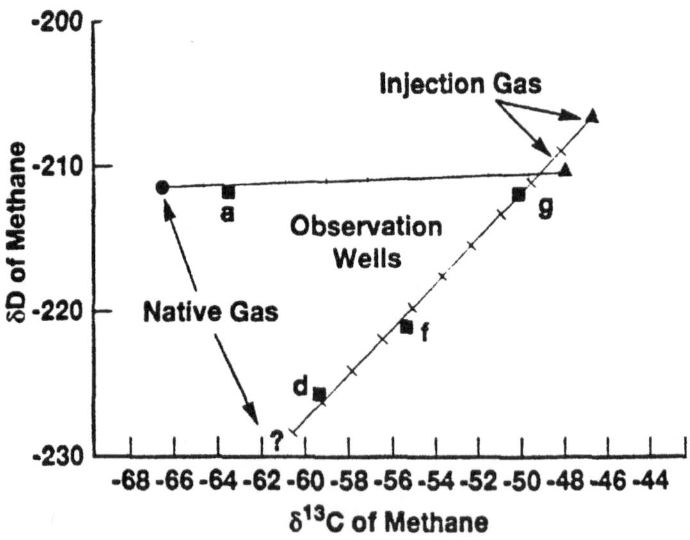

Figure 15-10. Carbon and hydrogen isotopic compositions of gas samples from a gas storage field. The question mark is located where the δD is that of the native gas of different vintage (no measurement of this gas was available). *(Modified after Schoell et al., 1993, p. 342, fig. 7.)*

that have migrated up from greater depths. A common problem is the determination of the origin of gas present at the surface: did it migrate from greater depths or is it simply the product of biological reactions at the surface? This is particularly important if one is attempting to determine whether gas is migrating from a petroleum reservoir or not. A cross-plot, such as the one shown in Figure 15-7, helps determine if the mixed gas is of thermogenic or biogenic origin. As the thermogenic (migrating) gas is mixed with the surface biogenic (native) gas, the isotope ratios will occupy a certain area of the diagram. The effect of oxidation of the gases is also shown in this figure. If the isotopic ratios of the native and migrating gas are known, the proportions of gases in a mixture can be directly calculated using a linear relationship (e.g., see Figures 15-9 and 15-10). For greater detail, refer to the classical works of M. Schoell, I. R. Kaplan, P. D. Jenden, A. W. A. Jeffrey, R. J. Poreda, G. E. Claypool, and D. D. Coleman.

CHAPTER 16

Typical Composition of Natural Gases

INTRODUCTION

Typical hydrocarbon components found in natural gases are shown in Table 16-1 and Figure 16-1. With few exceptions, natural gas consists of at least 95% hydrocarbons belonging to the paraffin series. The remainder are non-hydrocarbon gases such as nitrogen, carbon dioxide, and, sometimes, small quantities of hydrogen sulfide (and/or mercaptans). The presence of sulfur compounds, such as mercaptans and hydrogen sulfide, are responsible for odors sometimes associated with natural gases. In some areas, such as Texas, helium may also be present. Gases such as carbon dioxide and hydrogen sulfide are often referred to as acid gases as they can form acids on contact with water. Gases such as nitrogen and helium are referred to as inert gases. Natural gas may be partially or completely saturated with water vapor. Natural gas is a colorless and odorless mixture of gases, unless sulfur is present.

The principal hydrocarbons present (generally greater than 90% by volume) in natural gas are methane (CH_4), followed by ethane (C_2H_6), propane (C_3H_8), butanes (C_4H_{10}), and heavier components. The heavier hydrocarbons are present in a natural gas in decreasing proportions because of their high molecular weights. Upon migration, heavier hydrocarbons are preferentially adsorbed on rock minerals, mainly clays.

The composition of natural gases produced from an oil reservoir will vary with various operating oil/gas separator pressures. The example in Table 16-1 illustrates this variation of different components at various oil/gas separation pressures for the same produced gas. As shown, at lower oil/gas separator pressures, there is a lower relative

Table 16-1
Typical Composition of Natural Gases Expressed in Mole % (or Volume %)

	Type of Gas Field					
	Dry	Sour	Gas	Separation Pressures[d]		
Component	Gas[a] (mole %)	Gas[b] (mole %)	Condensate[c] (mole %)	400 psi (mole %)	50 psi (mole %)	vapor (mole %)
Hydrogen sulfide	0	3.3	0	0	0	0
Carbon dioxide	0	6.7	0.68	0.30	0.68	0.81
Nitrogen and air	0.8	0	0	0	—	2.16
Methane	95.8	84.0	74.55	89.57	81.81	69.08
Ethane	2.9	3.6	8.28	4.56	5.84	5.07
Propane	0.4	1.0	4.74	3.60	6.46	8.76
Isobutane	0.1	0.3	0.89	0.52	6.46	2.14
n-Butane	Trace	0.4	1.93	0.90	2.26	5.02
Isopentane	0		0.75	0.19	0.50	1.42
n-Pentane	0		0.63	0.12	0.48	1.41
Hexane	0	0.7	1.25	—	—	—
Heptane	0			0.15	1.05	4.13
Octane	0		6.3			
Nonane	0					
Total percent	100.0	100.0	100.00	100.00	100.00	100.00

[a] *Gas from Los Medanos, California*
[b] *Gas from Jumping Pound, Canada*
[c] *Gas from Paloma, California*
[d] *Gas from Ventura, California (oil and gas separator)*
After Chilingar et al., 1969, table 1, p. 89.

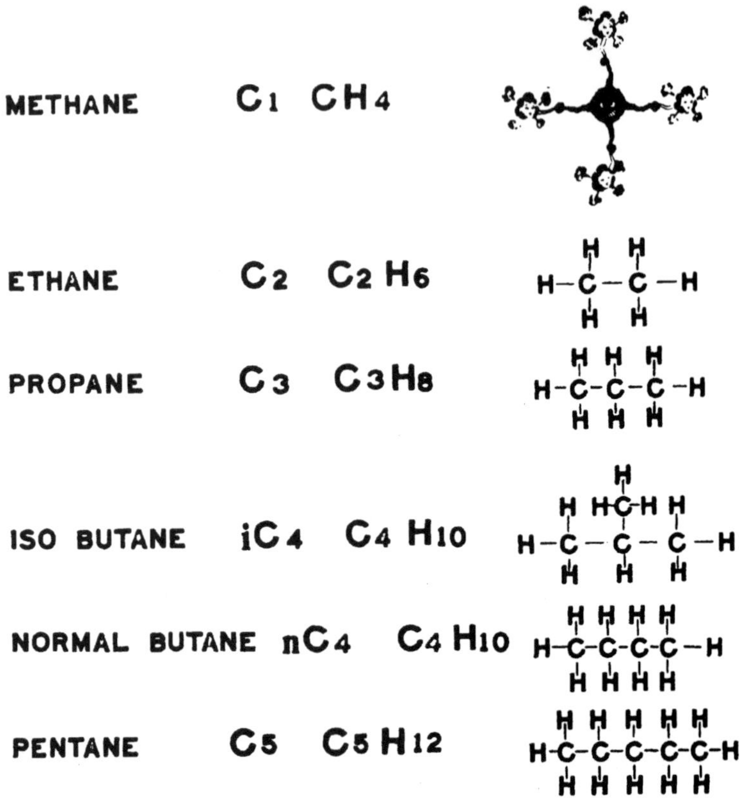

Figure 16-1. Structures of hydrocarbons commonly present in natural gases. *(After Chilingar et al., 1987, p. 180, fig. 7-3.)*

concentration of methane and higher concentration of heavier hydrocarbons in the produced natural gas. The actual composition of natural gas varies from field to field.

DENSITY OF NATURAL GASES

The ideal gas relationship as presented by Eq. 16-1 is convenient and usually acceptable when dealing with gases at less than 300 psia:

$$pv = nRT \tag{16-1}$$

Assuming 1 lb of gas,

$$\gamma_g = \frac{1}{v} = \frac{p}{nRT} = \frac{M}{RT} \tag{16-2}$$

where p = pressure, psia; v = specific volume, ft^3/lb; n = number of pound-moles; R = universal gas constant, 10.73 psi-ft^3/lb-mol°R; T = absolute temperature, °R = (°F + 460); γ_g = specific weight, lb/ft^3; and M = molecular weight.

In the situation of increasing pressure, Eqs. 16-1 and 16-2 become less accurate. Errors can be anticipated of about 2% at atmospheric pressure growing to over 500% at higher pressures. A simple method of compensating for such errors is the introduction of a "compressibility factor," Z, in Eqs. 16-1 and 16-2. This factor was determined empirically and is accurate within 5% for most gases. Modifying these equations to include the "compressibility factor," one obtains:

$$pv = ZnRT \tag{16-3}$$

$$\gamma_g = \frac{Mp}{ZRT} \tag{16-4}$$

The composition, temperature and pressure of the gas determine the "compressibility factor." For a single compound, Z is a function of reduced temperature, T_r:

$$T_r = \frac{T}{T_c} \tag{16-5}$$

and reduced pressure, p_r:

$$p_r = \frac{p}{p_c} \tag{16-6}$$

where T = absolute temperature of gas in degrees Rankine, °R = °F + 460; T_c = critical temperature of the compound, °R; p = pressure of gas, psia; and p_c = critical pressure of the compound, psia (see Table 16-2). The "compressibility factor," Z, can be obtained from Figure 16-2 if T_r and p_r are known.

(text continued on page 245)

Table 16-2
Physical Constants of Light Hydrocarbons and Other Components Associated with Natural Gas

Compound	Formula	Molecular Weight	Boiling Point at 1 atm. (°F)	Vapor Pressure at 100°F (psia)	Critical Temp. (°F)	Critical Pressure (psia)	Liquid Specific Gravity*	Gas Specific Gravity**
Hydrocarbons:								
Methane	CH_4	16.04	–258.68	—	–116.5	673.1	—	0.555
Ethane	C_2H_6	30.07	–127.53	—	90.09	708.3	0.375	1.046
Propane	C_3H_8	44.09	–43.73	190	206.26	617.4	0.5077	1.547
n-Butane	C_4H_{10}	58.12	10.89	51.6	305.62	550.7	0.5844	2.071
Isobutane	C_4H_{10}	58.12	31.10	72.2	274.96	529.1	0.5631	2.067
Isopentane	C_5H_{12}	72.15	82.1	20.44	370.0	483.	0.62476	2.4906
n–Pentane	C_5H_{12}	72.15	96.933	15.57	385.92	489.5	0.63116	2.4906
n-Hexane	C_6H_{14}	86.17	155.736	4.956	454.5	439.7	0.66405	2.9749

n-Heptane	C_7H_{16}	100.20	209.169	1.6199	512.62	396.9	0.68819	3.4591
n-Octane	C_8H_{18}	114.22	258.197	0.537	565.2	362.1	0.70677	3.9432
n-Nonane	C_9H_{20}	128.25	303.436	0.179	(613)	(345)	0.72171	4.4275
n-Decane	$C_{10}H_{22}$	142.28	345.2	0.073	(655)	(320)	0.73413	4.9118
Nonhydrocarbons:								
Nitrogen	N_2	28.02	–320.4	—	–232.8	492	—	0.9672
Oxygen	O_2	32.00	–297.4	—	–181.8	730	—	1.1047
Hydrogen	H_2	2.016	–422.9	—	–399.8	188	—	0.0696
Air	N_2+O_2	28.97	–317.7	—	–221.3	547	—	1.0000
Carbon dioxide	CO_2	44.01	–109.3	—	88.0	1073	0.8159	1.5194
Hydrogen sulfide	H_2S	34.08	–76.5	554.6	212.7	1306	0.790	1.1764
Water	H_2O	18.02	212.0	0.9492	705.4	3206	1.000	0.6220

** All measurements made with respect to water, Specific Gravity of water = 1.0*

*** All measurements made with respect to air, Specific Gravity of air = 1.0*

Data abstracted from Natural Gas Processors Suppliers Association, 1981.

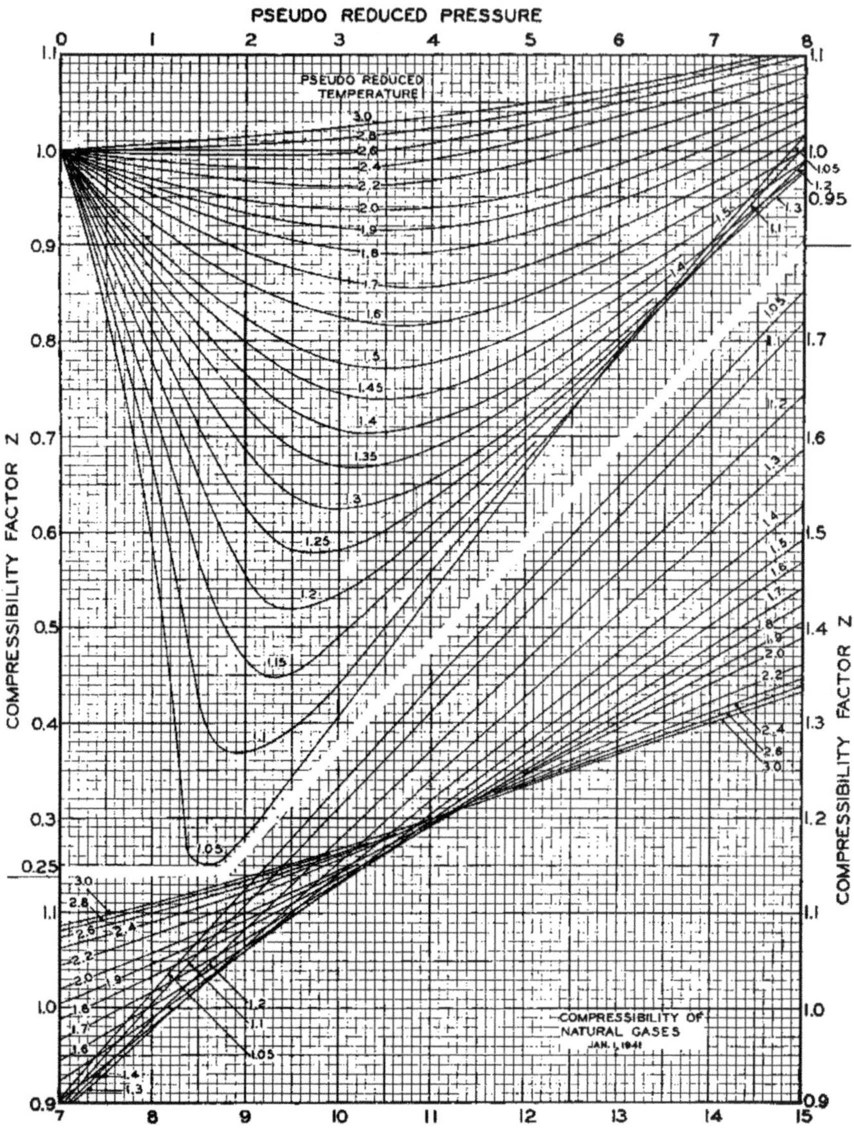

Figure 16-2. Compressibility factors for natural gases. *(Modified from Brown et al., 1941; in Natural Gasoline Supply Men's Association, 1957, and Natural Gas Processors Suppliers Association, 1981, fig. 16-3, p. 16-8.)*

(text continued from page 241)

If the gas represents a mixture of gases, the compressibility factor can be computed by first calculating the pseudo-reduced temperatures, T_{pr}, and the pseudo-reduced pressures, p_{pr}, for gases in the mixture and then obtaining an average value:

$$T_{pr} = \frac{T}{T_{pc}} = \frac{T}{\sum_{1}^{n} y_i T_{ci}} \qquad (16\text{-}7)$$

$$p_{pr} = \frac{p}{p_{pc}} = \frac{p}{\sum_{1}^{n} y_i p_{ci}} \qquad (16\text{-}8)$$

where y_i = mole fraction of individual gases in a mixture.

The pseudo-critical temperature and/or pressure of a mixture of gases may be calculated as the molecular average critical temperature and/or pressure of the components (volume % is equal to mole % for gases).

Sample Problem

Calculate the density of a gas mixture containing 90% methane, 8% ethane, and 2% propane at 55°F and 1,100 psig.

Composition	Mole Fraction, y	Molecular Weight, M	yM	Critical Temperature, °R	yT_c	Critical Pressure psia	yp_c
Methane	0.90	16		343		668	
Ethane	0.08	30		550		708	
Propane	0.02	44		666		616	
	1.00		17.7		366		670

Solving for T_{pr} by using Eq. 16-7:

$$T_{pr} = \frac{(55 + 460)}{366} = 1.41$$

Solving for p_{pr} by using Eq. 16-8:

$$p_{pr} = \frac{(1100 + 14.7)}{670} = 1.66$$

From Figure 16-2: $Z = 0.81$
Using Eq. 16-4:

$$\gamma_g = \frac{Mp}{ZRT} = \frac{(17.7)(1100 + 14.7)}{(0.81)(10.73)(55 + 460)} = 4.41\,\text{lb}/\text{ft}^3$$

The molecular weight of a natural gas can be determined from its composition. Assuming the following analysis of the natural gas:

Component	Composition, %
Methane (CH_4)	92.16
Ethane (C_2H_6)	4.88
Propane (C_3H_8)	1.85
i-Butane (C_4H_{10})	0.39
n-Butane (C_4H_{10})	0.55
i-Pentane (C_5H_{12})	0.17

The molecular weight can be calculated as follows:

Component	Mole Fraction, *y*	Molecular Weight, *M*	$y \times M$
Methane	0.9216	16.03	14.80
Ethane	0.0488	30.05	1.47
Propane	0.0185	44.06	0.815
i-Butane	0.0039	58.08	0.227
n-Butane	0.0055	58.08	0.319
i-Pentane	0.0017	72.09	0.123
		Average molecular weight	17.754

The specific gravity of gas (*SG*), with respect to air, can be calculated as follows:

$$SG = \frac{\text{Molecular weight of gas}}{\text{Molecular weight of air}} = \frac{\text{Molecular weight of gas}}{28.966} \tag{16-9}$$

$$SG = \frac{17.754}{28.966} = 0.612$$

The specific weight (lb/ft^3) of gas is equal to:

$$\frac{\text{Molecular weight of gas}}{379.498 \text{ ft}^3 \text{/lb-mole}} = \frac{17.754}{379.498} = 0.0468 \text{ lb/ft}^3$$

CHAPTER 17

Mechanisms of Gas Migration

INTRODUCTION

Gas migration from the source to the surface can be either continuous or discontinuous. Continuous gas migration occurs when there is an unbroken saturation of gas from the source to the surface. This type of gas migration is found in the layer of soil from the surface to the top of the water table. Discontinuous gas migration occurs in the form of globules/bubbles of gas migrating through the water-filled porous rock. This can occur below the water table, where the mechanism of gas migration is diffusion or mechanical, which may work either separately or in combination with each other.

TYPES OF GAS MIGRATION

Continuous Phase Gas Migration

The continuous phase migration of gas is similar to that of the flow in a pipeline. The cross-sectional area of flow (A) and the velocity of migration (v) control the volumetric rate of flow, Q:

$$Q = Av \qquad (17\text{-}1)$$

The flow rate of gas is directly related to the pressure drop along the migration path. A naturally occurring pathway of this type could be either a fracture or a fault. An example of a man-made pathway would be a wellbore drilled for either oil or water in a reservoir or where fractures were generated in an existing formation. The force moving the gas is the pressure drop between the beginning point and the end of the migration path. Gas flows from a higher-pressure area to a lower-pressure area. There are several types of pressure losses that can resist the movement of gas, such as frictional resistance to flow and capillarity.

The flow rate of gas in a sub-vertical or vertical column of gas saturating permeable rocks at any depth can be estimated as follows (Gurevich et al., 1993):

$$v = \frac{k}{\mu}(-\text{grad}\, p + \gamma_g) \tag{17-2}$$

where k is the permeability coefficient and γ_g is the specific weight of gas. The value of the pressure gradient (grad p) corresponds to the specific weight of water (1 g/cm^3) (i.e., 10^{-3} kg/cm^2/cm). Assuming a density for gas, ρ, at 0.015 g/cm^3 (at a pressure of 20 kg/cm^2), permeability, k, at 10^{-2} D, and a gas viscosity, μ, of 1.5×10^{-2} cP, the velocity, v, would be 7×10^{-4} cm/s or 60 cm per day. The actual rates of upward gas migration will depend on the vertical permeability along the path of migration.

The volumetric rate of migrating free gas along the continuous gas-filled paths is dependent on the cross-sectional area of the gas-filled path and flow velocity. In general, gas migration through gas-filled, as compared to water-filled, paths is capable of producing large quantities of gas because there is much less resistance to flow. The volume of gas arriving at the surface is dependent on the pressure gradient. Petrophysical features of the flow path, such as fracture width, length, water saturation, tortuosity, absolute and relative roughness of the fracture walls, etc., have a great effect on the volumetric rate of flow. Head loss due to friction, h_{lf}, is equal to:

$$h_{lf} = f \frac{l}{d} \frac{v^2}{2g} \tag{17-3}$$

where f is the friction factor, l is the length of channel, d is the equivalent diameter of cross-sectional area, v is the velocity, and g is the gravitational acceleration.

The volume of gas migrating through open fractures is much greater than that through the pore channels because porous media provide much more frictional resistance to the flow than the open space of a fracture. Boussinesque's formula allows making an order of magnitude estimation of the velocity of gas migration through fractures:

$$v = \frac{w^2}{12\mu}(-\text{grad}\, p + \gamma_g) \tag{17-4}$$

where w is the fracture width. At the same magnitudes of pressure gradient, density, and viscosity as in the example above and assuming w to be 1 cm, v will be equal to 5 m/day.

Wellbores penetrating gas/oil reservoirs are filled with gas, oil, and water. If the wellbore is filled with gas, the pressure distribution in the gas column can be described by the following formula:

$$p_h = p_b \exp\left(-\frac{0.000341\, H \rho}{ZT}\right) \tag{17-5}$$

where p_h is the pressure at any distance H from the wellbore bottom (or top of the fluid level), kg/cm^2; p_b is the pressure at the bottom of the wellbore (or top of the fluid level), kg/cm^2; ρ is the gas density relative to that of air (0.7 for methane); Z is the average gas supercompressibility (deviation from ideal gas behavior); and T is the average absolute temperature, °K.

In the case of a 3,500-ft (1,068 m) deep wellbore filled with gas in an oil/gas reservoir with a formation pressure (depleted) of 200 psia (14 kg/cm^2), a reservoir temperature of 307°K, and gas supercompressibility of 0.98, the wellhead pressure, p_h, is 185 psia (13 kg/cm^2). If the pressure distribution in the water-saturated formation surrounding the well is hydrostatic, then at a depth of 426 ft (130 m) pressure in the borehole will exceed the pressure outside the wellbore. Above this point, gas can escape into the surrounding formation. In the case where holes are present in the casing (due to corrosion) and the surrounding cement sheath, the gas could leave the wellbore and enter the surrounding formation above a depth of 426 ft ("escape zone"). The gas would then migrate toward the surface outside of the casing. Chemically aggressive corrosive formation waters often give rise to holes in the casing of older wells.

Discontinuous Phase Gas migration

Molecular Diffusion Mechanism

The diffusion of gas is a molecular process. The direction of gas flow is from areas with greater concentrations of gas to those of lower concentrations. The molecular diffusion mechanism (migration force) acts upon the gas at a molecular level. The driving force for this mechanism is the concentration gradient of gas molecules (i.e., there are more molecules of gas in one area than in another; see Figure 17-1). Over a period of time, the system will seek equilibrium through a more uniform distribution of gas molecules in all directions.

In general, gas molecules move from areas of higher to lower gas concentrations. The largest value for diffusion flow is in the direction of lower gas concentrations. The diffusion coefficient, D, is strongly affected by the type of rock porosity and the presence of fractures. Many other petrophysical properties of rock (i.e., tortuosity, surface chemistry, size of pores and pore throats, and width of fractures) also affect the diffusion coefficient. The exact values for the flow rate by diffusion in rocks cannot be easily established because the many variables that control the rate of diffusion are either unknown or cannot be physically determined. The total gas diffusion can be approximated by the following equation:

$$I = D\frac{C}{h} \tag{17-6}$$

where I = the diffusion flow intensity (cm^3/cm^2/s); D = the diffusion coefficient for gas moving through water-saturated porous rock (cm^2/s); C = the concentration of gas dissolved in water at the upper boundary of the oil or gas reservoir (m^3/m^3); and h = depth of this boundary (cm). As this equation is only an approximation, the diffusion coefficient, D, represents the average value for the rock through which the gas diffuses and is assumed to be uniform within the whole path of travel, h.

Although it is difficult to find an exact value of the diffusion flow intensity for a particular geologic formation in the field, an approximate value can be determined. Below the water table, formation pores are filled with water. The term C represents the solubility of gas in

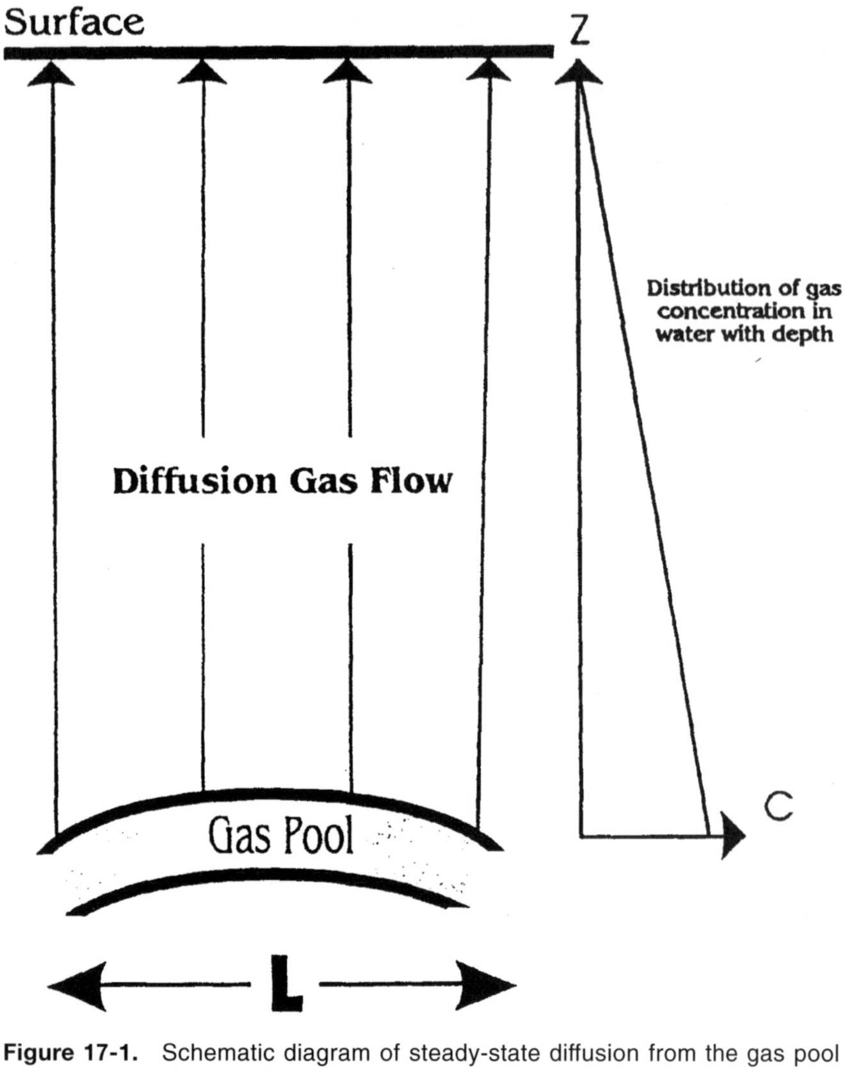

Figure 17-1. Schematic diagram of steady-state diffusion from the gas pool to the surface. C = concentration of gas in water (After Gurevich et al., 1993, p. 224).

water and the term C/h is, in fact, the solubility coefficient. A typical value for solubility of natural gas in water would be 0.3 $(m^3/m^3)/MP_a$, or 0.003 $(m^3/m^3)/m$ for a column of fresh water. Assuming a diffusion coefficient, D, of 10^{-6} cm^2/s, the calculated value for I is approximately 3×10^{-11} cm/s for gas diffusion flow through a horizontal cross-sectional area of $1\,cm^2$. This value is probably much higher than the actual values of diffusion flow intensity encountered in nature due to the assumption of maximum values that were made to calculate it.

Gurevich (1970) measured gas-saturated values of formation water samples taken at various depths above the caprock of oil reservoirs. At a distance of several meters above the caprock, he found that the gas content of water dropped to one-half or one-third of the maximum gas solubility value in water.

Several authors have studied the solubility of natural gas in water. Dodson and Standing (1944) prepared graphs that show the effects of water salinity and pressure on the solubility of natural gas in water. Figure 17-2 can be used to estimate the maximum solubility of natural gas in formation waters at various temperatures and pressures.

Often, the upward diffusing gas encounters moving water, associated with the near-surface fresh-water aquifers. If the moving water stream is not 100% saturated with gas, the gas moving upward into this body of water can be absorbed and carried off to other areas. The quantity of migrating gas absorbed by a unit column of the fresh-water stream can be expressed as follows:

$$Q = It = I\frac{L}{v} \tag{17-7}$$

where Q = the quantity of gas absorbed by the moving water (cm^3); I = the diffusion flow intensity ($cm^3/cm^2/s$); t = time necessary for the water stream to pass above the oil or gas pool (s); L = length of travel path of water through the zone above the oil or gas pool (cm); and v = water velocity (cm/s).

Assuming typical values for a path *(L)* of 10 km and a flow rate (v) of 1 m/day for the water stream and substituting in Eq. 17-7:

$$Q \approx \left[3\times10^{-11}\,cm^3/s\right]\left[\frac{10{,}000\,m}{1\,m/86{,}400\,s}\right][0.20]$$

$$\approx 5.2\times10^{-3}\,cm^3$$

Consequently, although some writers have given the diffusion process credit for migration of gas to the earth's surface from the underground gas sources, as demonstrated by the above calculations,

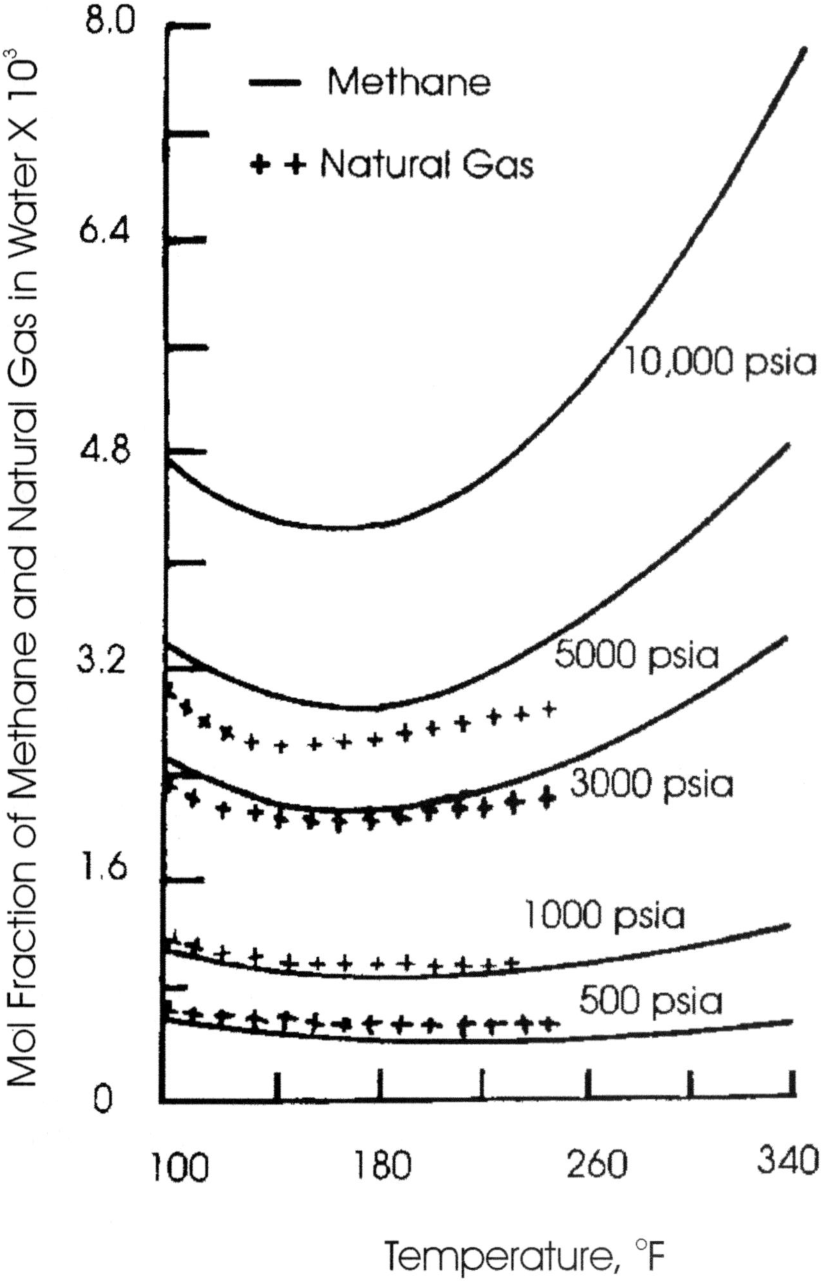

Figure 17-2. Comparisons of solubilities of methane and natural gas in water. *(Modified after Culberson and McKetta, 1951, fig. 4.)*

molecular diffusion of gas through rocks yields very low quantities of gas at the surface.

Hunt (1979, p. 433) also pointed out that vertical molecular diffusion of hydrocarbons above oil and gas reservoirs is too low to account for surface hydrocarbon anomalies found in the soil. Smith et al. (1971) estimated total hydrocarbon losses by molecular diffusion from a gas accumulation at a depth of 1,740 m (5,709 ft), temperature of 70°C, and pressure of 172 atm, at 2.8×10^8 m^3 (99×10^8 ft^3). These appear to be typical values for oil and gas reservoirs in the United States. The values shown in Table 17-1 indicate that it would take 140 million years for methane and 170 million years for ethane to migrate to the surface from a depth of 1,740 m. This calculation assumes that source rock gases replaced the diffusion losses in order to maintain a gas gradient at a constant concentration and composition. This is an optimistic assumption since the actual quantities of gases migrating to the surface are likely to be even less than indicated. Clearly, vertical molecular diffusion of gas from petroleum reservoirs buried at depths around 1,700 m (5,600 ft) is not the mechanism responsible for the volumes of gas observed in surface soils over oil fields (see Gurevich et. al., 1993).

In summary, diffusion, as a mechanism working on a molecular level, can move free gas toward the surface. The process, however, is a very slow one, occurring over a geologic time. Gas held in solution also does not move freely toward the surface. Although the process is slow, all gas eventually migrates to the surface. Given sufficient

Table 17-1
Time Period for Gas Migration Rate and Rate of Loss to the Surface for Vertical Steady-State Diffusion from a Depth of 1,740 m

	Time to Reach Surface (10^6 y)	Rate of Loss to Surface (cm^3/m^2/y)
Methane	140	2
Ethane	170	0.16
Propane	230	0.06
Butane	270	0.02

Modified after Smith et al., 1971; in Hunt (1979), table 9-2, p. 427.

time, all gases associated with the oil/gas reservoirs will diffuse through the caprocks and overlying formations to the surface. The thickness and integrity of the caprocks (e.g., shales) has a direct effect on the rate of gas diffusion for a particular oil/gas field.

INTERGRANULAR FLOW: GAS GLOBULES LARGER THAN THE PORE THROAT DIAMETERS

Discontinuous-phase Gas Migration

The rate of discontinuous gas migration in the form of gas bubbles depends on whether or not a gas bubble is smaller than the pore throat diameter of the porous rock through which the globule is migrating. It also depends on the pressure difference between the beginning of the flow path and that at the end. Here, the term "porous media" refers to rocks with intergranular and micro-fracture porosity.

Aschenbrenner and Achauer (1960) and Gurevich (1969), for example, have investigated the migration of oil and gas bubbles through water-filled porous rocks. In the situation where the diameter of the gas bubble is greater than that of the pore throat, the discontinuous buoyancy-type mechanism (see Figure 17-3) consists of gas or oil bubbles migrating through the channels of water-filled rocks having intergranular or micro-fracture porosity. The forces pushing the bubble through the porous rock are (1) the differential pressures resulting from differences in the specific weight between the gas bubbles and water, and (2) any pressure drop along the migration path. The forces resisting the flow (i.e., capillary forces) are subtracted from the upward forces. The bubbles will move in the direction of lower pressures, which generally is toward the earth's surface.

The bubbles move slowly through porous rocks with small voids. As a result of the very low velocity, the pressure losses to overcome friction (resistance to flow) are often negligible. The capillary forces are the primary forces necessary to overcome. The pressure within the gas bubble is almost the same at all points within the bubble (Figure 17-3). The resistance the bubble meets to movement will control the maximum size of the bubble. The bubble pressure cannot exceed the pressure of formation. The pressure within the bubble is almost the same as the water pressure at the bottom of the bubble. As a result, the pressure within the bubble will be greater than the water pressure

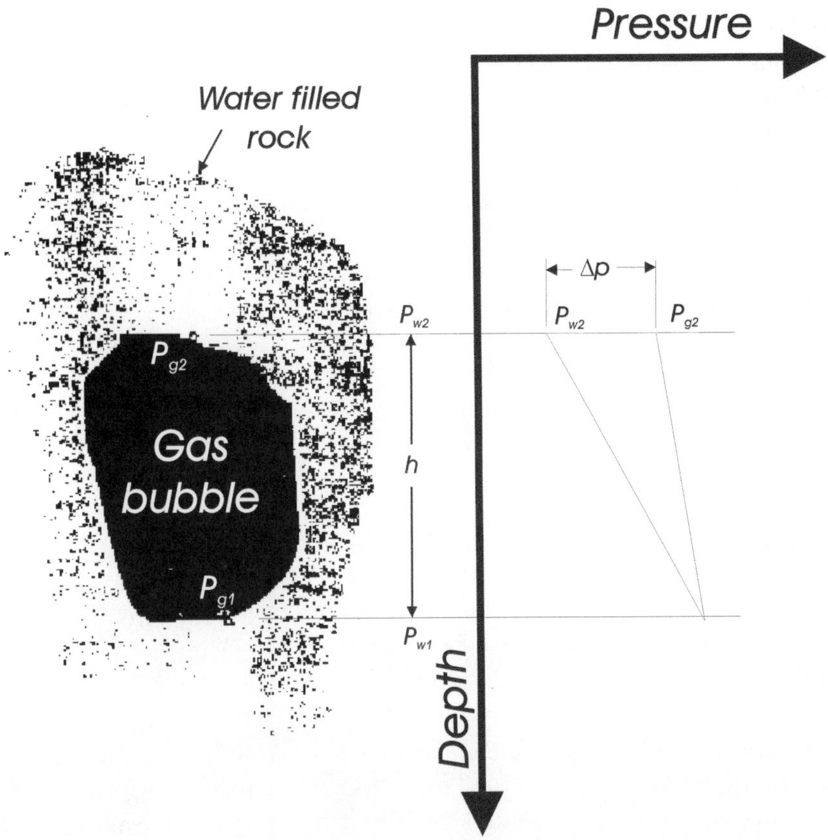

Figure 17-3. Schematic diagram of gas bubble migrating upward. *(After Gurevich et al., 1993, fig. 3, p. 226.)*

at the top of the bubble. Except for some very special cases of intensive underground water flows, it is the higher pressure at the top of the gas globule that overcomes the capillary pressure at the upper gas-water interface, where gas displaces water and initiates the upward movement of the bubble.

Whatever the bubble's shape, this excess pressure Δp, shown in Eq. 17-8, is equal to the vertical height of the bubble (h) times the difference in specific weights. This excess pressure may be calculated as follows:

$$\Delta p = (\gamma_w - \gamma_g)h \qquad (17\text{-}8)$$

where γ_w and γ_g are the specific weights of water and gas, respectively, and h is the vertical height of the gas column. In the case where the specific weight of water, γ_w, is 1.0 g/cm^3, the specific weight of a gas, γ_g, is 0.2 g/cm^3, and the height of the gas column, h, is about 2,000 m, the pressure difference would be about 200 kg/cm^2.

The capillary pressure, P_c, at the upper portion of the gas (gas-water interface), which resists its upward movement, depends on the values of interfacial tension, contact angle, and pore channel radius (Figure 17-4). The capillary pressure, P_c, may be determined by the following equation:

$$P_c = 2\sigma \frac{\cos\theta}{r} \tag{17-9}$$

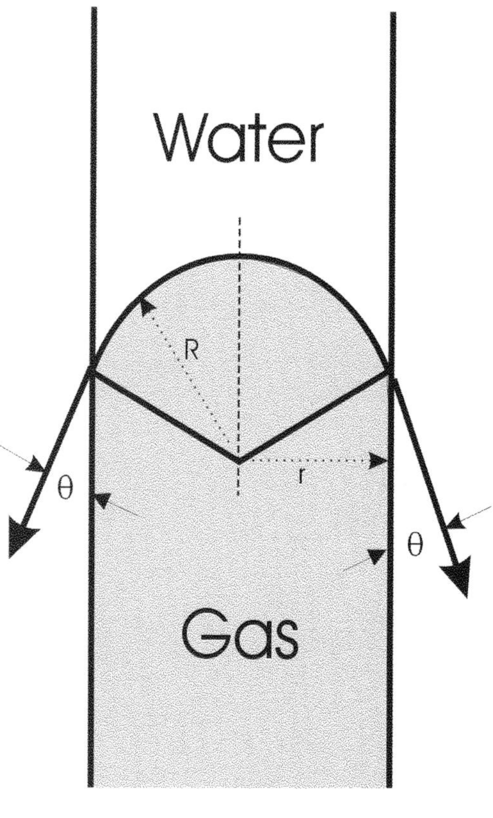

Figure 17-4. Gas-water interface in water-wet pore throat (capillary channel).

where the interfacial tension is represented by σ, the contact angle by θ, and the radius of the pore channel by r. To enable the bubble to move upward in the pore channel, the excess pressure (Eq. 17-8) must exceed the capillary pressure (Eq. 17-9):

$$\Delta p > P_c \tag{17-10}$$

To determine the minimum vertical height of the bubble, h, necessary to initiate the upward movement of the gas, Eqs. 17-8 and 17-9 can be rewritten in the following form:

$$h > \frac{2\sigma\cos\theta}{r(\gamma_w - \gamma_g)} \tag{17-11}$$

The value of interfacial tension at the gas-water interface depends on the temperature and pressure of the rock/water system. At a pressure of 200 kg/cm^2 and a temperature of 60°C, the interfacial water-gas tension is about 40 dynes/cm. The angle θ may be assumed to be 60° if the rock is water-wet. The radii of pore throats in coarse, medium, and very fine sands are about 0.02, 0.005, and 0.001 cm, respectively. The values of bubble height, h, necessary for a gas bubble to overcome the capillary pressure and begin moving upward in these types of sands will range from 2.5 cm for coarse sand, 10.2 cm for medium sand, and 51.0 cm for fine sand. Shales have smaller pore openings and, thus, the gas bubble height necessary to overcome capillary forces will be greater. The diameter of pore canals 10 μm and lower is common for shales. Thus, initial heights, h, of at least 5 m are necessary before the resistance of capillary pressure is overcome.

In gas pools or gas caps, the height of a continuous gas body almost always exceeds 5 m; often it is 20–30 m. Therefore, upward migration of gas through a caprock is possible if the thickness of the caprock is greater than the minimum height of gas column necessary to overcome the capillary pressure in water-wet rocks.

In recent, unconsolidated clays that have not lost their colloidal properties, pore channels can be either partially or completely blocked with adsorbed water at depths at which temperatures are below 50°C. In such a case, the force required to remove the adsorbed water from the pore channel so that gas can migrate is so great, that for all practical purposes, gas cannot mechanically penetrate these rocks as a free phase. In this case, the gas must seek fracture systems as pathways for migration.

After gas enters the caprock, its migration rate depends on the rate of water displacement. For a caprock of low permeability, the rate of gas migration is very low.

In summary, gas migration occurs when the upward force generated by the height of the gas column is greater than the capillary-force resistance of the rock through which the gas migrates. Some oil/gas reservoirs are composed of thin layers of alternating shale and sandstone. Only in gas-wet rocks, the gas migrates upward from layer to layer until it reaches the top of the water table, and then diffuses to the surface.

Minimum Height of Gas Column Necessary to Initiate Upward Gas Movement

As the gas bubble moves upward through the water in a capillary, several forces act on this column of gas of height h: (1) weight of column of gas (lb) acting downward, $W_g = (\pi r^2 h)\gamma_g$, where r is the radius of capillary (ft) and γ_g is the specific weight of gas (lb/ft^3); (2) upward force of water on gas (lb), $B = (\pi r^2 h)\gamma_w$, where γ_w is the specific weight of water (lb/ft^3); and (3) capillary force (lb), F_c, pulling the gas downward (in water-wet rocks), $F_c = 2\pi r(\sigma_{wg}\cos\theta_{wg})$, where σ_{wg} is the water-gas interfacial tension and θ_{wg} is the water-gas-rock contact angle. Thus, $W_g + F_c = B$, or $(\pi r^2 h)\gamma_g + 2\pi r(\sigma_{wg}\cos\theta_{wg}) = \pi r^2 h\gamma_w$. Solving for h (ft), one obtains:

$$h = \frac{2\sigma_{wg}\cos\theta_{wg}}{r(\gamma_w - \gamma_g)} = \frac{2\sigma_{wg}\cos\theta_{wg}}{r\Delta\gamma} \tag{17-12}$$

and inasmuch as $(2\sigma_{wg}\cos\theta_{wg})/r = P_c^{wg}$, which is the capillary pressure in the water-gas system:

$$h = \frac{P_c^{wg}}{\Delta\gamma} \tag{17-13}$$

In gas-wet systems (possibly some carbonates), the gas starts moving without reaching the minimum height because the capillary pressure aids the upward movement.

Buoyancy: Gas Bubbles Smaller than Pore Throat Diameters

Where the diameter of the gas bubble size is less than that of the migration channel, the gas bubble/globule is able to float upward in the channel. In this case, the capillary pressure plays little or no part in the resistance to gas flow as the gas globule does not need to displace the water from the pore channel to migrate upward (a bubble is a small globule).

Figure 17-5 shows an example of a gas bubble floating upward due to buoyancy forces. For a Reynolds number, N_R, below about 0.4

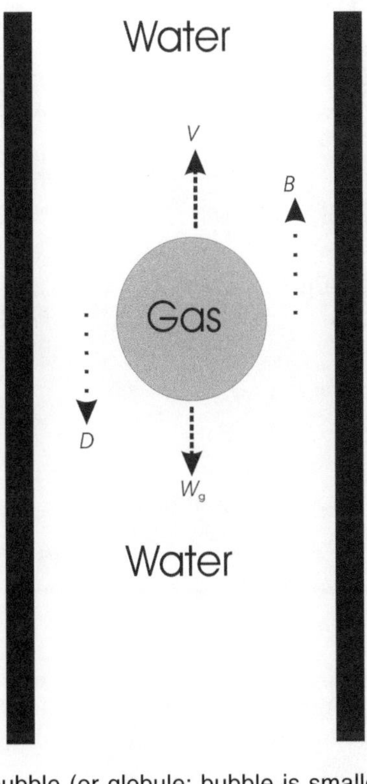

Figure 17-5. Gas bubble (or globule; bubble is smaller than globule) floating up through water in a channel where the diameter of the bubble/globule is smaller than that of the channel. W_g = weight of bubble/globule, B = buoyant force, D = frictional drag force, and v = upward velocity of the bubble/globule.

the drag coefficient, C_D, for a bubble of spherical form is equal to $24/N_R$. Thus, for a laminar or viscous flow of the gas bubble the drag force, D, in lb is equal to:

$$D = C_D\rho\frac{v^2}{2}A = \left[\frac{24}{\left(\frac{\rho v d}{\mu}\right)}\right]\times\rho\times\frac{v^2}{2}\times\left(\frac{\pi d^2}{4}\right) = 3\mu v\pi d \tag{17-14}$$

where ρ is the mass per unit volume in slug/ft^3, v is the upward velocity of the bubble in ft/sec, d is the diameter of the spherical gas bubble in ft, A is the largest projected area is ft^2, and μ is the viscosity of the fluid in lb-sec/ft^2 (or slug/ft-sec).

Inasmuch as the buoyant force, B, is acting upward, whereas the drag force, D, and the weight of the gas bubble, W_g, are acting down:

$$D + W_g = B$$

$$\left(\frac{\pi d^3}{6}\right)\gamma_w - \left(\frac{\pi d^3}{6}\right)\gamma_g = 3\mu_w v\pi d$$

$$\left(\frac{\pi d^3}{6}\right)\gamma_w = 3\mu_w v\pi d + \left(\frac{\pi d^3}{6}\right)\gamma_g$$

solving for upward velocity of bubble, v:

$$v = \frac{d^2(\gamma_w - \gamma_g)}{18\mu_w} \tag{17-15}$$

Assuming a temperature of 25°C, a pressure of 20 kg/cm^2, a specific weight of methane gas of about 0.015 g/cm^3, a specific weight of water of 1 g/cm^3, a water viscosity at this condition of approximately 10^{-2} poise, and a migrating gas bubble having the radius of 0.1 cm, the critical velocity is approximately equal to 0.22 cm/s.

Sample Problem

At laminar flow, calculate the rise velocity of a hydrocarbon gas bubble 0.1 mm in diameter through fresh water.

Given:

Gas bubble diameter, $d = 0.1\text{mm} \times \dfrac{1\text{ft}}{304.8\text{mm}} = 3.28 \times 10^{-4}\,\text{ft}$

Specific gravity of water, SG = 0.998

Specific weight of water, γ_w = (0.998 × 62.4 lb/ft^3)
= 62.27 lb/ft^3

Water viscosity, μ_w = 2.089 × 10^{-5} lb-sec/ft^2
(1 Poise = 2.089 × 10^{-3} lb-sec/ft^2)

Specific weight of gas, γ_g = 0.0422 lb/ft^3; specific weight of air at 59°F and 14.7 psia = 0.07651 lb/ft^3; molecular weight of hydrocarbon gas = 16; molecular weight of air = 29

The velocity of buoyant rise of the gas bubble is equal to:

$$v = \frac{d^2(\gamma_w - \gamma_g)}{18\mu_w} = \frac{(3.28 \times 10^{-4})^2(62.27 - 0.0422)}{18(2.089 \times 10^{-5})}$$

= 1.78 × 10^{-2} ft/sec or 0.54 cm/sec

CHAPTER 18

Paths of Gas Migration

INTRODUCTION

Faults and fractures form channels that allow large volumes of gas to migrate. The cross-sectional area of pore throats in rocks having intergranular porosity is generally several magnitudes lower than that for fractures and faults. As a result, the volumes of migrating gas that travel through faults and/or fracture systems are much greater than those that can migrate through porous rocks.

Natural fractures in rocks are the result of: (1) tectonic movements; (2) diagenetic processes, e.g., dolomitization process; (3) seismic activity; and (4) deformations caused by compaction of reservoir rocks due to fluid withdrawal and subsequent subsidence of overlying formations.

Man-induced fracturing of rocks occurs during oil field drilling operations around the wellbore itself by action of the drill bit. Inadequate sealing by cementing of the annular space between the wellbore and the well casing or improperly filling the fractures that surround the wellbore, results in "flues" or paths through which gas can migrate toward the surface. Many oil field production operations, such as fracturing or acidizing, also generate fracture systems. Man-induced fracturing also occurs when liquids are pumped from the formation and subsidence occurs. Fractures found due to fracturing operations (to increase production) tend to be vertical, cutting across bedding planes, and creating new paths for gas to migrate upward through layered rocks. Often, either water or gas is injected into reservoir rocks to maintain formation fluid pressure as fluids are produced. If fluids are injected at an equal rate to the fluid withdrawal, subsidence is either reduced or stopped. In faulted reservoirs, especially those that are divided into blocks by faults, stresses across the boundary zones (faults) of adjacent blocks due to water injection can be high enough to cause fracturing and breakdown of these fluid barriers (faults). Whenever the original hydrostatic formation pressure is exceeded, the brittle bedded formation can fracture, creating vertical paths for upward gas migration.

The most important paths or channels for gas migration are fault zones because they constitute wider and longer continuous conduits through the rock layers. As a result, usually more gas can migrate through a fault than through a porous or fractured medium. Many existing faults consist of a zone of fractured and broken material rather than a clean simple break in the geologic structure. One product of two rock blocks grinding against each other is rock flour (often referred to as gouge). This fine-grained material has a low permeability, which limits the flow of fluids across the fault zone. Our studies indicate that when the pressure drop across faults exceeds 300 psig, fluids can move across the fault zone. Thus, when differential pressure exceeds these values, the fault loses its integrity and becomes a "leaky" barrier. The pressure drop across faults becomes significant when reservoirs are repressured by water/gas injection. Many operators that believe faults act as barriers (seals) to gas and liquid flow, are not aware that several hundred psi of differential pressure across the fault is sufficient to overcome the resistance to liquid and gas migration across the fault. The liquid and/or gas may enter the fault and then travel upward toward the upper portions of the reservoir and, eventually, to the surface.

Zones of fracturing are often associated with geologic structures where beds are intensely folded. Fractured zones are also associated with localized tectonic and seismic activity and faults. Proper evaluation of fault zones is essential in establishing the integrity of an oil field and in preventing gas migration. This evaluation is even more important in those instances where older oil fields are converted to underground gas storage fields. Usually, normal faults appear to be more permeable than reverse faults.

Fractures also form in rocks as a direct consequence of oil field operations, such as drilling, cementing, acidizing, and repressurizing of wellbores and depleted reservoirs. These fractures are potential paths for the upward migration of gas.

Wellbores of idle or improperly abandoned wells often result in literal pipeline flow of large volumes of gas to the surface. This is particularly important where the abandoned wells were not properly sealed. During the cement squeezing operations, the wellbore may be even further hydraulically fractured. These vertical fractures may extend for tens of feet from the wellbore depending on the characteristics of the formation and the injection pressures used for placement of the cement. The cement will fill some of the larger fractures surrounding the casing, but cannot enter the smaller fractures away from the wellbore. The end result is that a new fracture system is created around the wellbore forming a path for the migration of gas.

Acidizing is a common practice in revitalizing the production of oil wells. Acid enters the pore channels and fractures, and enlarges them by dissolving the carbonates. Thus, there is widening of pre-existing fractures and pore channels. If the acid is injected into the rock at a high pressure exceeding fracturing pressure, new fractures are generated. Acidizing operations can damage the ring of cement around the well casing, which was placed there to prevent vertical migration of fluids and gas outside the wellbore. Tectonic and seismic activity along with movements due to subsidence (owing to production of oil field fluids) may also damage the casing and cement shield around it. Chemical attack by aggressive formation waters may also break down the cement seal. Corrosion may give rise to holes in the casing with consequent leaks. Oilwells drilled and completed many years ago were often abandoned according to different standards than those used today. Idle and/or abandoned wells are likely candidates for becoming vertical channels of gas migration toward the surface. The cement used in completion of older oilwells was of poor quality, whereas for wells drilled prior to the 1920s cement was not used at all. There are many examples of oilwells, abandoned according to modern standards, in which casing and cement bonds have broken down after a short time due to corrosive waters, repressurization of formations, local seismic activities, or other causes.

A fractured zone may be formed, or enhanced, by nonuniform compaction of sediments. Additional deformation can also occur as a result of differences in compaction rates of clays and sands. Differential movements occur over many oil fields because of differences in production or injection of fluids within various portions of the reservoir. Compaction can occur as a result of fluid withdrawal from the lower producing intervals with consequent subsidence at the earth's surface. Subsidence owing to production in the Wilmington Field, California, in some areas has been estimated to be as great as 20 feet.

Subsidence due to water withdrawal from water aquifers is also common. Subsidence of even a few feet can result in the creation of faults and fractures or cause widening and enlargement of existing faults and fractures, which facilitates gas migration to the surface. Nonuniform subsidence has the added problem of causing cracks in sidewalks, streets, and buildings. Reinjection of fluids into an oil reservoir, as shown in the Wilmington Oil Field, California, not only can stop subsidence but can also result in some rebound. Injection of water under pressure (waterflooding) can create fractures in the reservoir and the formations above the reservoir. In general, injection of fluids into a reservoir under pressure can cause fracturing and also break down the cement bonds in the existing wellbores. In some cases, this may create micro-quakes, which physically reflect fracturing of the formation. Continuously changing tension may widen the fractures, creating even larger channels for gas and, possibly, oil and water migration. In older, idle and/or abandoned wells, repressurization of the oil reservoir often forces gas, oil, and water toward the surface along these man-made migration channels (see Part IV for details).

Prior to production, many petroleum reservoirs do not have a gas cap because the gas is held in solution within the oil. When held in solution, this gas is not free to migrate toward the surface. During the course of oil field production operations, as a result of pressure drop, some of the gas held in solution is liberated. This free gas (gas cap) becomes available for upward migration along the faults, for example. As gas is freed, it can migrate upward toward the gas cap and gather under the caprock, where it forms a "free gas zone." This gas may escape across the caprock through (1) natural fractures, (2) man-induced fractures, and (3) diffusion. The caprocks of most oilfield reservoirs do not have sufficient sealing capacity to stop the migration of free gas (providing the rock is water-wet), but can increase the time of migration to thousands of years. Although the migration of gas through caprocks is very slow, over geologic time the gas can escape to the surface.

PRESSURE DETERMINATION

Water Columns

Fluids (liquids and gases) flow from a location with higher pressure to a location with lower pressure. Pressure is defined as follows:

$$\text{pressure} = \frac{\text{force acting perpendicular to an area}}{\text{area over which force is distributed}} \tag{18-1}$$

Pressure can be measured in pounds per square inch (psi) or pounds per square foot (psf). Consider two different-sized containers as presented in Figure 18-1. The larger container has a cross-sectional area of 1 sq ft, whereas the smaller container has a cross-sectional area of 1 sq. in. If the containers are filled with water at 60°F, the weight for each one-foot section of large container is 62.4 lb. The weight of each one-foot high by 1-square inch base container is 0.4333 lb. The pressure in this case is equal to the weight of water in one-foot sections of containers. The total weight of two one-foot sections of container would be 124.8 lb. The pressure at a depth of two feet would be 124.8 lb/ft^2. The total weight at a depth of 10 ft would be 624 lb. The pressure at this depth would be 624 lb/ft^2. The pressure gradient for this system is equal to:

$$\begin{aligned} \text{Pressure Gradient} &= \frac{\text{total weight/unit area}}{\text{height of column}} = \frac{624\ \text{lb/ft}^2}{10\ \text{ft}} \\ &= 62.4\ \text{lb/ft}^2/\text{ft} \end{aligned} \tag{18-2}$$

This value can be converted into psi by dividing it by 144 square inches per foot. In the case of a smaller container, the weight at a depth of 10 feet would be 4.33 lb. This would represent a total pressure of 4.33 lb/in^2. If one converts the pressure gradient in Eq. 18-2 to psi, the pressure gradient in each set of containers (large and small) is identical:

$$\text{Pressure Gradient} = \frac{62.4\ \text{lb/ft}^2/\text{ft}}{144\,\text{in}^2/\text{ft}^3} = 0.433\,\text{psi/ft}$$

Regardless of the cross-sectional area, the pressure gradient is the same in both containers.

Sample Problem

What is the pressure in a swimming pool at a depth of six feet?

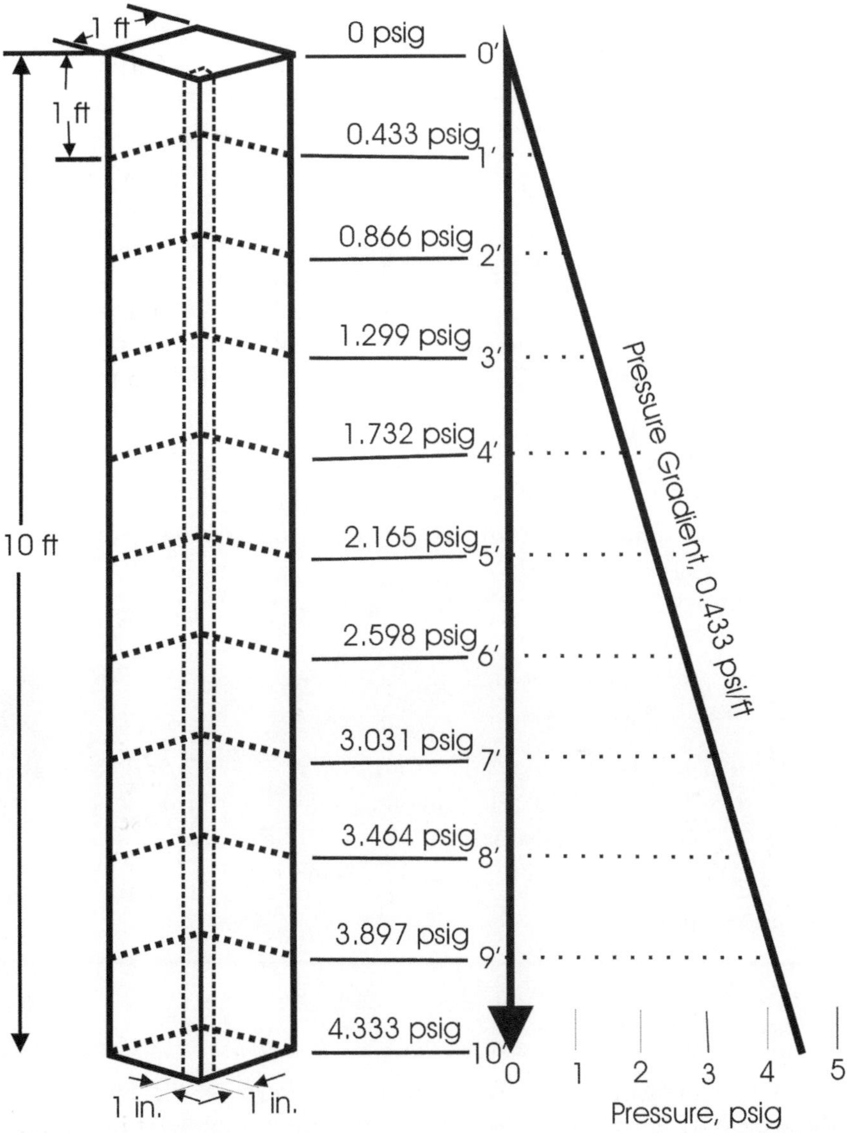

Figure 18-1. Pressure versus depth for a 0.433 psi/ft gradient. *(Modified after Brown et al., 1967, fig. 3.11, p. 28.)*

Solution:

$$p = \gamma_w \times h$$

where p is the pressure in lb/ft^2, γ_w is the specific weight of water in lb/ft^3, and h is the depth of water in ft. Thus:

$$p = 62.4 \text{ lb/ft}^3 \times 6 \text{ ft} = 374.4 \text{ lb/ft}^2 \text{ or } (374.4/144) \text{ lb/in}^2 = 2.6 \text{ psi}$$

The specific weight of a fluid is equal to:

$$\text{specific weight} = \frac{\text{weight}}{\text{volume}} \tag{18-3}$$

For example, the specific weight of water is 62.4 lb/ft^3 at 60°F and atmospheric pressure.

The specific weight of water must take into account the quantity of dissolved salts, volume of dissolved gas, oil content, temperature, and pressure.

Density (ρ) is commonly defined as the specific weight, γ, divided by the gravitational acceleration (g):

$$\rho = \gamma / g$$

Oil field brines commonly have a gradient of 0.45 psi/ft or higher due to dissolved salts. To determine the gradient for brine, a correction factor (C_b) for the specific weight of brine due to dissolved salts can be determined using Figure 18-2. The brine correction factor can be determined as follows:

$$C_b = \frac{\text{brine specific weight, lb / ft}^3}{62.4 \text{ lb / ft}^3 \text{ (pure water)}} \tag{18-4}$$

With increasing depth from the earth's surface, the temperature increases. An average temperature of the column of water should be used and the specific weight corrected to 60°F.

Inasmuch as the weight of dissolved gases in water is small, it may be neglected; however, gas in a column of water will reduce the specific weight of water. To correct for the effect of dissolved gas in

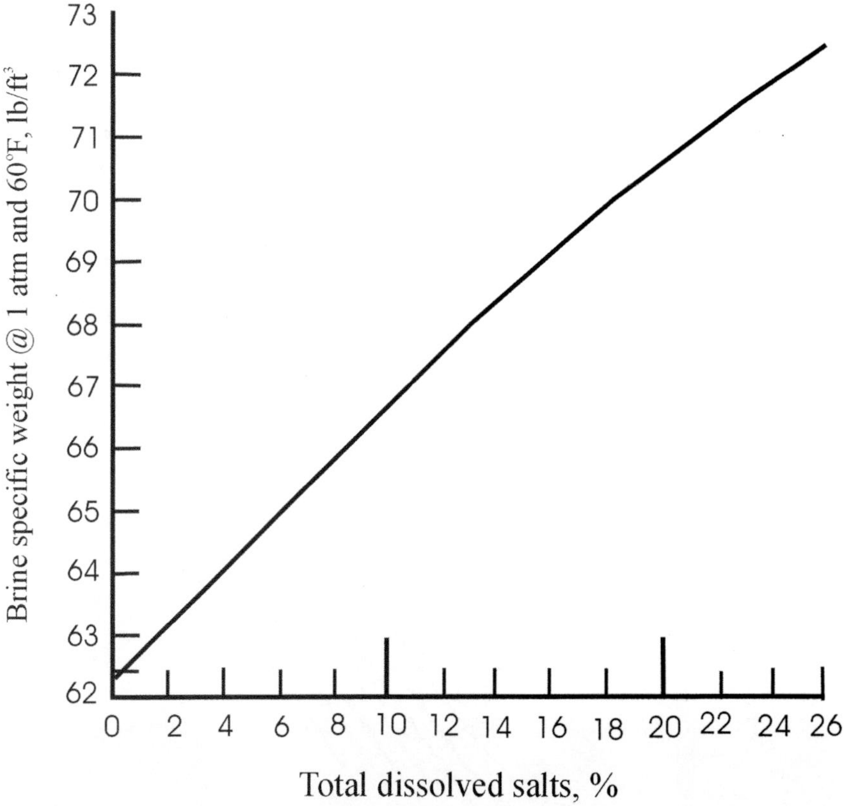

Figure 18-2. Density of brines as a function of total dissolved solids (TDS). *(Modified after Frick, 1962, fig. 22-21, pp. 22–24.)*

the brine, the following correction factor can be used using Figures 18-3 and 18-4. In these figures *FVF* = formation volume factor, ft^3/ft^3; i.e., the volume occupied at reservoir conditions divided by the volume occupied at standard conditions at the surface (60°F and 1 atm pressure).

Oil Columns

The petroleum industry classifies oil on the basis of specific weight and °API gravity. The °API gravity of oil may be obtained from the following relationship:

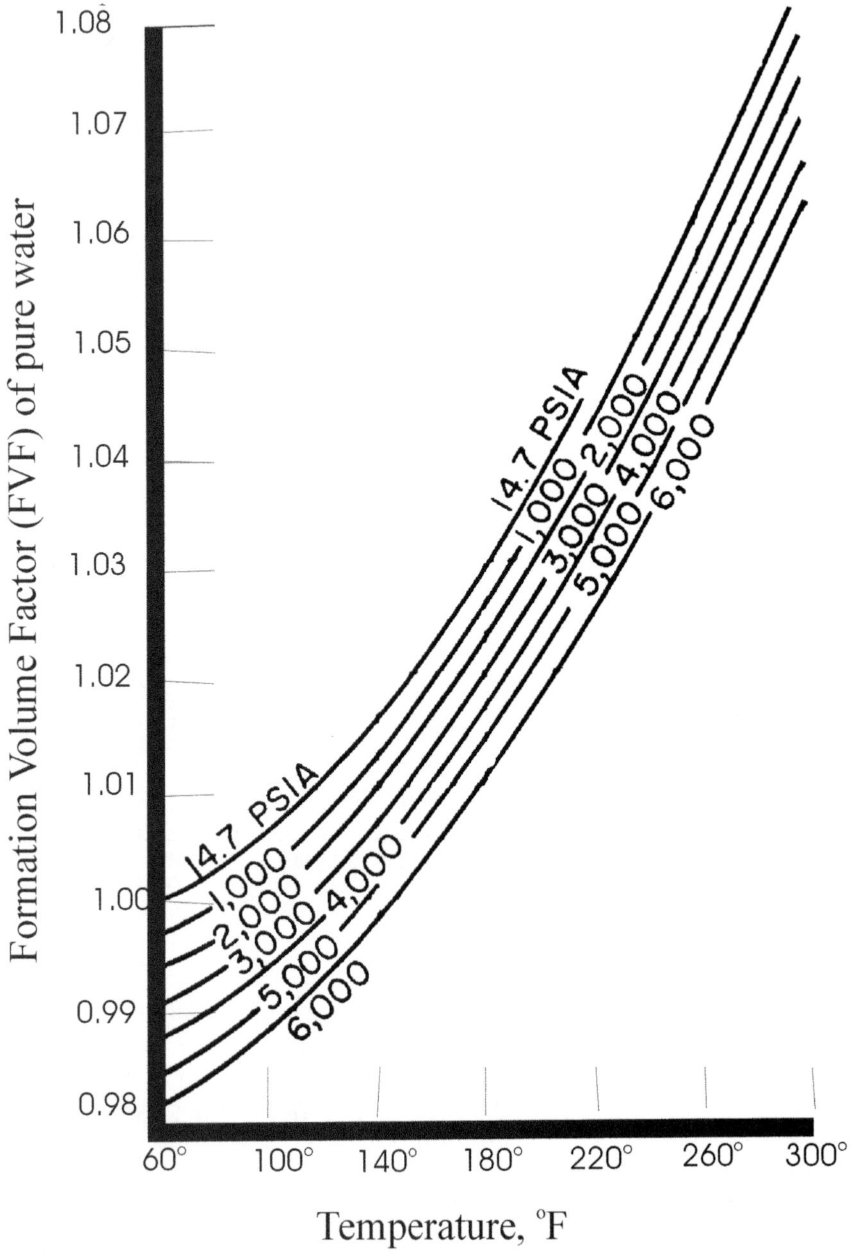

Figure 18-3. Formation volume factor of pure water without gas. *(Calculated from data of Keenan and Keyes, 1936; modified after Frick, 1962, fig. 22-15, pp. 22–21.)*

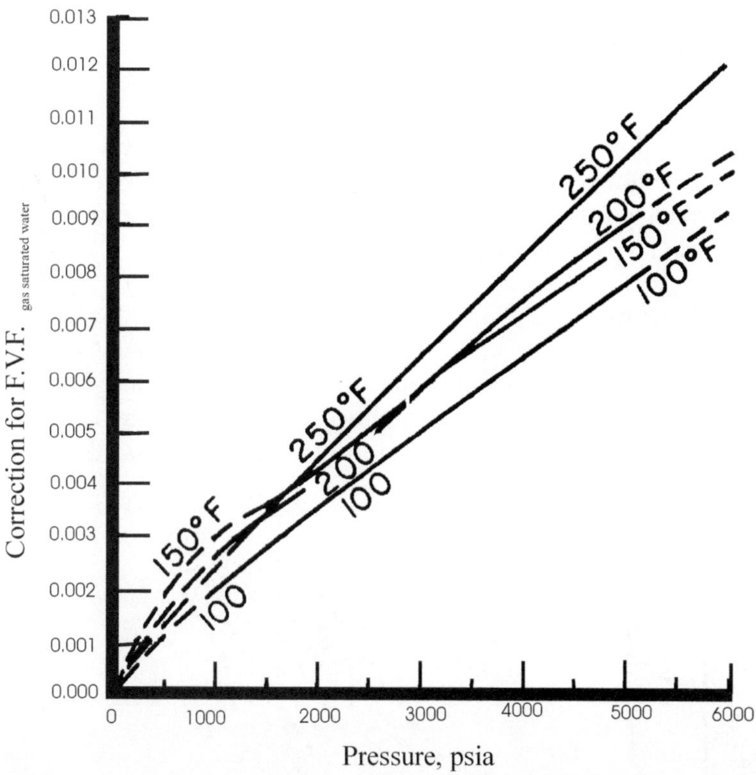

Figure 18-4. Difference between the formation volume factor of gas-saturated pure water and of pure water at various temperatures. Correction for FVF = $FVF_{gas\ saturated\ pure\ water} - FVF_{pure\ water}$. *(Modified after Frick, 1962, fig. 22-16, pp. 22–21.)*

$$°API = \frac{141.5}{SG\ @60°F/60°F} - 131.5 \qquad (18\text{-}5)$$

Higher °API gravities denote lighter and less viscous oils. Lower values denote thicker, heavier oils. A gravity of 10 °API is equvalent to a specific gravity (*SG*) of one (i.e., water).

Oil/Water Columns (Mixtures)

Often there is oil in the water column and it is necessary to determine the pressure gradient of a mixture of these two liquids. The average value of gradient, G_{ave}, may be determined as follows:

$$G_{\text{ave}} = \frac{(h_o\gamma_o + h_w\gamma_w)}{h_t \times 144}, \text{ psi/ft} \tag{18-6}$$

where h_t is the total height (ft) of the fluid column (oil and water); h_o is the height of the oil portion of the column, ft; h_w is the height of the water portion of the column, ft; γ_o is the specific weight of oil (lb/ft^3); and γ_w is the specific weight of water (lb/ft^3).

Gas Columns

The pressure at any depth (point) in a gas column can be calculated. In a situation, where point 1 is above point 2 in a gas column (see Figure 18-5), Eq. 18-7 can be used to calculate the pressure at point 2:

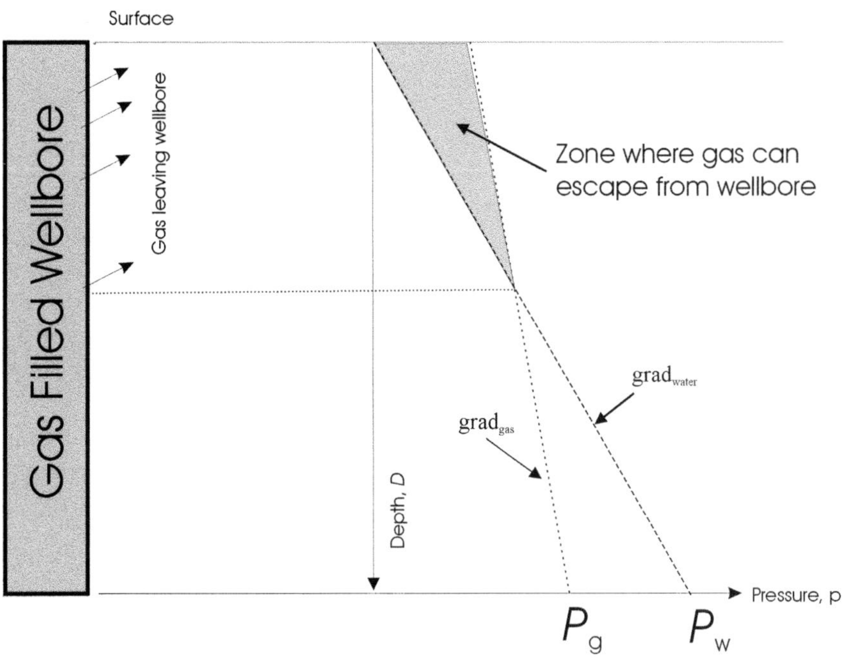

Figure 18-5. Schematic diagram of the pressure distribution in a gas-filled wellbore and in surrounding water-saturated formations. Arrows indicate potential escape of gas through corrosion holes in the casing. P_w = water pressure gradient; P_g = gas pressure gradient. *(Modified after Gurevich et al., 1993, fig. 6, p. 230.)*

$$p_2 = p_1 + \frac{(\gamma_g \times \Delta L)}{144} \tag{18-7}$$

where ΔL is the distance between points 1 and 2; p_1 is the pressure at point 1 in psi; and γ_g is the specific weight of gas in lb/ft^3.

The equation of state, considering 1 lb of gas, can be expressed as follows:

$$pv_g = Z_{ave} NRT_{ave} \tag{18-8}$$

where p = absolute pressure in lb/ft^2, v_g = specific volume in ft^3/lb, Z_{ave} = gas compressibility factor, N = number of moles of gas, R = universal gas constant, and T_{ave} = average absolute temperature in °R. Thus:

$$v_g = \frac{Z_{ave} NRT_{ave}}{p} \tag{18-9}$$

Inasmuch as:

$$v_g = \frac{1}{\gamma_g} \tag{18-10}$$

$$\gamma_g = \frac{p}{Z_{ave} NRT_{ave}} \tag{18-11}$$

Substituting Eq. 18-11 in Eq. 18-7:

$$p_2 = p_1 + \left(\frac{p}{144\, Z_{ave} NRT_{ave}} \right) \Delta L \tag{18-12}$$

Rearranging:

$$(p_2 - p_1) = \left(\frac{p}{144 Z_{ave} NRT_{ave}} \right) \Delta L \tag{18-13}$$

when $\Delta L \rightarrow 0$, Eq. 18-13 can be rewritten in differentials:

$$\frac{dp}{p} = \frac{dL}{144 Z_{ave} NRT_{ave}} \tag{18-14}$$

In integral form (when point 2 is at the bottom of the gas column and point 1 is at the surface) Eq. 18-14 becomes:

$$\int_{p_s}^{p_{bc}} \frac{dp}{p} = \int_0^L \frac{1}{144 Z_{ave} NRT_{ave}} \mathrm{d}X \tag{18-15}$$

where X is a variable of integration.

Thus, after integration,

$$\ln\left(\frac{p_{bc}}{p_s}\right) = \frac{L}{144 Z_{ave} NRT_{ave}} \tag{18-16}$$

and

$$\frac{p_{bc}}{p_s} = e^{\frac{L}{144 Z_{ave} NRT_{ave}}} \tag{18-17}$$

where Z_{ave} = average gas compressibility factor computed at the average temperature and pressure, T_{ave} and p_{ave}; N = number of moles of gas; R = universal gas constant (equal to 10.7 for one lb mole of gas); p_s = pressure at the surface, psia; p_{bc} = pressure at the bottom of gas column; L = vertical depth of gas column, ft; and T_{ave} = average temperature in gas column, °R. On simplifying the exponent and considering only one pound of gas:

$$p_{bc} = p_s e^{\frac{LG_g}{53.3 Z_{ave} T_{ave}}} \tag{18-18}$$

where G_g = gravity of gas as compared to that of air (SG of air = 1).

The average temperature, T_{ave}, in the gas column can be computed from the following equation:

$$T_{ave} = T_s + \left[\left(\frac{L}{2}\right)\left(\frac{dT}{dL}\right)\right] + 460° \quad (18\text{-}19)$$

where T_s = surface temperature, °F; L = vertical depth of gas column, ft; and (dT/dL) = geothermal gradient, which is commonly assumed to be about 2°F/100 ft. It should be noted, however, that the geothermal gradient varies with locality.

Sample Problem

Above what depth would gas leave the casing through holes formed owing to corrosive formation waters, given the following information:

Specific gravity of gas, G_g = 0.86 (compared to air); Casing depth, D = 4,000 ft; Supercompressibility factor, Z = 0.95; Casing diameter = 7 in; Surface casing pressure, p_{tc} = 180 psig; Water table depth = 70 ft below surface; hydrostatic gradient = 0.433 psi/ft; Temperature at surface, T_s = 55°F; Geothermal gradient = 2°F/100 ft.

Solution:

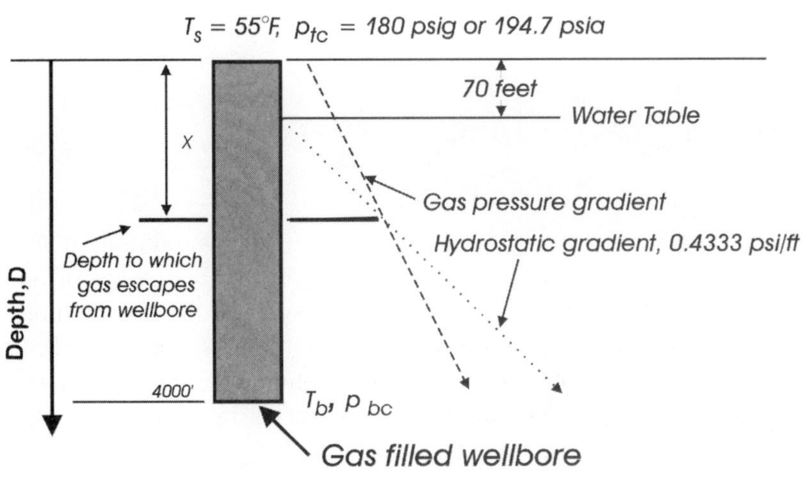

1. $T_b = 55 + \left[(4000)\left(\frac{2}{100}\right)\right] = 55 + 80 = 135°F$

$$T_{ave} = \frac{T_s + T_b}{2} = \frac{(55 + 460) + (135 + 460)}{2} = 555°R$$

$$p_{bc} = p_{tc} \times \exp\left(\frac{LG_g}{53.3 \times Z \times T_{ave}}\right) = 194.7 \times \exp\left(\frac{4000 \times 0.86}{53.3 \times 0.95 \times 555}\right)$$

$$= 220.05 \text{ psia or } 205.35 \text{ psig.}$$

2. Gas pressure gradient $= \dfrac{p_{bc} - p_{tc}}{D} = \dfrac{(205.35 - 180)}{4000}$

$$= 0.00634 \text{ psi/ft}$$

Hydrostatic gradient = 0.433 psi/ft
Water table is located at a depth of 70 ft

3. Assume that at depth x, gas would leave casing through corrosion holes

   Gas pressure at depth x = 180 + 0.00634x, psig
   Hydrostatic pressure at depth x = (x − 70) × 0.433, psig
   Gas pressure at depth, x = hydrostatic pressure at depth x
   Equating gas pressure and hydrostatic pressure and solving for x, x = 492.92 ft

CHAPTER 19

Hazards Resulting from Migrating Gas

INTRODUCTION

The following case studies represent the various environmental problems that can occur as a result of the migration of gas. The methane gas can explode when 5% by volume is mixed with air. Natural gas can also kill plants when it saturates the soil.

CASE STUDIES OF GAS MIGRATION

Los Angeles, California—Fairfax Area

The Fairfax area of Los Angeles, California lies over the depleted Salt Lake Oil Field. This is an area with numerous environmental problems as a direct result of gas migration to the surface. It also includes the internationally known La Brea Tar Pits, in which the bones of numerous Pleistocene mammal and bird fossils have been preserved.

The La Brea Tar Pits represents an area of oil and gas seepage to the surface from subsurface hydrocarbon deposits (Figure 19-1). After a rain, bubbles of gas can be observed thoughout this area in the cracks of streets and other paved areas. In several ponds, located in depressed areas, a significant volume of oil migrates to the surface and floats upon the surface of water. Over time, there is a loss of lighter hydrocarbons from this oil and the remaining residue is degraded into a heavy, thick, tar-like substance that sinks to the bottom of the pond. This tar, which has gathered in depressions, has trapped and preserved many Pleistocene mammals and birds. Similar small accumulations of tar can also be found in surface cracks of paved parking lots, sidewalks, and streets throughout the area. Gas and oil shows are not

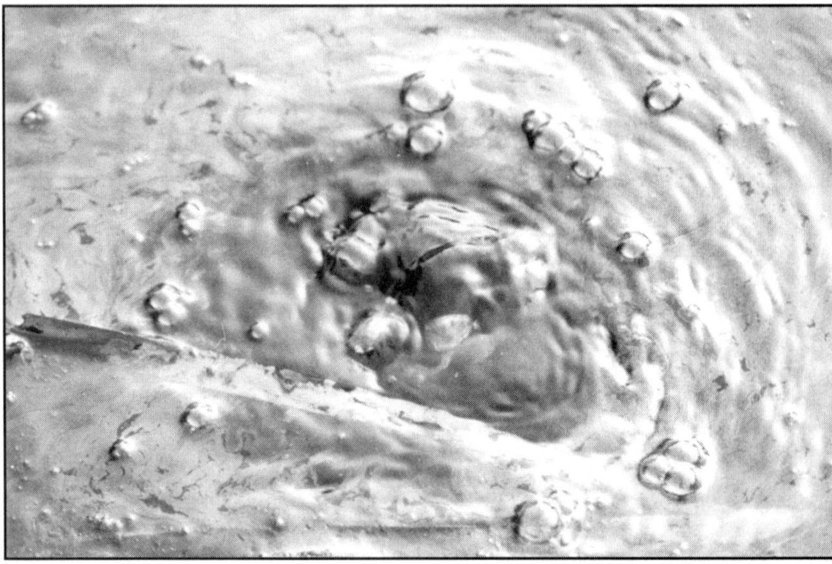

Figure 19-1. Bubbles of gas and oil breaking the surface at the La Brea Tar Pits, California.

limited to the pits alone: hydrocarbons also appear at the surface in several nearby underground parking areas and basements. In one case, the produced gas is collected and then burned to generate electricity for the commercial structure.

The existence and use of these seeps by local Indians led to the early exploration of the surrounding area for oil about 100 years ago. The authors of this book made a detailed study of the La Brea Tar Pits area. Geologic, well logging, and petroleum production data were gathered and analyzed. Gas and oil samples from the surface and underlying reservoir were analyzed to determine the origin of gas and oil migrating to the surface. Soil gas probes monitoring data gathered since 1985 were also examined to help determine the origin and paths of migration of gas to the surface.

The Salt Lake Oil Field, California, has been heavily exploited since the early 1900s The general geology is shown in Figures 19-2 and 19-3. By 1910, 325 wells had produced about 50 million barrels of oil. The first wells that had been drilled were abandoned to real estate development. In the 1980s, it was estimated that over 500 wells had

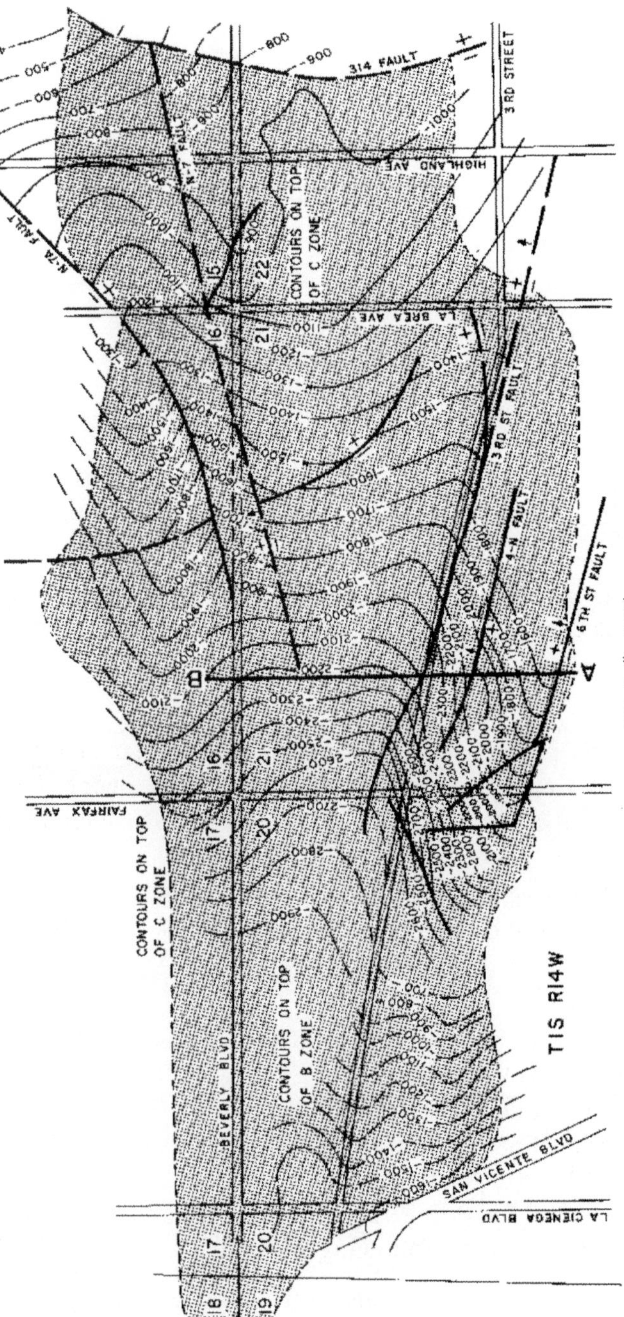

Figure 19-2. Horizontal contours on the Salt Lake Oil Field, California. *(Modified after California Division of Oil and Gas, California Oil and Gas Fields, 1974.)*

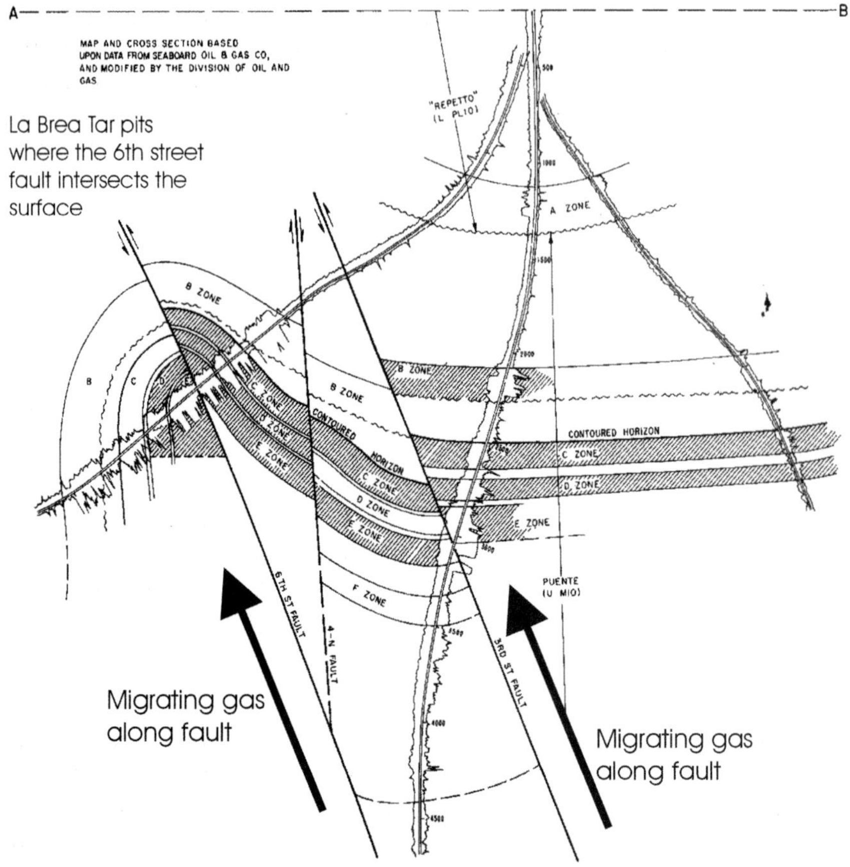

Figure 19-3. Structure of the Salt Lake Oil Field, California. *(Modified after California Division of Oil and Gas, California Oil and Gas Fields, 1974.)*

been abandoned in the area. Most of these wells were not abandoned according to today's standards.

As a result of this gas and oil production, the reservoir pressure has substantially declined. Additionally, there has been subsidence in the area as a direct result of fluid production from the reservoir. This has increased vertical fracturing. The volume of free gas in the reservoir has increased due to the following reasons:

1. When the fluids were produced from the reservoir, dissolved gas in the fluids was liberated.

2. A portion of the reservoir was used to store this free gas.
3. The reservoir dips steeply and, as a result, fluids within the reservoir migrate due to gravity (in gravity drainage: gas moves upward, whereas oil moves downward). Consequently, free gas is placed next to the fault zone and becomes available for migration.
4. Due to water injection into various portions of the reservoir, large volumes of free gas have been displaced to the upper parts of the reservoir. The vertical fracturing in the reservoir has occurred due to waterflooding operations when injection pressures have exceeded the fracturing pressure of the reservoir. This also occurred during cement squeezing operations.

Free gas is escaping from the Salt Lake Oil Field along several faults and wellbores that penetrate the reservoir. The 3rd and 6th Street faults are good conduits for gas and oil migration (Figure 19-3). Gas and fluid migration is occurring from the reservoir. The production history of the Salt Lake Field demonstrates that gas, oil, and water have crossed from one fault block to another, where the pressure differential across the fault blocks exceeded 300 psig.

Ross Dress-For-Less Department Store Gas Explosion

At a location approximately one-half mile north of the La Brea Tar Pits, a gas explosion and fire destroyed the Ross Dress-for-Less Store on March 24, 1985 (Figure 19-4). Twenty-three people were injured. After investigating, we found that gas had migrated up along the 3rd Street fault and certain abandoned wellbores. After the gas explosion in 1985, one vent well, 80 ft in depth, was drilled to bleed off the natural gas accumulated below the hard clay layer, which acts as a caprock at a depth of 20–30 ft. Several shallow probe holes, approximately 13 ft in depth, were also drilled in an effort to monitor the gas concentrations underground. The possible paths for gas migration are shown in Figure 19-5.

In 1989, the gas monitor wells indicated that large volumes of gas were again building up under the caprock. The concentration of gas in the soil had accumulated above the gas explosive limits, at a pressure of about 30 psig. The reason for the buildup of gas in the soil was that the gas vent well near the Ross Dress-for-Less store had become plugged with debris, preventing the accumulated gas from venting to the atmosphere. This situation was corrected, resulting in

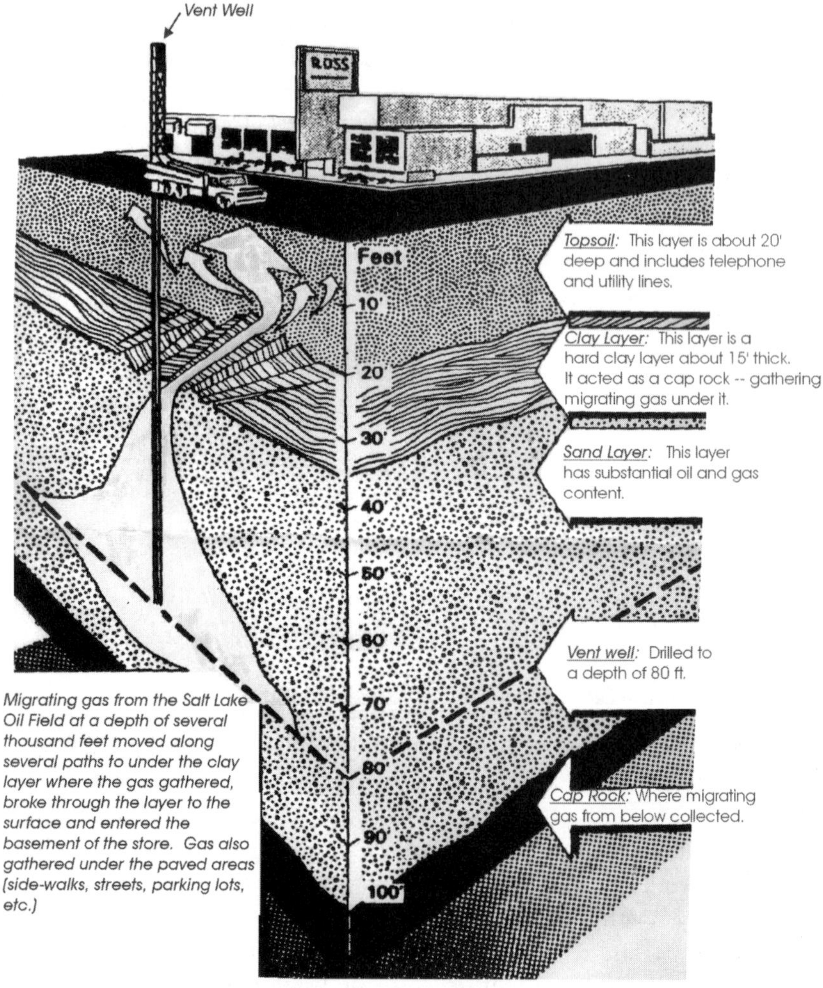

Figure 19-4. Schematic showing how gas entered the basement of the department store and the surrounding area. *(Modified after an article by George Ramos and Ted Thackrey in the* Los Angeles Times, *1985; illustration by Michael Hall.)*

a drop in gas pressure and concentration of gas. In 1992, the monitoring probes revealed an overall increase of gas concentrations in the subsurface near the 1985 explosion site.

Examples of gas problems at other properties surrounding the Fairfax area include the smell of gas and gas-oil-water leakages from

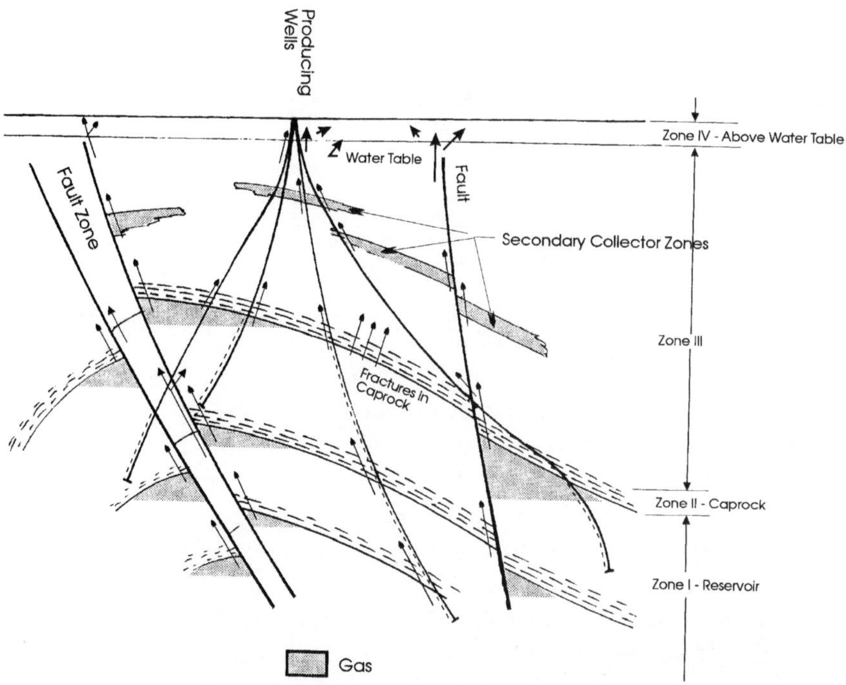

Figure 19-5. Schematic diagram showing migration paths of gas from the reservoir to secondary collector zones and then to the surface. The arrows show the various potential paths of gas migration. *(After Gurevich et al., 1993, fig. 17.)*

abandoned wells. In 1990, 5% of methane gas (by volume) was detected in a nearby bank, forcing its closure for several months until the gas seepage could be diverted from the structure. High methane gas concentrations also have been detected under the playground of a nearby elementary school. Since the 1970s, several older abandoned oil wells in this area have experienced gas and oil "blowouts." These wells have had to be reabandoned using modern techniques. Additionally, there is oil and gas leakage from several other abandoned wells in the area.

Gas analyses reported by the Los Angeles City task force investigating the incident showed that the Fairfax seeps consisted of 90% methane with less than 2% ethane (C_2H_4) content, 6% CO_2, and minor quantities of N_2 and oxygen of atmospheric origin. The gases reported from the Salt Lake field contained about 70% methane, 10%–12% ethane, and 15%–18% carbon dioxide (CO_2). The absence of helium and odorants ruled out the possibility of gas presence being caused by a local pipeline leak. Initially, the task force concluded that the origin of the gas was probably of bacterial origin, related to swamp deposits and tar accumulations in the shallow Quaternary alluvium cover of the area. However, in 1990 we showed that the methane gas migrating to the surface originated from the oil field below. Subsequent investigation by Schoell et al. (1993) also showed that the gases originated in the Salt Lake oil field and migrated to the surface mixing with surface gasses (see Figure 19-6). The 3rd and 6th Street faults and improperly abandoned oil wells appear to be the most likely paths of migration. During migration, the gases most likely lost some of the ethane due to adsorption and oxidation. Carbon dioxide was possibly lost due to precipitation of carbonates. Mixing of the thermogenic gases from the oil field with the near-surface biogenic gases further masked the origin of the gas. Fortunately, these processes often have only a minor effect upon the isotopic composition of the light hydrocarbons, especially methane. As shown in Figure 19-6, despite differences in composition of methane and ethane, $\delta^{13}C$ values of the seep and relief gases are similar to those of the Salt Lake oil field gases.

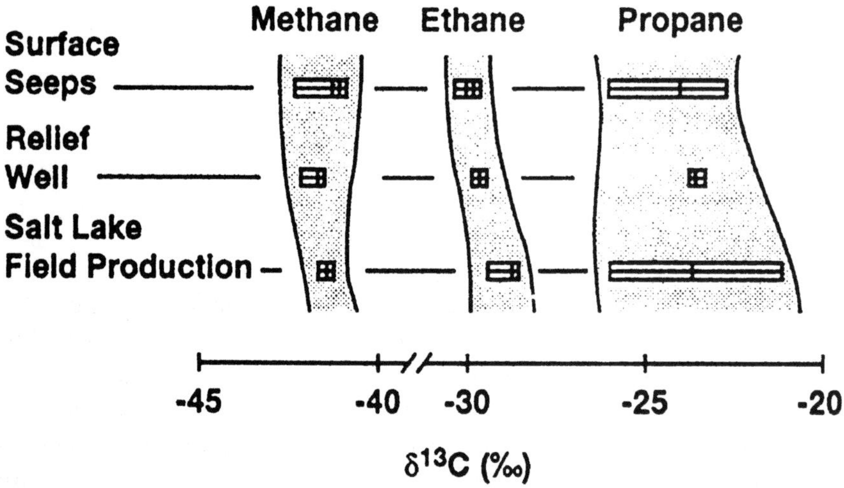

Figure 19-6. Carbon isotopic fingerprinting of gas leaking from a shallow reservoir to the surface. *(After Schoell et al., 1993, p. 7, fig. 8.)*

The isotopically heavy methane (> $-45^{o}/_{oo}$) precludes a bacterial origin for this gas. There are many potential paths of gas migration.

Throughout the area, there is evidence of gas migration along faults and abandoned wells. There is also evidence that the operations in the underlying oil fields have aggravated the problem by placing (moving) gases next to the fault due to the gravity drainage. Production operations can free gas that was once held in solution, providing a gas source for migration to the surface. Water and gas injection operations can force gas to migrate within the reservoir and seek paths of migration toward the surface. Wells abandoned in the 1920s through the 1940s were not properly sealed off and, as a result, often provide additional paths for gas migration to the surface. There is a risk of gas migration into homes, apartment buildings, and commercial structures, which have been built in the area of producing oil fields. The cost of reabandoning old wells would be very high, even though there is strong evidence of gas leakage in many cases. Much of the commercial area has been paved over, thus trapping gas migrating to the surface and not allowing it to dissipate into the atmosphere. This trapped gas under buildings, parking lots, and schoolyards may create dangerous situations and a potential explosion hazard. Detailed studies, surface gas monitoring, flaring of migrating gas, along with changes in oil field production practices must be made to prevent future explosions.

Santa Fe Springs, California—Methane Gas Accumulation in Soil

A high concentration of methane gas was discovered in the ground about 60 feet below the surface in Santa Fe Springs, California. The methane concentration in the ground was found to be 300,000 parts per million. Most of the area where this gas was discovered was vacant and a part of the Santa Fe Springs redevelopment agency area. This area overlies the Santa Fe Springs oil field, a field known for its high gas production (Figure 19-7). City officials felt that there was no danger as long as the methane gas did not accumulate in an enclosed area, such as the Fairfax area in Los Angeles, California. A methane barrier and vents were planned for all new constructions in that area to channel gas outside of the buildings and prevent natural gas from seeping into the buildings.

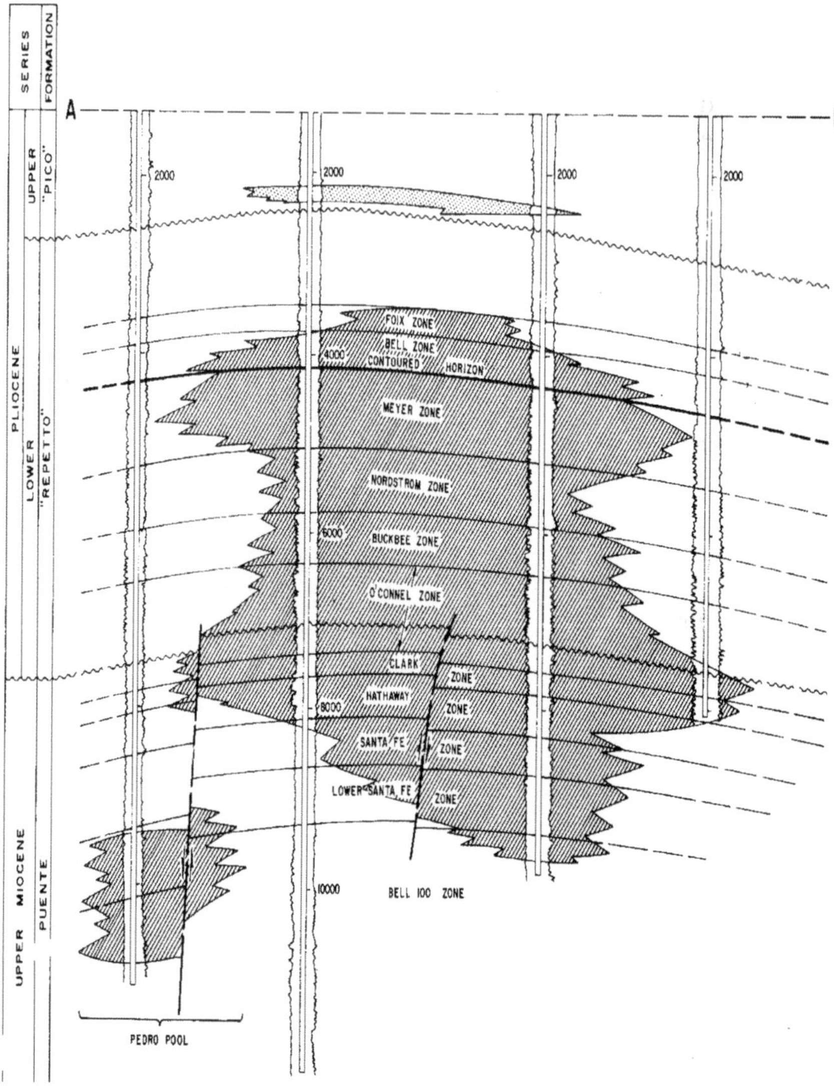

Figure 19-7. Structure of Santa Fe Springs Oil Field, California. *(Modified after California Division of Oil and Gas, California Oil and Gas Fields, 1974.)*

Sylmar, California—Tunnel Explosion

In 1971, a tunnel (approximately 5 miles long) in the Sylmar area of Southern California was being excavated for a future water project. The excavation had proceeded uneventfully until the tunneling machine encountered a fault plane. The fault was filled with natural gas that had migrated from a nearby oil field. Natural gas from that oil field had migrated along the fault plane and had filled the fault zone. The tunneling machine, upon contacting the gas-saturated fault zone, set off the explosion. Consideration of gas-saturated fault zones had not been a part of the design of this project. Seventeen workers were killed in the explosion.

Commerce City, Colorado—Landfill Explosion

On January of 1978, an explosion occurred in Commerce City, Colorado, killing two men. Methane gas (biogenic) from a landfill 35 feet away entered a water conduit and traveled uphill to a manhole. To correct the problem, gas extraction plants were proposed at the landfill to decrease the methane gas danger and produce revenue at the same time.

CHAPTER 20

Hazards of Gas Storage Fields

INTRODUCTION

The practice of using depleted oil fields for the purpose of storing large volumes of natural gas underground has become more common. The economics are such that it is more profitable to store gas in the underground reservoirs than construct an equivalent surface tankage. Over 300 underground natural gas storage projects are currently being operated in the United States. Underground storage facilities have demonstrated a long history of gas migration problems. Experience has shown that using depleted oil fields for gas storage facilities can create a serious risk of explosions and fires, especially when located in urban settings. One must consider the following in designing underground gas storage facilities: (1) the presence of faults and fractured zones, (2) lithology and structure of the overlying formations, and (3) breakdown of cement and the seals that block the vertical migration of fluids and gases in the wellbores and their annular space. In fact, it is impossible to assume that the vertical migration of natural gas to the surface from the gas storage reservoir will never occur.

In California, for example, gas is transported from out-of-state through pipelines. The gas is then stored in several underground reservoirs until needed by consumers. These storage fields operate under large pressure fluctuations, dictated by seasonal variations in the demand for (and usage of) natural gas. Sometimes these cyclic storage pressures exceed the original oil field reservoir pressure. This overpressuring of the reservoir can create vertical fractures, substantially increasing the risk of gas leakage to the surface. In many cases, the original rural settings for these gas storage areas have been changed to high-density housing developments. Today, the risks of injury from fire and/or explosions have increased because a greater percentage of the land is now covered, not permitting the gas to escape into the atmosphere.

All of the previous discussion relating to gas migration from oil fields to the surface is applicable to the concerns associated with underground gas storage facilities. The problems can be grouped into three areas: (1) problems related to the natural pathways for lateral and vertical gas migration (e.g., faults and fractured zones); (2) problems related to man-made pathways for vertical gas migration (e.g., improperly abandoned wells, fracturing of the reservoir by drilling and/or production operations, fracturing due to overpressuring and secondary waterflooding, and enhanced oil recovery operations); and (3) problems related to carcinogenic chemicals present in the natural gas (e.g., benzene and toluene, and added odorization agents).

It should be remembered that an explosion hazard exists with an air-natural gas mixture if the content of natural gas is 5% to 15%.

CASE HISTORIES OF GAS MIGRATION PROBLEMS

El Segundo Oil Field—California

The El Segundo Oil Field has a depth of about 3,000 feet. Gas that was stored here in the early 1970s, migrated into the adjoining geologic formations. Gases were detected in a nearby Manhattan Village, California, housing development that was under construction. As a result, the construction was stopped. To protect the housing development, a $750,000 passive venting system was installed to prevent the buildup of gases, which may cause an explosion, and the injection project was shut down.

Honor Rancho and Tapia Oil Fields—California

Castaic gas storage is located in the depleted Castaic Hills Oil Field (Figure 20-1) near the producing Honor Rancho (Figure 20-2) and Tapia (Figure 20-3) oil fields. Figure 20-4 shows the relationship of the three oil fields. The arrows show the direction of gas migration. The Tapia oil field producing zone has an average depth of 1,050 feet. The Honor Rancho has several producing levels ranging from an

(text continued on page 295)

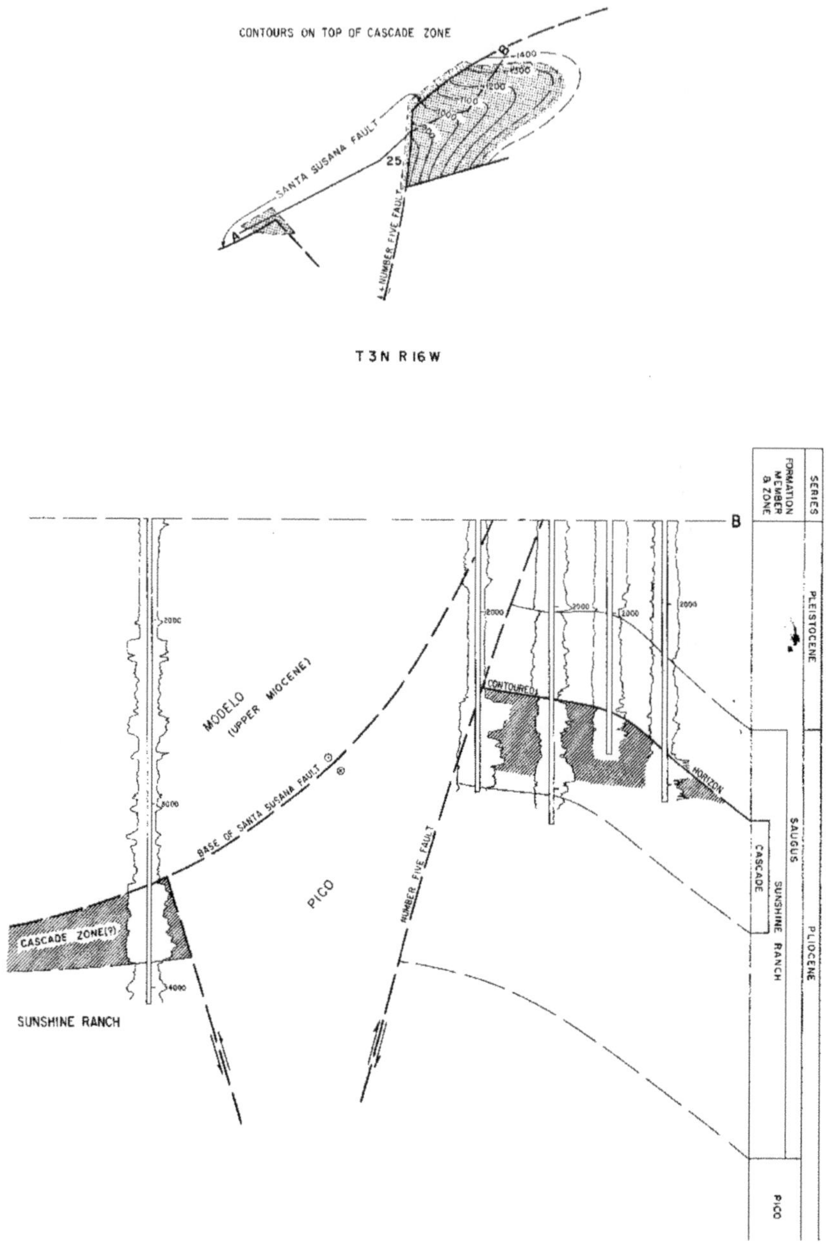

Figure 20-1. Horizontal contours and structure for the Castaic Hills Oil Field, California. *(Modified after California Division Oil and Gas, California Oil and Gas Fields, 1974.)*

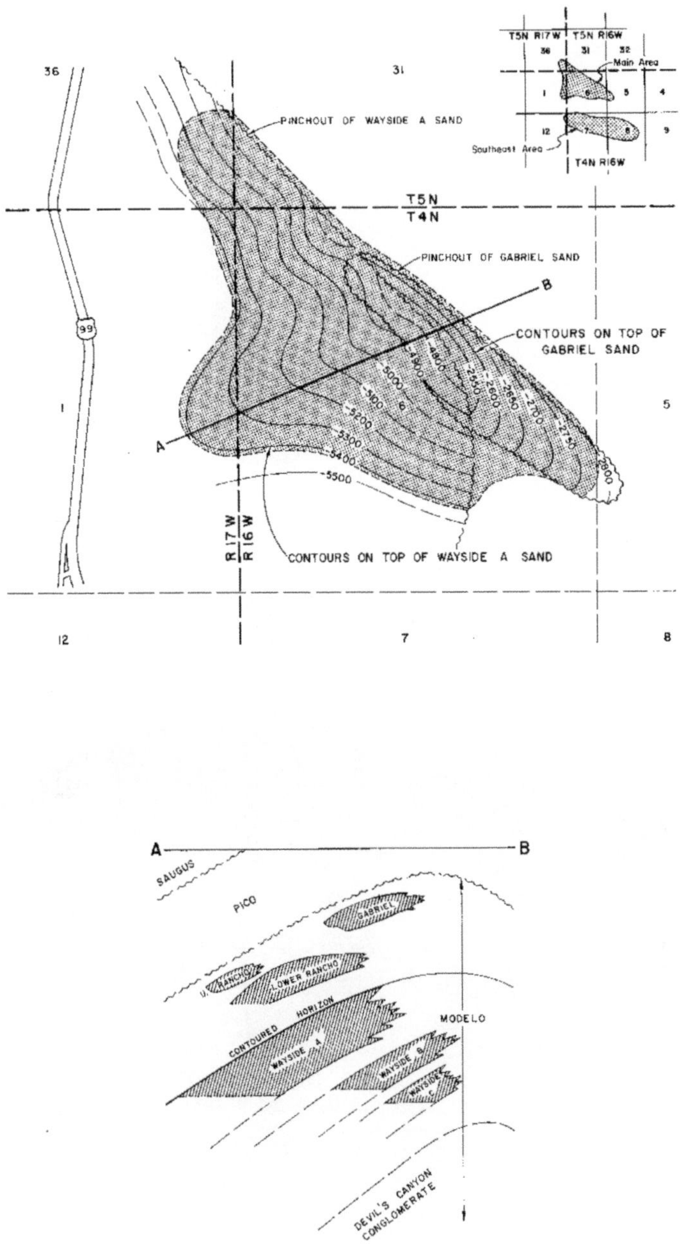

Figure 20-2. Horizontal contours and structure for the Honor Rancho Oil Field, California. *(Modified after California Division Oil and Gas, California Oil and Gas Fields, 1974.)*

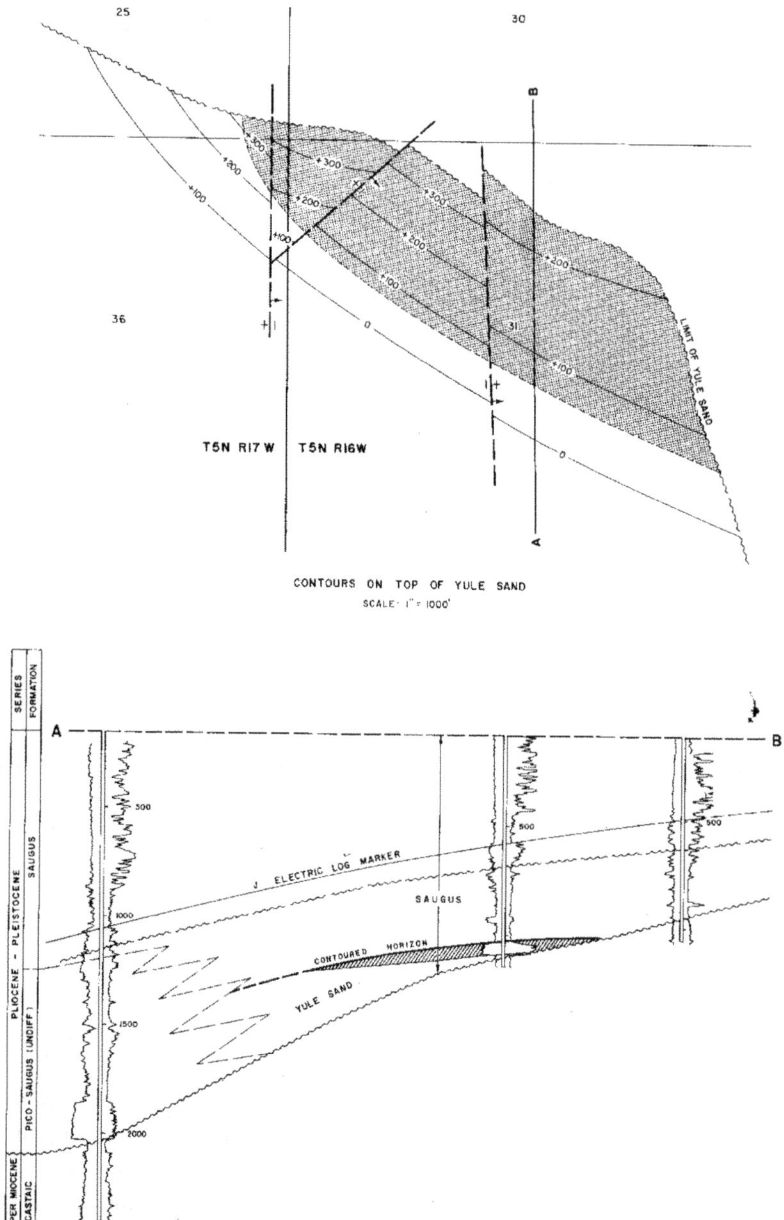

Figure 20-3. Horizontal contours and structure for the Tapia Oil Field, California. *(Modified after California Division Oil and Gas, California Oil and Gas Fields, 1974.)*

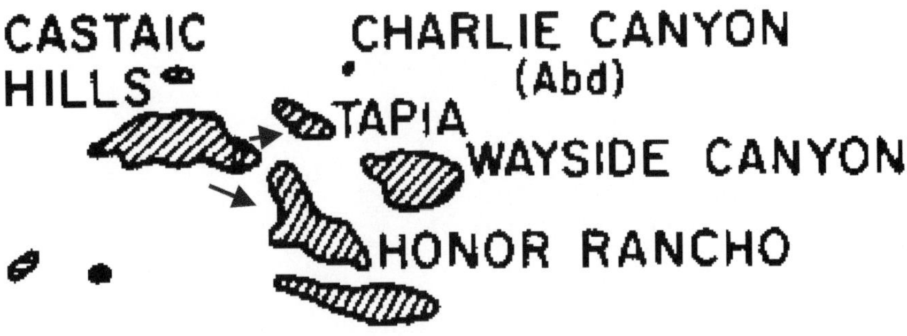

Figure 20-4. Physical relationship of Castaic Hills, Honor Rancho, and Tapia oil fields, California. Arrows show the prevailing direction of gas migration. *(Modified after California Division Oil and Gas, California Oil and Gas Fields, 1974.)*

(text continued from page 291)

average depth of 3,800–6,400 feet. Gas from the gas storage project broke into producing wells of the Honor Rancho and Tapia oil fields. Indications of gas migration along faults at the surface included the killing of oak trees along the surface trace of faults in the area. Gas bubbles were also noted in a nearby water reservoir. The helium content identified the migrating gas as originating from the gas storage project at the Castaic Field as the native natural gas has a very low helium content, whereas the imported gas from Texas does contain helium.

Montebello Gas Storage Field—California

The gas company stores large volumes of natural gas, which is transported into the Los Angeles basin, California, from out-of-state sources, using the Montebello Oil Field in California. The depth of this oil field ranges from 5,000–7,000 feet (Figure 20-5). The large storage facility is used to reduce the quantities of gas that would otherwise have to be shipped into the Los Angeles area by pipeline during periods of high demand, namely, by pumping in more gas than required during the periods of low demand. The gas storage facility also helps to even out the demands for importing gas during a typical year of operation. The gas is pumped into the reservoir under high pressure, which often exceeds prior oil field pressures, causing fracturing.

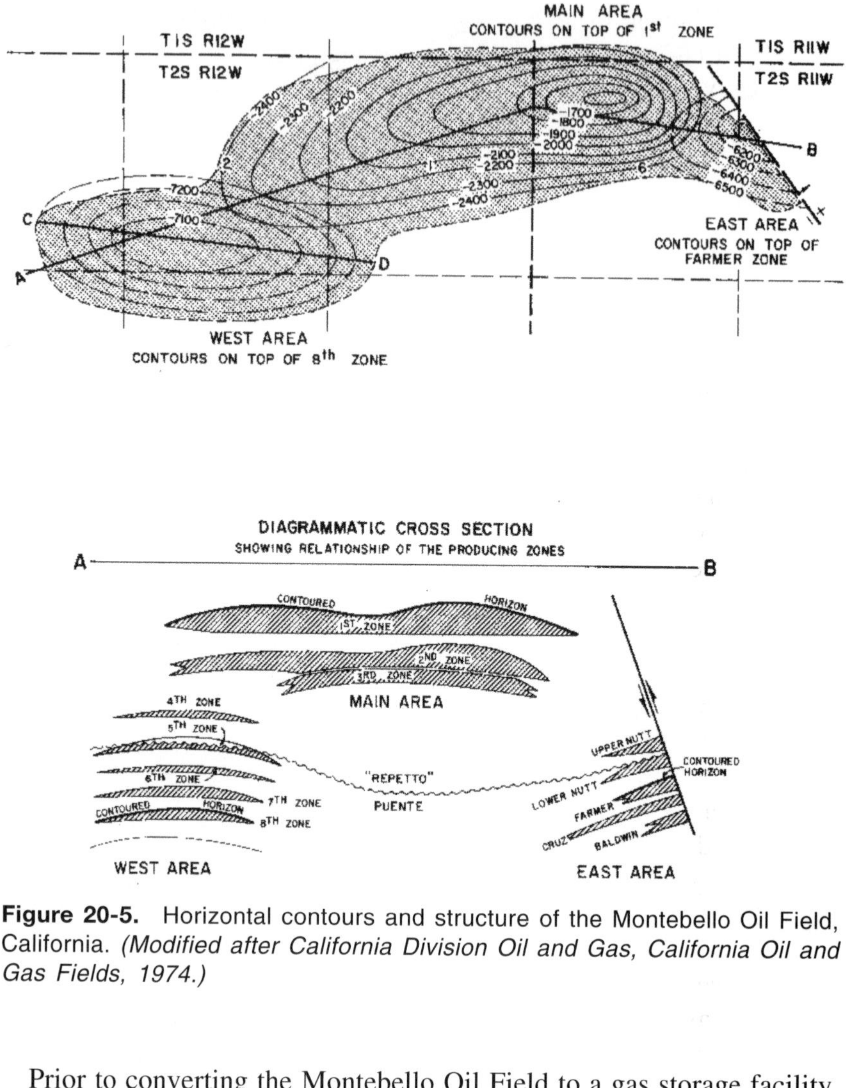

Figure 20-5. Horizontal contours and structure of the Montebello Oil Field, California. *(Modified after California Division Oil and Gas, California Oil and Gas Fields, 1974.)*

Prior to converting the Montebello Oil Field to a gas storage facility, many of the wells in this field had been abandoned using standards that were less stringent than those required at present. The oil field also contains several fault planes, which are potential paths for gas migration.

The gas company began storing gas in a portion of the Montebello Oil Field in the early 1960s. In the early 1980s, significant gas seepages were discovered at the surface within a large housing development above the gas storage facility. Many of these gas seepages appeared to be associated with poorly abandoned wells. The gas seepages endangered homes located on the surface and required evacuation of families on numerous occasions. To reabandon these wells and stop the dangerous vertical gas migration, the gas company purchased and tore down several houses in order to obtain access to the previously abandoned oil wells. These areas were then landscaped and left undeveloped. This gas storage project continues to operate today.

Playa Del Rey Gas Storage Field—California

The Playa Vista proposed surface development and gas storage project is located over a portion of the depleted Playa Del Rey Oil Field in California (Figure 20-6). This field was discovered in 1929 and was in production until the pilot storage project was initiated in 1942. The project was expanded to full-scale operations in 1943. The reservoir has a capacity of about 2.6 billion cubic feet. In the late 1950s, the gas company reabandoned many of the older wells in the wetlands area.

The State of California Division of Oil and Gas (D.O.G.) has reported lateral migration away from the produced area of several million cubic feet of gas to the north and west. There are vertical faults in the region and some investigators incorrectly believe that they act as seals. In Southern California oil fields, a differential pressure of about 300 psig across the fault can result in liquids and gas crossing the fault plane. It was estimated in 1953 by the D.O.G. that 25% of the injected gas migrated to an adjoining oil field. In this case, the oil company returned the gas to the gas company to keep the gas loss to a minimum. The D.O.G. estimated that the overall losses including the "cushion," fluid replacement, and entrapped gas was about 13% for the reservoir.

There have been numerous complaints by local landowners concerning noxious gas odors. In 1990, the gas company vented approximately 300,000 cubic feet of natural gas into the air each month. In 1991, this amount was reduced to about 50,000 cubic feet per month. There also appear to be additional fumes coming from abandoned wells and along faults in the area. It appears that gas is not only escaping from the surface production facilities, but also migrating up along faults and abandoned wells from the storage reservoir. Gas can be seen bubbling through water in the Ballona Channel, in the local marina, and offshore in the ocean. Analysis of this escaping gas showed that it is a dry gas of thermogenic origin.

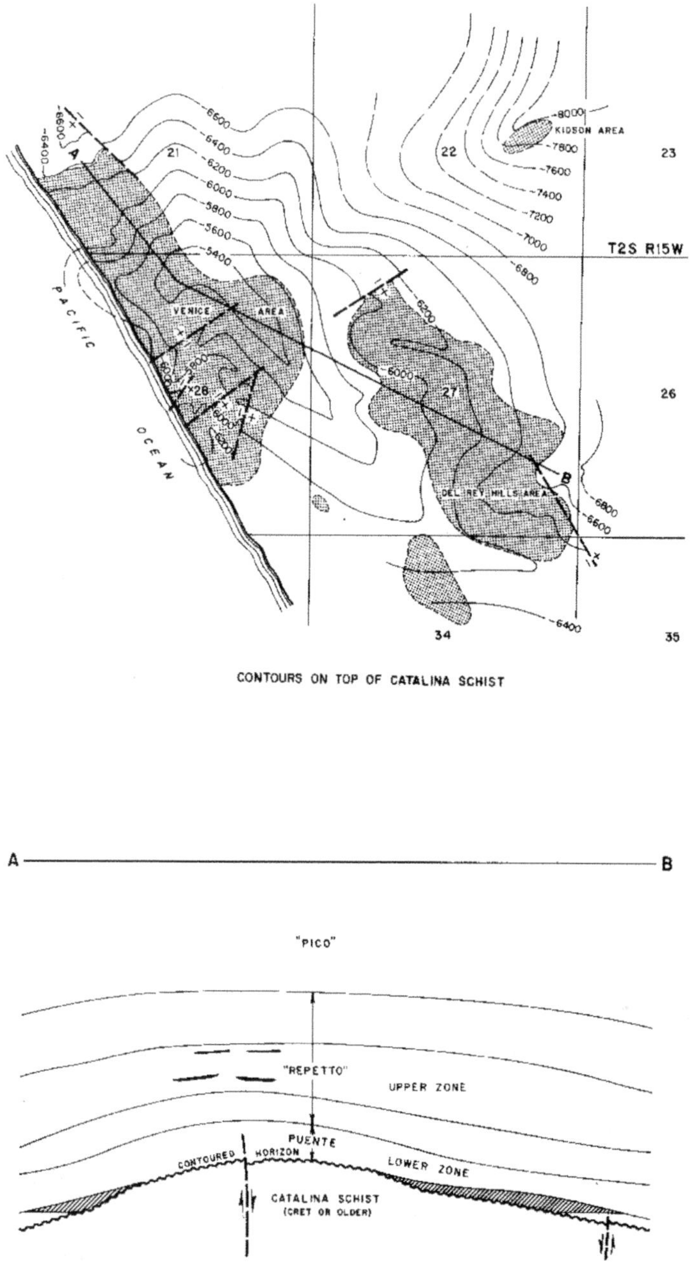

Figure 20-6. Horizontal contours and structure for the Playa Del Rey Oil Field, California. *(Modified after California Division Oil and Gas, California Oil and Gas Fields, 1974.)*

Huntsman Gas Storage—Nebraska

Located in southwestern Nebraska, the Huntsman Gas Storage Field was a depleted 4,800-feet deep gas field prior to its conversion to gas storage. Gas leakage occurred from this field into the adjoining oil- and gas-producing field several miles away. A different company handled production. In this case, a large lateral gas migration occurred through the barrier (fault), which several experts had thought to be impermeable. The gas company purchased gas from the oil company. In fact, they purchased the migrating gas several times over, as the gas was recycled between the two fields. Isotopic analysis proved that the gas company was purchasing its own gas.

Mont Belvieu Gas Storage Field—Texas

In October 1980, a serious gas leak developed in a salt dome gas storage field beneath Mont Belvieu, Texas, located 33 miles east of Houston. The gas seepage was detected when an explosion ripped through the kitchen in a house. This explosion occurred when the homeowner turned on the dishwasher and the spark of electricity ignited the gas mixture. More than 50 families were evacuated from their homes as a result of the explosion caused by the gas leak. A flash fire caused by the gas that had seeped into the home burned the housewife. The gas consisted of a mixture of ethane and propane with traces of butane. The concentration of these gases ranged from 2% to 14% by volume. The gas storage company had noted an "unexplained" drop in the reservoir pressures in September of 1980 after the explosion. The event caused severe financial difficulties for the city, which paid for housing and lodging of the displaced families.

Isotopic gas identification results showed that the source of gas was the gas storage facility.

Leroy Gas Storage Facility—Wyoming

The Leroy Gas Storage project lies about 100 miles northeast of Salt Lake City. Shell Oil Co. drilled the first well in the summer of 1951. After testing, it was decided that the Thaynes would be a good storage formation (Figures 20-7 and 20-8). Additional wells were drilled and completed in 1970 though 1972. The Federal Power Commission approved this storage project on November 17, 1972.

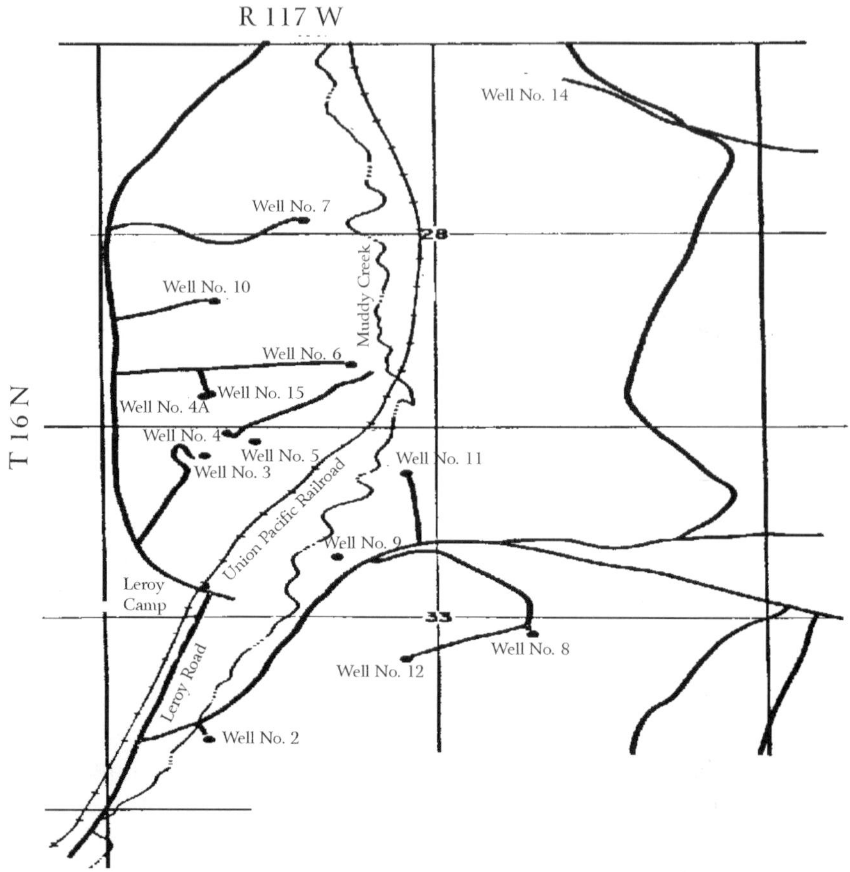

Figure 20-7. Surface map illustrating road (solid lines) and well locations for the Leroy Gas Storage Project, Wyoming. *(Modified after Tek, 1987, fig. 11-7, p. 316.)*

Gas migration to the surface was first confirmed during the later part of November 1978 through bubbling of gas in a nearby creek and pond. The gas migrated from the reservoir and was trapped in a secondary collector formation (Figure 20-9). The gas leaked from this storage to the surface and was, overall, a result of corrosion problems in wellbores and migration of gas along the fault plane. The gas leakage was confirmed by identification of the formation gas bubbles in the adjacent creek and pond. The rate of gas loss to the project was estimated to be (Tek, 1987, p. 323):

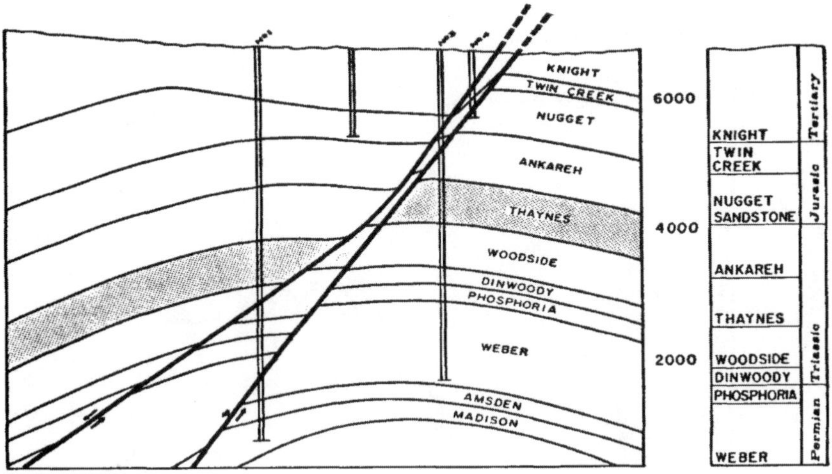

Figure 20-8. Lithologic cross-section and stratigraphic sequence of the Leroy Gas Storage Project, Wyoming. *(Modified after Tek, 1987, fig. 11-8, p. 317.)*

$$q_1 = 3.74 \times 10^{-7}\left(p_g^{\,2} - 1.600^2\right)^n \tag{20-1}$$

where q_1 is the daily leak rate in MMCF/D and p_g is the maximum gas bubble pressure in psia. The exponent n was assumed to be equal to 1.0. The variation of leak-rate versus formation pressure is shown in Figure 20-10.

RECOMMENDATIONS FOR GAS STORAGE

Underground reservoirs provide the most economic method of gas storage. The problem is that such storage areas can become a health and safety hazard if located in an urban area. It has been determined that gas will start leaking to the surface from an underground storage project within 50 years after initiating the project.

The following recommendations are suggested for selection of a gas-storage reservoir site:

1. No structures should be built over gas storage sites. The soil gas should be continuously monitored for the presence of natural gas.
2. No structures should be built over abandoned oil or gas wells.

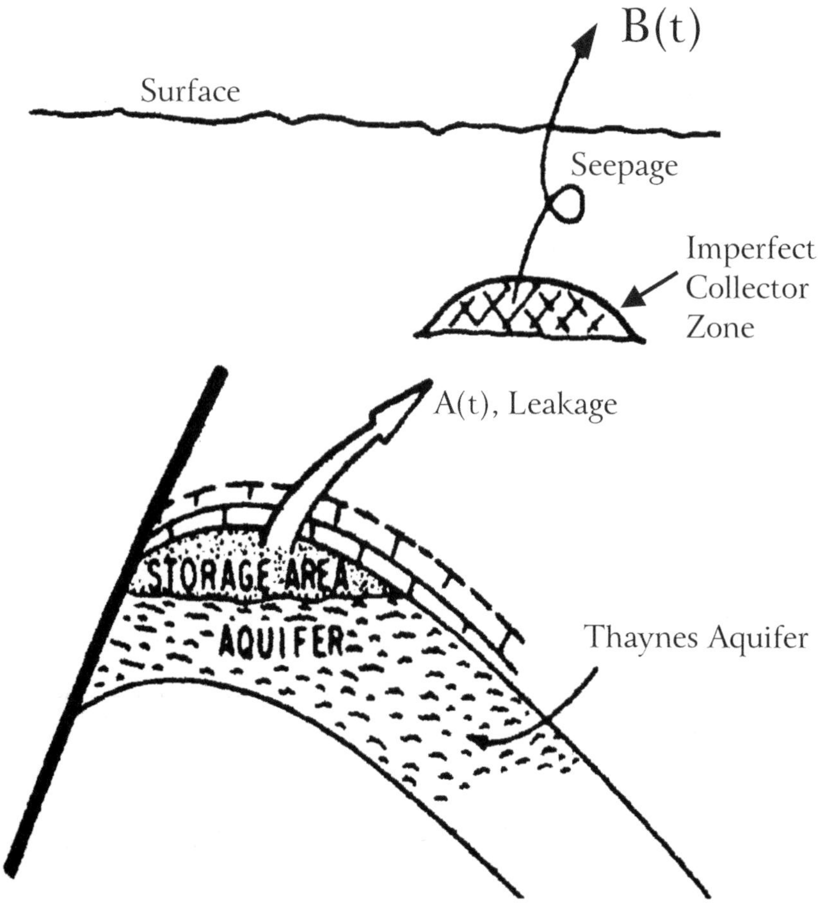

$A(t) = c_p\,(p_g^2 - p_o^2)^n$

Figure 20-9. Schematic representation of gas leakage from the Leroy Gas Storage Project. c_p = performance coefficient for leak, MMCF/D/psi^{2n}; p_g = maximum storage pressure, psia. *(Modified after Tek, 1987, fig. 11-12, p. 323.)*

3. Continuous gas monitoring should be conducted for the presence of natural gas around all wells penetrating the gas storage reservoir. This is particularly important if structures are within two times the radius of the storage area.

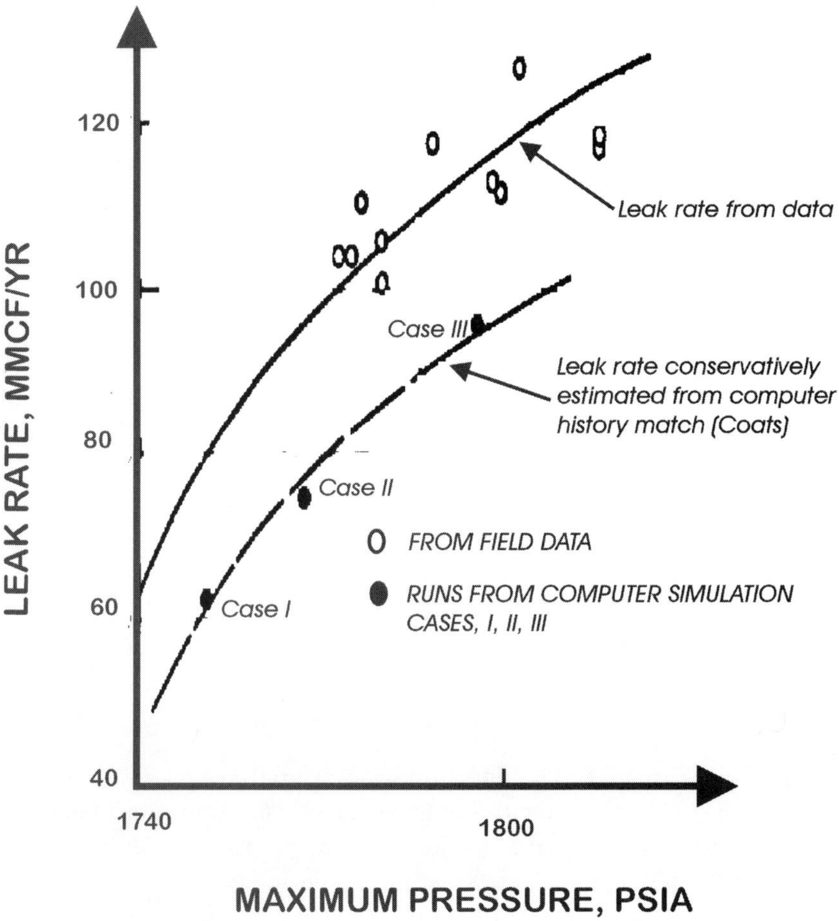

Figure 20-10. Gas leak rate for various maximum reservoir pressures for Leroy Gas Storage project, Wyoming. *(Modified after Tek, 1987, fig. 11-16, p. 326.)*

4. Health hazards upon exposure to natural gas must be evaluated annually, particularly if the gas contains benzene, toluene, and mercaptans.

Underground gas storage facilities, especially those using depleted oil fields, are subject to the same gas migration hazards as producing oil fields. From an environmental viewpoint, there are three areas of concern: (1) the existence of faults and fractures; (2) the presence of abandoned, idle, and producing wells; and (3) fractures formed as a result of gas injection pressures exceeding formation fracturing pressures and those previously existing in the producing reservoir. Gas storage projects require a thorough evaluation to establish whether or not the gas storage facility can be safely operated. This is particularly important in an urban setting where there are human activities. It should also be recognized that even if the reservoir can trap and hold fluids, it might not safely hold "pressurized" gas.

Generally, it is not advisable to locate an underground gas storage facility near an urban setting because of the potential danger of gas migration to the surface, which can create an explosion hazard. Continuous soil-gas monitoring is essential to maintain the safe operation of these facilities, especially in areas close to fault planes and abandoned wells. The anticipated life of a gas storage facility is approximately 50 years. However, even if the facility does not leak initially, it will with time. The important question is not "if" the storage facility will leak, but rather "when."

CONCLUSIONS

The following conclusions can be reached on the migration of gas:

1. The primary force controlling the migration of gas from an oil and/or gas reservoir to the upper layers and traps (secondary gas accumulations) and, eventually, to the earth's surface is the difference between the specific weight of water and that of gas.
2. Gas, oil, and water do migrate along faults. They are also known to migrate along fracture systems and behind wellbore casings to the surface. The volumetric rate of gas migration toward the surface is a function of the type and width of the path along which it migrates.
3. Gas migrating along existing faults, fracture zones, and man-made paths can create surface hazards if the gas is allowed to concentrate in localized collector zones (secondary traps), especially if they are near the surface. It is not advisable to build over abandoned or idle wellbores. In some cases, geophysical methods must be used to establish their presence. Over time, the cement and well casing deteriorate, resulting in the creation of paths for gas migration to the surface. Also, structures should not

be built over fault planes, especially those that intersect oil field reservoirs, which often contain free natural gas. In urban areas, the migrating gas can cause fires and/or explosions when it accumulates in the confined areas of the structure.

4. Underground gas storage facilities have demonstrated a long history of gas migration problems. Gas migration hazards are aggravated in those facilities where pressures are increased to levels beyond the original oil field pressures. Experience has shown that underground gas storage facilities can create a serious risk of explosion and fire and should not be placed under urban settings. It is virtually impossible to ensure that the gas will not migrate to the surface. Especially in the case of earthquakes, the faults may become more permeable and there could be an upsurge of gas to the surface. The potential fires, especially in the presence of winds, could be disastrous and uncontrollable by local fire departments. In addition, the soil saturated with gas may lose its cohesive properties and convert to "quicksands."

To avoid the possible catastrophic events described in this chapter, a fundamental awareness and understanding of gas migration, paths of migration, and associated hazards is required. It is necessary to monitor and evaluate:(1) soil gas content, (2) the rate of gas migration to the surface, (3) oil and gas production from the underlying oil fields, (4) older improperly abandoned oil wells, and (5) all oil field operations that produce fractures and free gas, and predict possible environmental complications associated with these operations.

PART IV

Interrelationships among Subsidence, Gas Migration, and Seismic Activity

Collapse of bridge due to subsidence in Venice, Italy, (modified from photograph by G. Lotti).

CHAPTER 21

Interrelationships among Subsidence, Gas Migration, and Seismic Activity

INTRODUCTION

Changing rates of gas migration to the earth's surface may signal the development of ground subsidence. Subsidence caused by withdrawal of ground water, oil, and gas has been observed and studied for more than 100 years (Poland and Davis, 1969; Strehle, 1989). Some of the earliest best-known examples of subsidence due to ground water withdrawal are Osaka, Japan (first noted in 1885), London, England (first noted in 1865) and Mexico City, Mexico (first noted in 1929). Subsidence often occurs due to man's need for fresh water, oil, and gas to satisfy domestic, industrial, and agricultural demands. In urban areas, the large-scale withdrawal of subsurface fluids, when accompanied by subsidence, can become an environmental hazard as natural gases may migrate upward along newly-formed faults and fractures and accumulate in underground structures.

SUBSIDENCE DUE TO FLUID WITHDRAWAL

One of the earliest examples of subsidence caused by the withdrawal of oil is Goose Creek oil field in Texas (first noted in 1918), described by Pratt and Johnson (1926). Subsidence is caused by compaction of reservoir rocks due to an increase of the effective stress

[$p_e = p_t$ (total overburden stress) – p_p (pore pressure)] as a result of the reduction of pore pressure. Pratt and Johnson (1926) added one more factor—the extraction of sand and other unconsolidated sediments—which is important in formations with loose sands and other unconsolidated sediments.

Withdrawal of the fluids from a petroleum reservoir or an aquifer reduces the pore pressure and results in the increase in effective (grain-to-grain) stress. Thus, compaction of reservoir rocks occurs and to balance the existing overburden pressure, the overlying formations and the land surface subside (Figure 21-1). Bending of rock formations due to subsidence results in compression, extension, and shear, which lead to fracturing of rocks. Fractures tend to occur at the periphery of a subsiding basin. Fracturing of rocks damages wellbores and sometimes induces technological earthquakes. The process of subsidence begins gradually in beds immediately overlying a reservoir or aquifer (from which fluids are withdrawn), while the land surface and shallow formations are unaffected. During this process, different beds and thin strata can separate from one another, opening existing bedding joints and creating new ones. Gradually, the deformation process expands upward. Deformation is not uniform in a vertical section and contains elastic, plastic, and brittle components in proportions that can change with time and with the rate and total amount of deformation (Gurevich and Chilingarian, 1998).

The surface subsidence may not begin until several years after commencing withdrawal of fluids from large subsurface deposits of unconsolidated or poorly-consolidated sediments. It may also proceed as a relentless slow process over a wide area. Endres and Chilingar (1999) have estimated that the area of subsidence on the surface is about twice the area of the reservoir or aquifer from which the fluids are withdrawn.

The primary cause of subsidence is rock compaction. Compaction is the result of several simultaneously acting phenomena that are induced by the increase of effective stress in the rocks in which the pore-pressure is reduced due to fluid withdrawal. High compressibility of unconsolidated (or semi-consolidated) rocks, confinement of the reservoir, presence of interbedded undercompacted clays or shales (e.g., aquitards), and high porosity, all contribute to the amount and regional distribution of the surface subsidence. Confinement of the reservoir is an integral part of the mechanism because natural replace-

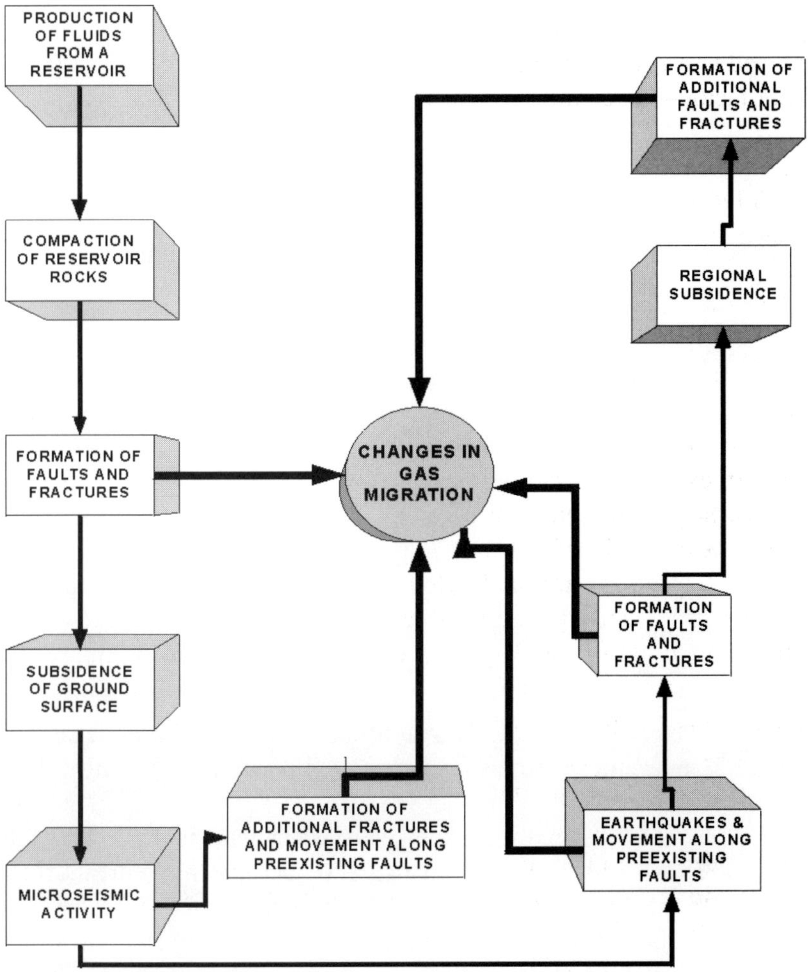

Figure 21-1. Schematic diagram of system relationships among the production of fluids, compaction, subsidence, and seismic activity. (*Modified after Chilingarian et al., 1995, fig. 1, p. 41.)*

ment of the withdrawn fluids may be severely restricted. Stratigraphic confinement of the reservoir may occur after deposition of the sediments and lithification, whereas structural confinement can also occur during deposition due to contemporaneous faulting or intrusion of extraneous bodies such as salt diapirs (Donaldson and Chiligarian, 1997).

Sinkholes are large caverns often filled with unconsolidated sediments, which are supported by the pore (fluid) pressure (shallow groundwater level). This support is weakened by the withdrawal of water (water table is lowered) after which the ability of sediments to support the total overburden is decreased. This can result in several stages of sudden subsidence for a small area, generally less than 50 m in diameter, around the sinkhole. Abandoned mining shafts (or tunnels) that have been filled with unconsolidated sediments, can give rise to similar surface subsidence.

COMPACTION

The phenomenon of compaction due to fluid withdrawal is more generally associated with shallow reservoirs, less than 1,000 m deep. The geologic history of the specific formation is an important factor. The compressibility of a shallow formation may be low if it has been previously subjected to high stress and consolidation during deep burial, prior to the later uplift and erosion. In such a case, the rocks will retain most of their prior low compressibility and high degree of compaction. As a result, fluid withdrawal may not give rise to appreciable compaction. Younger sediments of the Cenozoic Era (less than 1,000 m in depth), on the other hand, are in general compaction prone and are the major areas of concern regarding the surface subsidence.

The formations supporting the overburden tend to have low intergranular stress and high pore pressure if tectonic movements during deposition of younger sediments develop a confined reservoir or aquifer. As the depth of burial increases, the fluids in the pores of the formation will support a greater part of the overburden load. The Gulf Coast (U.S.) geo-pressured formations, which are undercompacted in relation to the depth of burial, represent such examples. Withdrawal of fluids, therefore, increases the effective stress leading to the grain rearrangement and compaction (Donaldson and Chilingar, 1997).

SUBSIDENCE BOWL

Shallow reservoirs will undergo viscous creep, deforming horizontally as well as vertically if the overburden pressure is principally supported by the grain-to-grain stress over a long period of time. A subsiding bowl develops slowly, almost imperceptibly, over a very large area with structures inclining toward the center of the subsiding bowl and fractures developing at the periphery.

Presence of shale and clay beds within the productive zone can significantly contribute to the overall reservoir compaction. The gradual loss of water from such beds may be responsible for a large portion of the land subsidence. In relatively young, shallow aquifers, these shale and clay beds are commonly undercompacted and exhibit high compressibility. Compaction takes place over a long period of time because the process involves the slow expulsion of water. The aquitards and reservoir rocks undergo inelastic compaction and, therefore, such subsidence is mostly irreversible (Geertsma, 1973; Dusseault, 1983). Petroleum reservoir rocks will rebound to a certain extent if waterflooding is started early.

CHAPTER 22

Subsidence Mechanics

INTRODUCTION

Pore pressure produces two separate effects: (1) the compression of solid matter, thus reducing the volume of the solid matrix; and (2) creation of body-forces proportional to the variations of the pressure drop, Δp (Jaeger, 1979). The term "body-force" is rather confusing in this application. It is physically correct to define the latter forces not as body forces (like gravity acting on every particle of a body), but as resultants of non-uniform surface forces (pore pressures) acting against the surfaces of solid grains. Fluid-solid force interaction only acts at the fluid-solid contacts on the surfaces of mineral and rock grains. The force acting against this surface is pore pressure.

There are two different approaches regarding the role of pore pressure in deformations of fluid-filled rocks. The first is the poro-elasticity approach (Biot, 1941; Geertsma, 1957, 1973; Jaeger and Cook, 1969; Jaeger, 1979; Fjaer et al., 1992). Geertsma (1973) proposed the following formula for calculating a component of the stress tensor, σ_{ij}:

$$\sigma_{ij} = 2G\left[e_{ij} + \frac{v e \delta_{ij}}{(1-2v)}\right] - (1-\beta)p\delta_{ij} \tag{22-1}$$

where:

σ_{ij} = stress component, related to the bulk stress system
e_{ij} = strain component
$e = \Sigma e_{ii}$ = dilatation or relative volume change of the bulk material
G = bulk shear or rigidity modulus

ν = Poisson's bulk ratio
p = pore-fluid pressure
β = ratio of rock matrix compressibility to the bulk compressibility
δ_{ij} = Kronecker's delta

This approach can only be used for elastic, reversible deformations. The majority of deformations due to natural and engineering processes, however, contain irreversible components that often prevail.

Terzaghi introduced a different approach—the effective stress concept—in 1923 (Terzaghi, 1943; Terzaghi and Peck, 1967). The effective stress concept was first empirically formulated on the basis of laboratory experiments. This approach assumes that the load, L (overburden pressure), applied to a unit of fluid-filled soil or rock, is supported by the solid frame (grain-to-grain stress, σ_e) and the pore pressure increase, Δp. The support provided by the solid frame, within this scheme, was also called effective-stress, σ_e. Effective stress does not correspond to any actual stress in a rock, but is the stress in the model medium and is an average stress (computed value) on a horizontal plane:

$$\sigma_e = L - \Delta p \tag{22-2}$$

In the Terzaghi model, physical meanings of measured values are quite definite: L represents a new, additional load applied to a physical body in mechanical equilibrium and Δp is the elastic response of the pore fluid to the total (elastic, reversible, and plastic, irreversible) deformation of the specimen. Terzaghi assumed that effective stress, $\sigma_{e,}$ is the stress in the solid frame under these conditions (see Gurevich, 1980).

In Terzaghi's equation ($\sigma_e = L - \Delta p$), although Δp is called the pore pressure, it actually represents the pressure change. It is an elastic response of the pore water to deformation of a sample, caused by the applied load and not the total pore pressure itself. In the laboratory, the height of a specimen is, usually, about one inch. Thus the hydrostatic pore pressure, before load application, is negligible in such a specimen and the excess pressure above the hydrostatic pressure was taken for the whole pressure value. Numerically, this was mathematically correct, but physically it was not, leading to unavoidable confusion. Thus, the excess pore pressure and additional stress in the solid frame support the load.

Later, Terzaghi's concept was extended to the relation between the total load, including the weight of rock column, and the total pore fluid pressure acting through a buoyancy mechanism. Whereas in the previous model fluid pore pressure existed only dynamically, in the course of the deformation process in this extended case pore fluid pressure exists even at equilibrium. At equilibrium, pore fluid pressure due to its gradient provides buoyancy of grains, which results in the reduction of their weight (see Rieke and Chilingarian, 1973, p. 6). However, owing to the fact that grain contacts are not point contacts, the relation:

$$\sigma_e = L - p \tag{22-3}$$

where the total value of pressure is used, is probably incorrect for the majority of cemented rocks (Jaeger, 1979; Gurevich, 1980). Eqs. 22-2 and 22-3 are physically different.

In a dynamic situation, hydrostatic uplift and elastic response act together. Thus, the physical meaning of the effective stress concept being applied to deformations of rocks in situ is rather obscure physically and often leads to confusion. This is especially true because the effective-stress concept completely ignores both the nature of deformations and mechanical properties of rocks that are deformed. This concept does not take into account the fact that, not just a small piece, but rather a large mass of rocks are being deformed as a single unit. Thus, when deformation can not be reduced to a one-dimensional model, some additional problems arise. For example, the generation of a vertical tension and strain of rocks in the course of subsidence of formations above a compacting reservoir is not compatible with the effective-stress concept: the overburden weight is not fully transmitted to the reservoir but, nevertheless, compaction continues (Gurevich and Chilingarian, 1995, p. 196).

Additional problems with the effective-stress concept arise because of heterogeneity of the rock and its mechanical properties, and presence of fractures. Owing to heterogeneity, some scale effects arise and should be taken into account (Bell and Dusseault, 1991; Enever et al., 1990; Ito et al., 1990; Li, 1990; Ratigan, 1990).

Another source of confusion is the disagreement on whether or not compressibility of sediments and rocks obtained from compaction tests in the laboratory and those in situ differ. There is a dependence of compressibility on loading history and according to some investigators, the measured degree of compaction in situ is lower than those predicted from laboratory tests. Radioactive bullet surveys in the Groningen gas field showed that the actual compaction values were three times lower than the amount predicted (Mess, 1979). It is important to know, however, whether uniaxial or hydrostatic compaction equipment was used in the laboratory. For example, when using hydrostatic compaction apparatus, the compressibilities of unconsolidated sands are about twice as high as those obtained when using uniaxial compaction apparatus.

The effective-stress concept is attractive because of its simplicity. It is believed to work well in simple cases when deformation is a one-dimensional compression. Rieke and Chilingarian (1974) relied heavily on uniaxial compaction apparatuses because they believed that "as the overburden load becomes large enough, the pressures are probably uniaxial."

The overburden pressure, L, is equal to the specific weight of the overlying water-saturated rock (γ_b) multiplied by the depth (D):

$$L = \gamma_b \times D \tag{22-4}$$

The effective stress (σ_e) continuously increases with decreasing porosity (ϕ) and is a function of either porosity or the remaining fluid content:

$$\sigma_e = f(\phi) \tag{22-5}$$

$$\sigma_e = f(S_g, S_w, S_o)$$

where S_g = gas saturation, % of pore space; S_w = water saturation, % or pore space; and S_o = oil saturation, % of pore space.

$$S_g + S_w + S_o = 100\% \tag{22-6}$$

During sedimentation, the stresses (p_e, p_t, p_p)[1] attain temporary equilibrium with different degrees of support assumed by the rock-matrix skeletal-structure and the fluids occupying the pores ($p_t = p_p$

[1] p_e = effective or grain-to-grain stress or σ_e; p_t = total overburden stress or σ_t; and p_p = pore or fluid pressure.

+ p_e). Gravitational stress (mass of the overburden) is transmitted vertically through the grain-to-grain contacts. Hydrostatic stress from the mass of the interstitial water above the compacting structures is transmitted through the water column. Another stress that aids grain redistribution is a viscous drag caused by movement of water downward and toward producing wells.

There are several types of recoverable and non-recoverable deformations that may occur in response to the imbalance of the p_t, p_p, and p_e forces. Increase in the grain-to-grain stress can result in recoverable elastic and visco-elastic (time dependent) deformation of the grains, or the grains may permanently deform by structural yield (crushing under the increased load) and plastic deformation of shape.

Compaction in natural environments is accompanied by other processes and occurs at a much slower rate (Gurevich, 1969, 1980). Compaction in nature depends both on acting overburden or tectonic load and on rock strength. Continual tectonic movements influence the rock strength. These movements sporadically break grain contacts and, thus, result in the rearrangement of grain packing and compaction without additional load. Periodic changes in temperature, seismic waves, and other factors have the same impact on compaction. Shaking (seismic waves) facilitates compaction. These mechanisms of compaction are missing in the laboratory tests. In situ loading and compaction are sometimes six to eight orders of magnitude slower than in laboratory tests. This means that molecular processes of slippage along the grain-crystal boundaries also play a more significant role in nature than in the laboratory.

CHANGES IN THE AQUIFER

Rapid sedimentation accompanied by tectonic movements leads to confined, undercompacted reservoirs. A decrease of the pore pressure of an undercompacted reservoir by removal of interstitial fluids increases the grain-to-grain pressure, which results in one, or a combination of several, types(s) of compaction.

In addition to the grain response to increased stress, the subsurface reservoir as a whole undergoes several types of changes adjusting to the imbalance of forces caused by fluid withdrawal. Loosely cemented rocks undergo permanent rearrangement of grains and pore spaces, resulting in the loss of porosity and permeability. This type of compaction induced by the grain mobility is only partially reversible if the fluid pressure is increased by water injection or natural recharge. Thus, the indiscriminate fluid production can lead to permanent loss of the capacity of an aquifer or a reservoir to contain, and to conduct, fluids.

Subsidence due to fluid withdrawal occurs when: (1) reservoir fluid pressures are lowered, (2) reservoir rocks are compactable and/or are unable to effectively resist deformation upon the transfer of load from the fluid phase to the grain-to-grain contacts, and (3) the overburden lacks internal self-support and the formations can easily deform downward (see Donaldson and Chilingarian, 1997, p. 253).

The principal lithological and structural characteristics of the subsiding areas include the following:

1. Sediments are unconsolidated and lack appreciable cementation.
2. Porosity of the sands is high (20–40%).
3. The sediments are of Miocene age or younger.
4. Producing formations are located at a shallow depth (300–1,000 m).
5. The aquifer is thick (>50 m).
6. Overburden is composed of structurally weak sediments.
7. Tension-type faulting, often with a graben, is present.
8. Sands are interbedded with clays, fine silts, and/or siltstones and shales.
9. Pore pressure is greatly reduced by voluminous production
10. Overburden is composed of structurally weak rocks.

CHAPTER 23

Fracturing Due to Subsidence

RATES OF SUBSIDENCE

Information published on the total and annual rates of subsidence is abundant but quite incomplete. Little information, however, has been published on the dependence of the subsidence rates on pressure decline rates, depths of reservoirs, and characteristics and thickness of the reservoir and overlying rocks. Some information on subsidence has been published by the following investigators: Deflache (1979), Carbognin et al. (1979), Holzer and Thatcher (1979), Kumar (1979), Lofgren (1979), Scott (1979), Meyer and Powly (1988), Strehle (1989), Andronopoulos et al. (1991), Balestri and Villani (1991), Bravo et al. (1991), Esaki et al. (1991), Gambolati et al. (1991), Holdahl et al. (1991), Morales et al. (1991), Murria (1991), Pottgens and Brouwer (1991), Prokopovich (1991), and Rivera et al. (1991).

The total subsidence due to withdrawal of fluids has been known to reach as much as 10 m. By 1970, the largest subsidence in the San Joaquin Valley, California reached 8.5 m (28 ft). In the Wilmington oil field total subsidence was 8.8 m. Lesser values of subsidence of the order of 0.9–1.5 m (3–5 ft) are commonplace. The annual rate of subsidence depends on the rate of pore pressure decline and has been recorded as high as 0.6 m (2 ft) per year. In the San Joaquin Valley, the average annual rate of subsidence increased to 2.4 m (8 ft) for 10 years (1959–1969), an average rate of 24 cm (0.8 ft) per year. In the Wilmington oil field, the annual rate for a three year period (from 1951 to 1954) was 70 cm/yr (2.3 ft/yr) (Figure 23-1). The Taipei basin, Taiwan, experienced subsidence at the rate of 25 cm/yr (0.84 ft/yr) in 1960, whereas in Tokyo, Japan, the annual rate of 32 cm/yr (1.07 ft/yr)

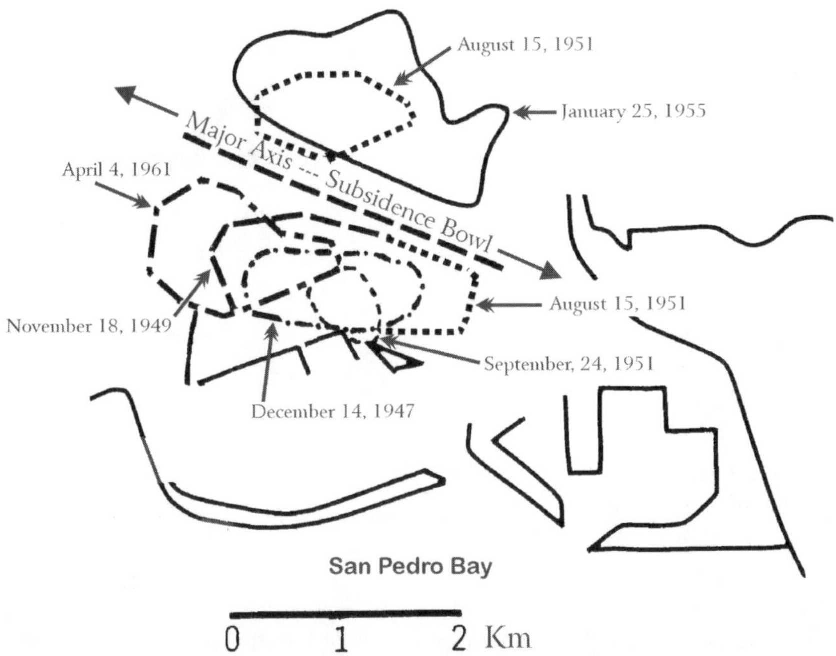

Figure 23-1. Locations of epicenters and slip planes of subsidence earthquakes, Wilmington oil field, Long Beach, California. *(Modified after Kosloff et al., 1980; in Gurevich and Chilingarian, 1993, fig. 3, p. 245.)*

were recorded in 1960. Throughout the world in areas of water aquifers and producing oil fields, rates of subsidence on the order of 0.3 m/yr (1 ft/yr) are common.

STRESS AND STRAIN DISTRIBUTION IN SUBSIDING FORMATIONS

Redistribution of stresses and strains within the rock mass above the compacting reservoir is the most important feature of subsidence from the viewpoint of fracturing the rocks, increasing the vertical permeability and enhancing the upward migration of gases (see Gurevich and Chilingarian, 1993).

Horizontal and vertical tensions are the highest surrounding the central core of the subsidence bowl (Figure 23-2), whereas compression predominates in the central core. Figure 23-3 shows the

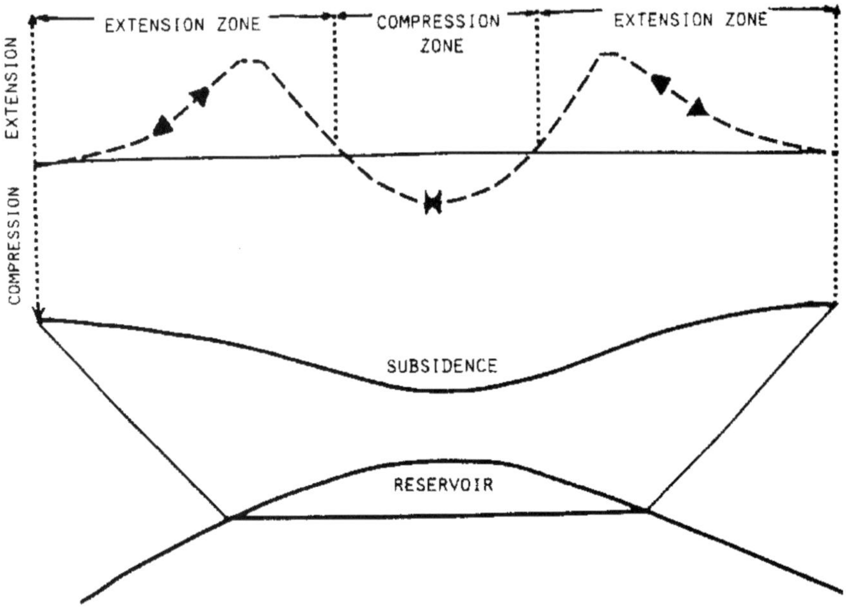

Figure 23-2. Schematic diagram of compressive and tensile stress distribution in subsiding formations. *(Modified after Gurevich and Chilingarian, 1993, fig. 1, p. 244.)*

measured elongation at five successive times. The most important feature of this graph is the demonstration of the existence of vertical strains above the petroleum reservoir. Horizontal displacement in the Wilmington oil field reached a maximum of about 3.66 m (Allen, 1973; Lee, 1979; Kosloff et al., 1980). This extension can be presented, for example, as four new 0.5 mm-wide fractures per each meter in the tension zone. This combination of horizontal shear stress with vertical tension caused several small earthquakes (Kosloff et al., 1980). Kosloff et al. (1980) emphasized that the hypocenters were at shallow depths, between 450 and 550 m, in bedded shale formation. The fault planes were always close to horizontal (Richter, 1958; Mayuga, 1965; Kovach, 1974). Locations of the epicenters are shown in Figure 23-1. (1980). These deformations, which released tensile stress and caused earthquakes, also formed new fractures, both lateral and vertical.

The very fact of the existence of vertical tensile strains and elongations above the compacting reservoir is direct evidence that the weight of the overburden is not fully transmitted to the compacting rocks due

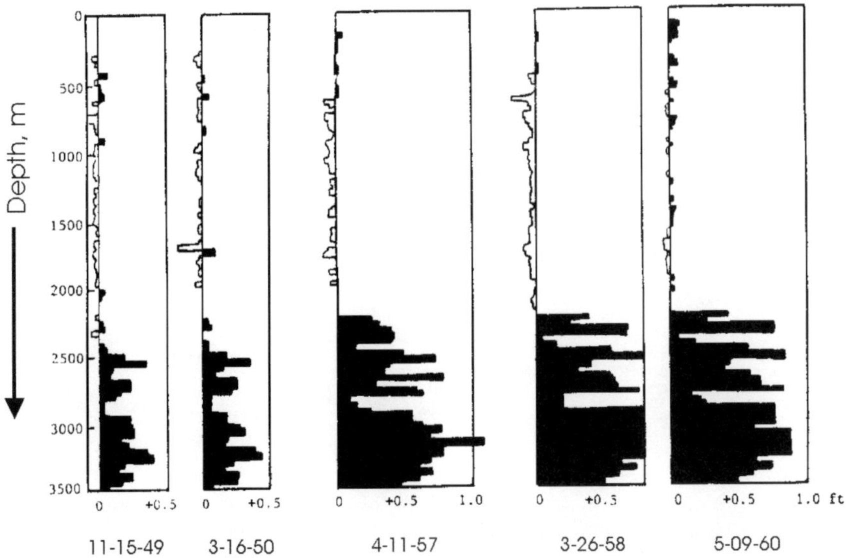

Figure 23-3. Schematic of casing count surveys of a typical well in the Wilmington oil field. *(Modified after Poland and Davis, 1969; in Gurevich and Chilingarian, 1993, fig. 2, p. 244.)*

to the bridging-effect of overlying formations. The effective-stress concept and models based on it do not exactly fit this type of deformation. In a sense, the existence of vertical tensile strain means that a reservoir can compact faster than the overlying beds bend. Thus, beds will break apart vertically. This process is especially intense in the peripheral (extension zone of the subsidence bowl), which makes this zone more permeable for the migration of gases. It is worth noting that in the case of sand production, when extraction of sand forms a cavity around the borehole, this effect will be enhanced and deformations similar to those shown in Figure 23-4, may be possible.

The role of strength for formations above a compacting reservoir is also indicated in the time lag in subsidence. Meyer and Powly (1988) noted that surface subsidence commonly lags behind cumulative production. This means that for some time compaction of the reservoir is compensated not by equal subsidence, but by vertical extension of the formation above the reservoir, whereas the deformation zone slowly expands upward. Thus, the first fractures form in the caprock of the depleting reservoir. This impairs the reliability of the caprock in preventing the upward migration of gas from the reservoir.

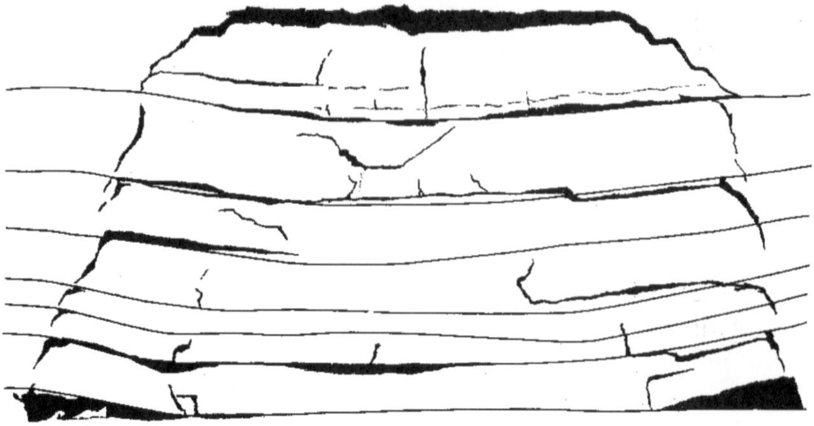

Figure 23-4. Deformation pattern of strata above longwall extraction with strong overburden (physical model). *(Modified after Whittaker and Reddish, 1989; in Gurevich and Chilingarian, 1993, fig. 4, p. 245.)*

SURFACE FISSURES CAUSED BY SUBSIDENCE

Tensile horizontal strain can cause fissures on the earth's surface (Guacci, 1979; Beckwith et al., 1991; Holzer, 1984; Lister and Secrest, 1985; Love et al., 1987; Pewe et al., 1987; Pampeyan et al., 1988; Contaldo and Mueller,1991; Haneberg et al., 1991; Keaton and Shlemon, 1991). In shallow aquifers (alluvial), large surface fissures form as a result of water withdrawal. Surface deformation (mostly horizontal displacements and fractures) as a result of the production of oil and water from oil fields and aquifers, has been observed where there is a substantial formation pressure decline (Pratt and Johnson, 1926; Strehle, 1989).

Origin and occurrence of earth fissures with dewatering of phreatic aquifers in Arizona, California, and Nevada have been studied by Holzer (1984). A common mechanism of fissure generation is localized differential compaction of unconsolidated aquifer material over bedrock irregularities. Association of earth fissures with zones of variable aquifer thickness suggests the occurrence of differential compaction (Jachens and Holzer, 1979). Differential compaction is the dominant source of horizontal tension causing earth fissures in the Picacho Basin, Arizona. An analysis of tensile strains at fissures at the times of their formation ranged from 0.1% to 0.4%.

The deepest reported fissure [more than 16.8 m (55.1 ft)] was located near Pixley in the San Joaquin Valley, California (Guacci, 1979). It was 0.8 km (0.5 mile) long and 2.4 m (8 ft) wide. The fissure was open to a depth of 1.8 m (6 ft) (Figure 23-5). Reported cracks attributed to hydrocompaction extended up to 4.4 m (14.5 ft) deep and were no more than 1 cm (0.4 in) wide (Beckwith et al., 1991). Contaldo and Mueller (1991) studied 13 discrete locations in the Mimbres Basin in southwestern New Mexico. They found that measurable fissure depths range from less than 0.3 to 12.8 m, whereas the width of fissures ranged from an incipient hairline to 9.7 m. Little is known about the full depth of the cracks and fissures, which is important from the gas migration viewpoint.

Kreitler (1977), Gabrish and Holzer (1978), Holzer and Thatcher (1979), and Van Sickle and Groat (1981) have studied the effect of subsidence on faulting. Holzer and Thatcher (1979) investigated

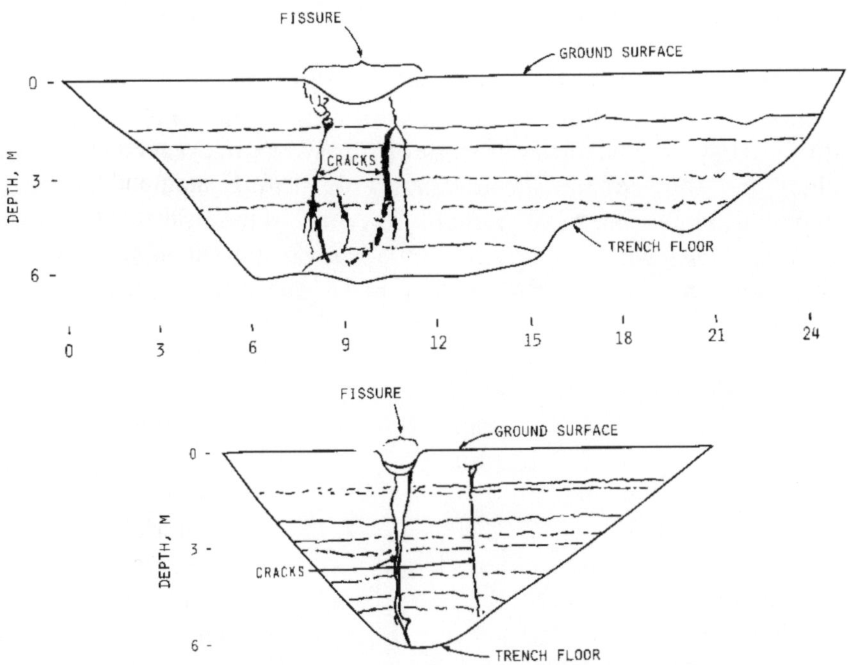

Figure 23-5. Trench logs of the Pixly fissure, San Joaquin Valley, California. *(Modified after Guacci, 1979; in Gurevich and Chilingarian, 1993, fig. 5, p. 246.)*

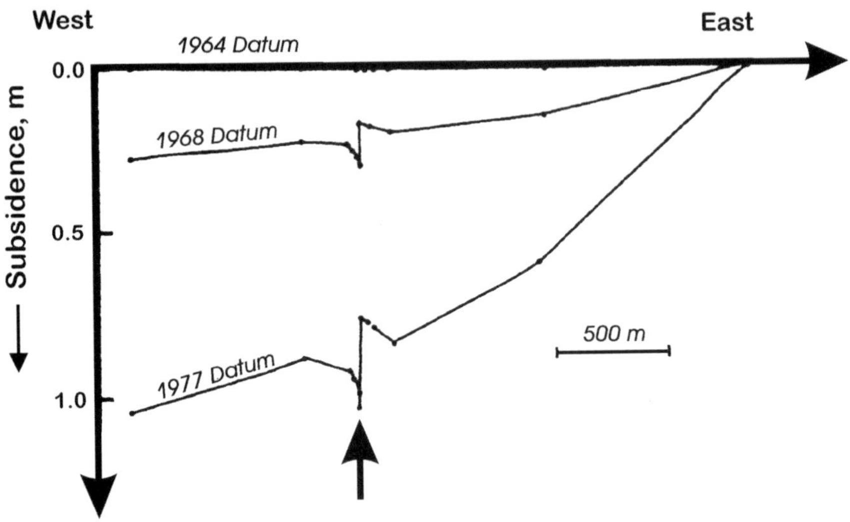

Figure 23-6. Subsidence profiles across the Picacho fault, Arizona. *(Simplified after Holzer and Thatcher, 1979; in Gurevich and Chilingarian, 1993, fig. 6, p. 247.)*

changes of surface altitudes on both sides of the Picacho fault (Figure 23-6). They simulated the process of differential movements of fault sides and showed that the difference in altitude changes of sides depends on the angle of a fault. Physically, it is quite clear that existing faults will provide differential movements of their sides if a shear stress exists, relative to the fault plane. The widths of old and new fractures should be studied from the upward gas migration viewpoint.

CHAPTER 24

Water Aquifer Subsidence

INTRODUCTION

The following review of the global problem associated with subsidence due to withdrawal of fresh groundwater shows that in all cases the sediments are of Miocene age or younger. Compaction occurs due to fluid (water) depletion (1) of the fine-grained clay/silt aquitards (that are interbedded with sand/gravel aquifers), and (2) of aquifers. Subsidence is caused by grain rearrangement and compaction when the overburden load is transferred from water to the grain-to-grain contacts. Little information is recorded on the fracture systems generated by the fluid withdrawal and its effect on the generation of fractures.

China

Land subsidence in China occurs in cities and industrial complexes located along the coastal plains of eastern China, and in the alluvial plains along the big rivers. Young sediments (Miocene age and younger) are unconsolidated and interspersed with aquitards of fine-grained clay and shale. Unregulated withdrawal of water from these areas resulted in subsidence that was noted in the early 1960s. Before the land subsidence could create severe environmental problems, studies of the rates and amounts of water withdrawal (together with the evaluations of the mechanisms of compaction and recharge of the various reservoirs) led to the development of control measures. Inasmuch as the majority of water withdrawal took place during the summer months (for industrial usage and cooling) several regulations were imposed:

1. Water production for industrial purposes was limited to the summer months only; this, in itself, reduced the rate of land subsidence considerably.
2. During the winter months, the wells are used for artificial recharge of the aquifers using surface water that is trapped and injected.
3. Injection of cold surface waters into deeper fresh-water aquifers. This cools the groundwater and makes it suitable as industrial cooling water in the summer months.

These regulations, as well as restrictions on rates of withdrawal, have allowed the continued use of the groundwater aquifers for municipal and industrial purposes with control of land subsidence (Guangxiao and Yiaogi, 1984). (Also see Donaldson and Chilingarian, 1997.)

Tokyo, Japan

The great cities of Japan (e.g., Tokyo) are located on coastal plains where the subsurface aquifers are composed of Miocene age, or younger, unconsolidated sands with interspersed layers of clay and shale. Therefore, it is not surprising that Japan has experienced land subsidence in some 50 areas, in and around its most important cities, due to water production at rates exceeding the rates of recharge. In some areas, excessive rates of water withdrawal led to salinification of the aquifers from the incursion of seawater and increased pumping costs due to decrease of permeability around the production wells as compaction of the reservoir rocks took place (see Donaldson and Chilingarian, 1997). A computer program for estimating the loss of porosity and permeability, as a function of distance from the producing wells where compaction is taking place, was presented by Chilingarian et al. in 1995.

Venice, Italy

The subsurface beneath Venice is a system composed of six sand aquifers separated by clay/silt layers, one above the other. Withdrawal of water from this system leads to rapid subsidence as the overburden stress is transferred from the fluid to the grain-to-grain contacts. The pressure difference established across the aquitards leads to slow (diffusion controlled) compaction of the clay/silt layers. As industrial withdrawal of water increased in the 1950s and 1960s, the piezometric levels under the industrial zone declined from 7 m above to 16 m below the surface. This water level decline was transferred to the area under Venice as a drop in piezometric level to about 7 m below the surface, resulting in rates of subsidence of 5 mm/y in areas within the city; this translates to 9 cm of subsidence in the city by 1970. It was estimated that a 1 m lowering of the piezometric level under the city causes subsidence of 2 cm. About 70% of the total subsidence occurred in the period from 1952 to 1969 during industrial expansion that was accompanied by increases in water withdrawal (Carbognin et al., 1976, Carbognin and Gatto, 1984; Brighenti et al., 1995). Measures taken after 1969, however, ended the subsidence and resulted in a 2 cm rebound.

Mexico City, Mexico

The portion of Mexico City that underwent subsidence due to water production is underlain by a thick bed of clay (80% montmorillonite), mixed with fine-grained sand and silt, ranging up to 50 m in thickness. The fresh-water aquifer is a thick stratum of sand and gravel, up to 50 m in thickness. The permeability of the aquifer is very high, delivering 350 m^3/hr of water. Records indicate that the first water well was installed in 1860 and water production increased throughout the city as the need for municipal water increased with increasing development and population growth. Quantitative measurements of the amount and rate of subsidence have been made since 1891 and the effects of subsidence on structures in the city were noticed by 1920. It was not until 1948, however, that subsidence within the city was attributed to withdrawal of groundwater for municipal and industrial use. The estimated subsidence in the older part of the city, from 1891 (when the first well was installed) to 1973, was about 8.5 meters. Research has shown that 75% of the subsidence occurred in the overlying clay layer and the other 25% within the aquifer (see Vega, 1976, 1984; Donaldson and Chiligarian, 1997).

Taipei City, Taiwan

Land subsidence occurred in Taipei City, Taiwan, from 1955 to 1975 creating a bowl-shaped depression with a depth of about 1.2 m on the east side of the city. A mandated reduction of the amount and rate of water withdrawal in 1974 gradually reduced this rate of subsidence. The depression shifted to the west side as pumping was moved to the western border, where the rate of subsidence reached 14 cm per year. Subsidence occurred in both areas because of reduction of the piezometric head, principally in the Linkou Formation at a depth of 50 to 100 m, which is composed of gravel, sand, and clay. The Linkou Formation is overlain by the Sungshan Formation composed principally of clay, silt, and fine sand under a layer of alluvial beds. Most of the compaction is occurring in the Linkou Formation with probable time-delayed compaction in the Sungshan Formation (Wu, 1976).

Bangkok, Thailand

Land subsidence due to water production has occurred to some extent in Bangkok, Thailand. Bangkok, however, cannot tolerate a great deal of surface subsidence because the average ground elevation above mean sea level is only 1.0–1.5 m and the average ground water table is only 1–2 m below the surface in areas outside of the pumped zones. The top sedimentary layer is a gray marine clay about 30 m thick (Bangkok clay), which is followed by sequences of unconsolidated sands and gravel, separated by thick layers of clay, down to a fractured basement rock more than 1,000 m deep. Water production is generally from the Nakhon Luang aquifer at 150 m depth and the Nonthaburi aquifer at a depth of 200 m, which have very high transmissibility, in the order of 50–125 m^3/hr-m (Piancharoen, 1976). The Phrapradaeng aquifer below the Bangkok clay has been generally abandoned because of saltwater incursion.

Bergado et al. (1984) reported that the Asian Institute of Technology campus, just north of Bangkok, experienced subsidence as evidenced by cracks in buildings, concrete structures, and roads due to differential vertical movement. Compaction was taking place at a depth of 200 m from pumping near the campus (see Brand and Balasubramaniam, 1976; Yamamoto, 1984; Donaldson and Chilingarian, 1997).

Orange County, California, United States

Subsidence occurred near the coast in Orange County, California, from compaction of sands within the aquifer from which fresh water was withdrawn. The principal cause of subsidence is attributed to compaction and shrinkage of organic peat deposits lying on top of the aquifer at a shallow depth of about 20 m and ranging in thickness up to about 60 m. Decline of the piezometric level caused invasion of saline water from the coastal aquifers. When the chloride ion level began to rise, a hydraulic barrier to seawater incursion was installed (Fairchild and Wiebe, 1976).

Santa Clara Valley, California, United States

Land subsidence in the Santa Clara Valley was first recognized in 1932. The area that underwent subsidence ranges from the southern end of San Francisco Bay to the city of San Jose (about 40 km). The Santa Clara valley is a trough located between the Hayward Fault on the east and the San Andreas Fault on the west forming a confined aquifer system. Subsidence reached 8 m at the center between the cities of Santa Clara and San Jose until arrested in 1965 by (1) a decrease of pumping rates, (2) the importation of water from rivers, and (3) a recharge of the aquifer from impounded local rain water runoff. Compaction of the aquifer and associated inclusions of clay/silt lenses was responsible for the surface subsidence, which has been stabilized by maintaining the piezometric level at 1972 levels using prudent pumping rates (Poland, 1984).

San Joaquin Valley, California, United States

The greatest area of land subsidence in California is the San Joaquin Valley, which occupies the southern two-thirds of the Central Valley. The fluvial deposits of lenticular layers of sand, gravel, and silt are interspersed with deposits of lacustrine origin. The eastern side of the valley consists of deposits from major streams, which form a series of commingled alluvial fans made up of coarse deposits of sand and gravel between layers of clay and silt. All of the deposits are younger than Miocene age.

There are two principal freshwater-bearing aquifers divided by a bed of clay ranging to 40 m in thickness: (1) the upper zone that extends from the surface to depths of 275 m, and (2) the lower water zone with a thickness range of 60 to 600 m. During World War II the rate of water production was increased from about 4,000 hectare-m to more than 12,000 hectare-m by 1966, keeping pace with the rapidly expanding agricultural enterprises in the valley. During this period the piezometric level declined by 60 to 180 m in the region. A combination of three types of subsidence, in response to the piezometric level decline, resulted in the spectacular amount of decline that occurred over a large area. Subsidence up to 10 m occurred along a narrow trough west of Fresno and extending 140 km south. Other areas throughout the valley subsided 1 to 4 m during the period of increased water withdrawal (Donaldson and Chilingarian, 1997).

Subsidence was attributed to: (1) hydro-compaction when water for irrigation was applied to moisture-deficient, near-surface deposits; (2) compaction of the aquifer system, which includes major compaction of the fine-grained clay/silt sediments and compaction of the aquifer sands, as the piezometric level was decreased, transferring vertical stress to the grain-to-grain contacts; and (3) compaction due to hydrocarbon production in some areas. After 1966, subsidence was stopped when large imports of surface water from the north, west, and south sides of the valley were used for irrigation in place of groundwater (Prokopovich, 1976; Poland et al., 1984).

Houston, Texas, United States

High rates of water withdrawal has led to subsidence of 2.3 m southeast of Houston (in the Pasadena City area) and about 0.5 m in the industrial area of Texas City south of Houston. It is estimated that the piezometric level in the principal aquifer declined by almost 100 m. The sand aquifers contain a large number of interbedded layers of clay, ranging up to 6 m in thickness, which retard vertical movement of water and cause large vertical differences in water pressure. Major amounts of compaction occur in the fine-grained clay bodies. The subsidence has caused inundation of homes across the bay (in Baytown) and there is speculation that future hurricane tides may cause considerable damage in the south Houston area (Gabrysch, 1976).

CHAPTER 25

Technologically Induced Earthquakes

Several authors have documented technologically induced earthquakes in the areas of petroleum production (Pakiser et al., 1969; Lee, 1978; Krestnikov et al., 1988; Wetmiller, 1986). The conditions of

Table 25-1
Aftershocks of a Technologically-induced Earthquake at the Starogroznenskoye Oil Field, Russia

Date	Depth, (km)	Earthquake Magnitude (Richter scale)
05-26-71	2.5	3.8
05-26-71	5.0	2.9
06-07-71	5.0	2.7
06-14-71	4.5	1.5
08-19-71	8.0	2.5
10-09-71	1.7	2.1
11-17-71	2.5	2.4
12-04-71	4.5	3.1
12-11-71	3.0	1.9
03-28-72	3.0	2.6
05-29-72	3.2	2.0
05-30-72	2.2	3.0
08-14-72	5.0	2.5
08-16-72	2.8	2.6
09-26-72	2.0	3.4
10-14-72	2.9	2.6
12-22-72	9.0	3.4
04-22-73	6.2	3.2
06-22-73	8.0	3.0

After Kouznetsov et al., 1994, table 1, p. 59.

ground subsidence over producing reservoirs were discussed by Allen (1968), Ammosov et al. (1987), Chilingarian and Wolf (1975, 1976), Erickson (1977), and Gurevich et al. (1993). Relationship between the acoustic fields and fracturing of the rocks was studied by Kouznetsov and Simkin (1990).

In seismically active areas, the scope of subsidence is generally higher compared to that of seismically quiet territories. This means that seismic activity combined with intense fluid withdrawal from the producing reservoir causes frequent redistribution of stresses (flickering stresses), which leads to the rapid development of additional vertical and subvertical fractures. This, in turn, facilitates the subsidence process (Gurevich and Chilingarian, 1993).

Fracture development and flickering stresses may induce tectonic movement, which can produce mid-size earthquakes in the area of petroleum production and its vicinity. There are a number of documented examples of production-induced earthquakes, some of which were described by Kouznetsov et al. (1994). The intensity of these man-induced earthquakes is comparable to that of natural seismic events and can be dangerous for people. Two-year records of time and magnitude of the production-induced seismic events at the Starogroznenskoye oil field in Russia are presented in Table 25-1.

CHAPTER 26

Gas Migration in Seismically Active Areas

Upward gas migration from petroleum reservoirs to the earth's surface depends on the distribution and types of vertically and subvertically oriented faults and fracture zones and the permeability along them. There are two groups of factors facilitating gas migration to the surface in seismically active areas: (1) seismic activity can form vertical and subvertical faults and fractures above petroleum reservoirs, and (2) seismic vibrations can considerably increase the permeability due to desorption of gas from the rock matrix (Kouznetsov et al., 1994).

The higher upward gas mobility in the field of seismic energy was corroborated by the repeated geochemical mapping over petroleum reservoirs before and after applying seismic vibrations to the oil-bearing formation (Kouznetsov et al., 1994). The experiments showed a noticeable increase in the content of heavy hydrocarbons in the samples taken near the earth's surface as a result of applying vibro-energy. To monitor a number of environmental parameters related to hydrocarbon production, seismic activity, geochemical processes, and tectonic activity over the Tengiz oil field in Russia, a special system of environmental monitoring was designed and implemented. It presents an important example of the monitoring system that allows controlling and predicting environmental hazards associated with gas leakage to the earth's surface in the seismically active areas.

Due to increasing rates of upward gas migration over petroleum reservoirs after earthquakes in seismically active areas, gas leakage may present more serious environmental problems compared to those in the relatively quiet (seismically) areas. This is an important reason for carefully monitoring the gas concentrations (and rate of migration) in the vicinity of faults over petroleum reservoirs in seismically active areas. In addition, as has been repeatedly emphasized in this book, this monitoring can provide valuable information for the purposes of earthquake prediction.

To enhance oil and gas production, artificial sources of seismic waves can be applied to the petroleum reservoirs (Kouznetsov et al., 1994). In such cases, due to the release of additional seismic energy, the rates of the upward gas migration become considerably higher. In addition, the induced seismic vibrations result in fracturing and compaction of the rocks above the petroleum reservoir, leading to ground surface subsidence. In turn, substancial subsidence may cause movement of the tectonic blocks and produce a technologically induced earthquake. Small earthquakes can trigger large ones; this may occur in Kuwait, for example.

CHAPTER 27

Joint Forecasting of Subsidence, Gas Migration, and Seismic Activity

The processes of subsidence, variations in gas leakage, and seismic activity are related to large-scale nonlinear effects such as fracture development, changes in faults' surfaces and their widths, and nonlinear deformations. This makes it extremely difficult to develop a methodology of forecasting these phenomena based on direct numerical modeling of underlying physical processes.

The methodology, described here, is based on the use of neural networks. They are especially efficient in working with data of complex structure. Neural nets have shown great potential in the identification of subtle patterns among the data parameters and in the prediction of phenomena that are linked to input data by nonfunctional relationships that cannot be approximated by the regression curves of predefined structure.

The neural net is usually trained to transform a set of input data vectors $\overline{X}_m$, $m = 1, 2, \ldots, M$ into desired outputs g_m, The structure of the neural net is automatically chosen (by the computer program) in such a way that the transformation error E is minimal:

$$E = \sum_m \left[g_m - \Psi\left(\overline{X}_m\right) \right]^2 \tag{27-1}$$

Here, $\Psi(\overline{X}_m)$ is the output of the neural net for the input data vector $\overline{X}_m$.

In Eq. 27-1, the values of g_m must be known and are normally obtained from observations. After the neural net is synthesized, it may be used to predict unknown (future) values $g_{z'}$ using as its input respective vector $\overline{X}_m$. This may be written as follows:

$$\Psi(\overline{X}_z) \rightarrow g_z' \tag{27-2}$$

where g_z' is the prognostic value of g_z.

In the considered case, a volume of input data may often be of the same order as the number of parameters describing the composed neural net. In this case, minimization of criterion (Eq. 27-1) may lead to unstable solutions. To avoid instability in the results of neural network-based prediction, this criterion must be modified to (Katz et al., 1994):

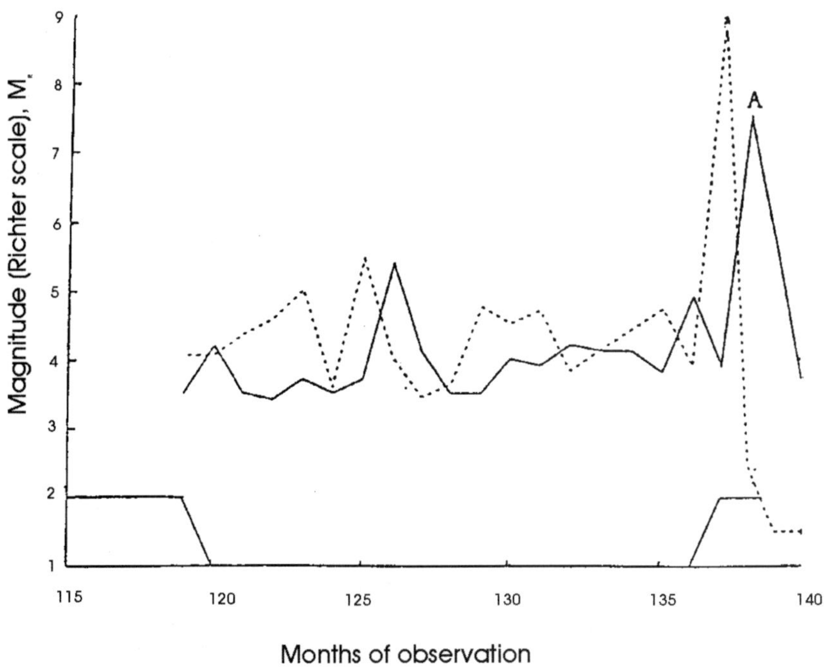

Figure 27-1. Neural network-based earthquake prediction. Integral characteristics of seismicity (A = actual, * = predicted) that included average depth and maximum magnitude as input parameters. *(Modified after Chilingarian et al., 1995, fig. 2, p. 45.)*

$$E = \sum_m \left[g_m - \Psi\left(\overline{X}_m\right)\right]^2 + \alpha^2 \|\overline{\mu}\|^2 \tag{27-3}$$

where $\overline{\mu}$ is a vector of parameters defining the structure of the neural net, $\|\cdot\|$ is the norm of the vector, and α^2 is the regularization parameter.

To define the input vector $\overline{X}_m$ (which represents past values of the parameters of subsidence, rates of gas migration, and seismic activity) in situ, it is necessary to develop a special monitoring system (systems) for the region of interest for continuous collection of processing data relevant to the three groups of processes. The following are important locations for monitoring gas leakage and gas concentration: (1) the vicinity of abandoned wells, (2) the vicinity of faults, and (3) the basement of buildings in the zones of petroleum production.

There is close correlation between seismic activity and subsidence (Bolt, 1988; Holtzer, 1984; Katz et al., 1994).

Monitoring of seismic activity includes two components: (1) integral characteristics describing features and behavior of seismic processes over large geographical regions, and (2) monitoring of localized phenomena related to the ground behavior at the location of monitoring. Examples of integral and local characteristics can be found in Katz et al. (1994). As shown in this publication, such characteristics can be used for earthquake prediction using neural network methodology. Figure 27-1 shows an example of this type of prediction.

CHAPTER 28

Conclusions

Some applications of the work by the writers are as follows: (1) Methodology for detection of increased danger of upward gas migration and surface gas leakage due to oil and gas production and variation in time of seismic-tectonic processes. (2) Formulation of new earthquake precursors directly linked to current seismic-tectonic activity in the area of oil and gas production based not only on variations in seismic activity, but also on the analysis of environmental phenomena related to oil production. (3) Increase in the gas leakage and gas concentration in soils may also indicate formation of vertically and subvertically oriented fractures over oilfields, which may be caused by reservoir compaction and consequent ground subsidence. Hence, monitoring of gas leakage at the earth's surface can be used as an additional indicator of possible increases in ground subsidence over oil fields and aquifers, and of movements along faults due to tectonic activity that could render faults more or less permeable.

In many instances, water is injected into the reservoir to increase the production of oil (waterflooding). Water supplies energy to drive oil toward producing wells and increases pore pressure in the reservoir. However, injected water also displaces the free gas in the reservoir, forcing it to migrate. When encountering faults and fractured zones, this free gas may migrate toward the surface.

Tectonic activity, ground subsidence, and gas leakage to the earth's surface over petroleum reservoirs can affect and considerably amplify each other especially in seismically active areas (Katz et al., 1994). Even small seismic disturbances of oil and gas production can cause environmental disasters, such as subsidence, earthquakes, and gas explosions. Gas or water injection can also result in fracturing of the subsurface formations due to overpressurizing the formations and earthquakes. A schematic diagram of system relationships among earthquakes, production of fluids, compaction, and subsidence, with resulting formation of faults and fractured zones, is presented in Figure 28-1.

Gas migration to the surface occurs in pathways along vertically and subvertically oriented fracture systems, faults, and fractures and cavities formed around the wellbores. Fractures in rocks over producing petroleum reservoirs form due to: (1) tectonic and diagenetic history, (2) current tectonic and seismo-tectonic movements, and (3) deformations caused by compaction of reservoir rocks with consequent subsidence of overlying formations. Fracturing modifies production, causes changes in the upward gas migration, and enhances subsidence at the ground surface.

In reservoirs that are divided into independent fault blocks, stresses at the boundary zones of adjacent blocks, increased by the injected fluids, can become high enough to cause massive fracturing of the rocks. These fractures increase the rate of upward gas migration. In seismically active areas, the increased upward gas migration can become a major environmental hazard. If a petroleum reservoir is intersected by an active fault, and an earthquake increases the permeability of the fault zone, then there may be a rapid increase in the rate of upward gas migration. This can result in massive explosions and fires.

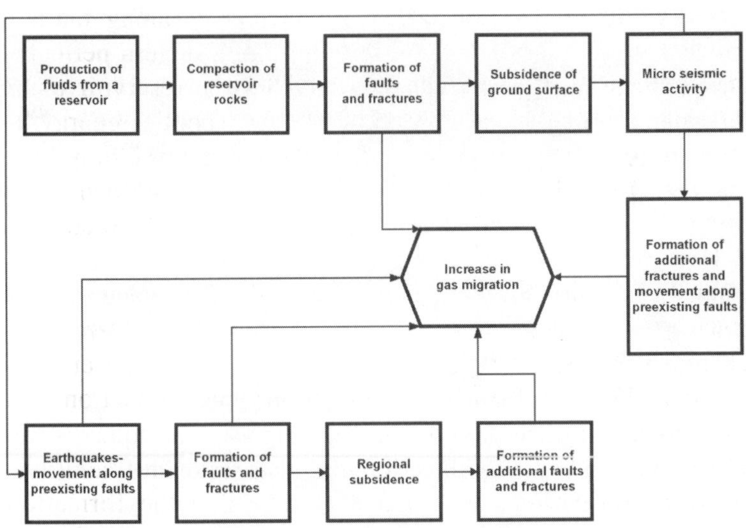

Figure 28-1. Schematic diagram of systems relationships among production of fluids, compaction, subsidence, and seismic activity *(modified after Katz et al., 1995).*

References

PART I

Albee, A. L. and Smith, J. L. 1969. Earthquake characteristics and fault activity in Southern California, pp. 9–33. In R. Lung and R. Proctor (eds.), *Engineering Geology in Southern California,* Assoc. Engr. Geologists, Arcadia, CA, 389 pp.

Allen, A. A. and Schlueter, R. S., 1969. Estimates of Surface Pollution Resulting from Submarine Oil Seeps at Platform "A" and Coal Oil Point: General Research Corp., Memo 1230. In Wilkinson, E. R. (1971).

Aschenbrenner, B. C. and Achauer, C. W., 1960. Minimum conditions for migration of oil in water-wet carbonate rocks. *AAPG Bull.*, 44(2):235–243.

Barends, F. B. J., Brouwer, F. J. J. and Schroder, F. H. (eds.), 1995. *Land Sub-sidence*, Proc. Fifth Internat. Symp. on Land Subsidence, The Hague, Netherlands, 16–20 Oct.

Barnes, I., Irwin, W. P., and White, D. E., 1978. Global distribution of carbon dioxide discharges and major zones of seismicity, scale 1:40,000,000. *Water Resour. Invest. WRI 78-39,* U.S. Geol. Survey, Washington, D.C.

Bernard, B. B., 1980. Geochemical prospecting for petroleum and natural gases. *U.S. Department of Energy Symp.*, St Petersburg, FL, Dec., pp. 15–17.

Blocker, P. C., 1964. Spreading and evaporation of petroleum products on water. *Proceedings of the Fourth International Harbor Conference*, June 1954, Antwerp, Belgium.

Bolt, B. A., 1978. *Earthquakes.* W. H. Freeman and Company, New York, NY, 282 pp.

Brown, K. E., 1967. *Gas Lift Theory and Practice*, The Petroleum Publishing Co., Tulsa, OK, 924 pp.

Carlisle, C. T., Bayliss, G. S. and Van Delinder, D. G., 1975. Distribution of light hydrocarbons in sea floor sediments: Correlations between geochemistry, seismic structure and possible reservoir oil and gas. In *Proc. 1975 Conf. Offshore Technology,* Dallas, TX, 3:65–70.

Chapman. R. E., 1983. *Petroleum Geology*. Elsevier, Amsterdam, pp. 245–251.

Chilingar, G. V. and Beeson, C. M., 1969. *Surface Operations in Petroleum Production*, Am. Elsevier Publ. Co., Inc., New York, NY, 397 pp.

Chilingarian, G. V. and Vorabutr, P., 1983. *Drilling and Drilling Fluids*. Elsevier, Amsterdam, pp. 71–79.

Chilingarian, G. V., Robertson Jr., J. O., and Kumar, S., 1987. *Surface Operations in Petroleum Production*, 1, Developments in Petroleum Science, 19A, Elsevier, Amsterdam, pp. 146–147; 416–420.

Chilingarian, G. V., Mazzullo, S. J. and Rieke, H. H., 1992. *Carbonate Reservoir Characterization: A Geologic Engineering Analysis*, Part 1, 30. Elsevier, Amsterdam, 639 pp.

Chiligarian, G. V., Katz, S. A. and Khilyuk, L. F., 1995. Neural network based upon prediction of subsidence, surface gas migration, and seismic activity. In F. B. J. Barends, F. J. J. Brouwer, and F. H. Schroder (eds.), *Land Subsidence, Proc. Fifth Internat. Symp. on Land Subsidence*, The Hague, Netherlands, 16–20 Oct., pp. 41–46.

Chilingarian, G. V., Donaldson, E. C. and Yen, T. F., 1995. *Subsidence Due to Fluid Withdrawal*, Developments in Petroleum Science, 41, Elsevier, Amsterdam, 498 pp.

Chilingar, G. V., Khilyuk, L. F. and Katz, S. A., 1996. Pronounced changes in upward natural gas migration as precursors of major seismic events. *J. Pet. Sci. Eng.*, 14:133–136.

Claypool, G. E. and Kaplan, I. R., 1974. The origin and distribution of methane in marine sediments. In I. R. Kaplan (ed.), *Natural Gases in Marine Sediments*, Plenum Press, New York, NY, pp. 99–140.

Conner, D. C. and Calvin, R. J., 1977. Development geology of Pineview field, Summit County, Utah, in Rocky Mountain thrust belt; geology and resources. *Wyoming Geol. Assoc. Guidebook*, no. 29:639–650.

Craig, H. 1953. The geochemistry of stable carbon isotopes. *Geochim. Cosmochim. Acta*, 3:53–92.

Culberson, O. L. and McKetta, J. J., Jr., 1951. Phase equilibrium in hydrocarbon-water systems. III. The solubility of methane in water at pressures to 10,000 psi. *Trans. AIME*, 192:223–226.

DeGolyer, E., 1940. Future position of petroleum geology in the oil industry. *Am. Assoc. Petrol. Geologists*, 24:1389–1399.

DeMets, C., Gordon, R. G., Stein, S. and Argus, D. F., 1990. Current plate motions. *Geophys. J. Int.*, 101:425–478.

Dickey, P. A. and Hunt, J. M., 1972. Geochemical and hydrgeologic methods of prospecting for stratigraphic traps. In *Stratigraphic Oil and Gas Fields—*

Classification, Exploration Methods and Case Histories. AAPG Memoir 16—Soc. Expl. Geophys. Spec. Publ., 10:136–167.

Dodson, C. R. and Standing, M. B., 1944. Pressure-volume-temperature and solubility relations for natural gas-water mixtures. *Drilling and Production Practices*, API, pp. 173–179.

Duncan, R. A., 1991. Oceanic drilling and the volcanic record of hotspots. *GSA Today*, 1:213–216, 219.

Endres, B., Chilingarian, G. V. and Yen, T. F., 1992. Environmental hazards of urban oilfield operations, *J. Pet. Sci. Engr.*, 6:95–106.

Eremeev, A. N., Sokolov, V. A., Sokolov, A. P. and Yanitskiy, I. N., 1973. Application of helium surveying to structural mapping and ore deposit forecasting. In *Geochemical Exploration*, 1972. *Proceedings Intl. Geochem. Exploration Symp.*, (4):183–190.

Follansbee, G. S., 1943. Lost Hills oil field. *California Division of Mines Bulletin*, 118:494–495.

Frick, T. C., 1962. *Petroleum Production Handbook, Volume 2*, Society of Petroleum Engineers of AIME, Dallas, Texas, 1,000 pp.

Galloway, J. M., 1992. *Natural Oil, Tar, and Gas Seeps of the Carpinteria Area*. Minerals Management Service, I.T.M. Field Trip, May 12.

Geertsma, J., 1973. Land subsidence above compacting oil and gas reservoir. *J. Petrol. Technol.*, 25(6):734–744.

Gold, T. and Soter, S., 1980. The deep-earth gas hypothesis. *Sci. Amer.*, 242: 154–161.

Gold, T. and Soter, S., 1982. Abiogenic methane and the origin of petroleum. *Energy Explor. Exploit.*, 1:89–104.

Gordon, R. G. and Stein, S., 1992. Global tectonics and space geodesy. *Science*, 256:333–342.

Gurevich, A. E., 1970. *Processes of Groundwater, Oil and Gas Migration*. Nedra Publ., St. Petersburgh, Russia, 112 pp.

Gurevich, A. E., 1970. Processes of groundwater, oil and gas migration in the petroliferous basins of Mangyshlak. Ph.D. diss., VNIGRI, St. Petersburgh, Russia.

Gurevich, A. E., Endres, B. L., Robertson, Jr., J. O. and Chilingar, G. V., 1993. Gas migration from oil and gas fields and associated hazards, *J. Petro. Sci. Engr.*, 9:223–238.

Gurevich, A. E. and Chilingarian, G. V., 1995. Possible impact of subsidence on gas leakage to the surface from subsurface oil and gas reservoirs. In G. V. Chilingarian, E. C. Donaldson and T. F. Yen (eds.), *Subsidence Due to Fluid Withdrawal.* Developments in Petroleum Science, 41, Elsevier Amsterdam, 498 pp.

Harden, D. R., 1998. *California Geology.* Prentice Hall, Inc., Upper Saddle River, N.J., 461 pp.

Hoots, H. W., Blount, A. L., and Jones, P. H., 1935. Marine oil shale source in Playa del Rey field, California. *Am. Assoc. Petrol. Geologists,* 19:172–206.

Howell, D., Cole, F., Fanelli, M., Wiese, K., 1993. Natural Gas. *Earth,* September, pp. 53–59.

Hunt, J. M., 1979. *Petroleum Geochemistry and Geology.* Freeman and Co., San Francisco, CA, 617 pp.

Igarashi, G., Saeki, S., Takahata, N., Sumikawa, K., Tasaka, S., Sasaki, Y., Takahashi, M., and Sano, Y., 1995. Groundwater radon anomaly before the Kobe earthquake in Japan. *Science*, 269:60–61.

Irwin, W. P. and Barnes, I. 1980. Tectonic relations of carbon dioxide discharges and earthquakes. *J. Geophys. Research*, 85(B6):3115–3121.

Ivanov, V. V., Ovchinnikov, A. M., and Yarotskiy, L. A., 1960. Map of underground mineral waters of the USSR, with explanatory notes, scale =1:7,500,000. *Ministry of Public Health, USSR Health Resort and Physiotherapy, State Res. Inst.*, Moscow.

Janezic, G. G., 1979. Biogenic light hydrocarbon production related to near-surface geochemical prospecting for petroleum. *AAPG Bull.*, 63:403.

Jeffrey, A. W. A, Alimi, H. M., and Jenden, P. D., 1974. Geochemistry of Los Angeles basin oil and gas systems. California Division of Oil and Gas, California Oil and Gas Fields, Vol. 2, *South, Central Coast and Offshore California*, Calif. Dept. Conservation, Div. Oil and Gas, Publ. TR12, 350 pp.

Jenden, P. D. and Kaplan, I. R., 1989. *Analysis of Gases in the Earth's Crust.* Final Report, January 1, 1982–May 31, 1988, Gas Research Institute. Global Geochemistry Corp., 6919 Eton Ave., Conoga Park, CA, 91303-2194.

Jenden, P. D., Kaplan, I. R., Poreda, R. J. and Craig, H., 1988. Origin of nitrogen-rich gases in the California Great Valley: Evidence from helium, carbon and nitrogen isotope ratios. *Geochimica et Cosmochimica Acta*, 52(4):851–861.

Jenden, P. D. and Kaplan, I. R., 1989. Origin of natural gas in Sacramento Basin, California. *AAPG Bull.*, 73(4):431–453.

Jenden, P. D., Drazan, D. J., and Kaplan, I. R., 1994. Mixing of thermogenic natural gases in Northern Applachian basin. *AAPG Bull.*, (6):980–998.

Jones, L. M., 1995. *Putting Down Roots in Earthquake Country.* United States Geological Survey, Southern California Earthquake Center, Los Angeles, California, 30 pp.

Jones V. T. and Drozd, R. J., 1983. Predictions of oil or gas potential by near-surface geochemistry. *A.A.P.G. Bull.*, 67(6)932–952.

Kamenskiy, I. L., Lobkov, V. A., Prasolov, E. M., Beskrovnyy, N. S., Kudryavtseva, E. I., Anufriyev, G. S., and Pavlov, V. P., 1976. Components of the upper mantle in the volcanic gases of Kamchatka. *Geochem. Int.*, 13:35–48.

Kaplan, I. R., 1994. Identification of formation process and source of biogenic gas seeps. *Isr. J. Earth Sci.*, 43:297–308.

Katz, D. L. and Williams, B., 1952. Reservoir fluids and their behavior. *Am. Assoc. Petrol. Geol. Bull.*, 36:342–358.

King, C. Y., 1986. Gas geochemistry applied to earthquake prediction: An overview. *J. Geophys. Res.*, 91:12269–12281.

King, C. Y., Walkingstick, C., and Basler, D., 1992. Radon in soil gas along active faults in central California. *U.S.G.S. Bull.,* no. 1971:77–78.

King, P. B., 1969. *Tectonic Map of North America*, scale = 1:5,000,000, U.S. Geol. Surv., Washington, D.C.

Kious, W. J. and Tilling, R. I., 1997. This dynamic earth: The story of plate tectonics. *U.S. Geological Sur*vey, CO, ISBN 0-16-048220-8, 78 pp.

Koltov, B. A., Chmirev, V. A., Polyakov, V. A., Maldov, I. I., Azimi, N., and Alam, K. H., 1977. Minor elements in carbonated waters of central Afghanistan (in Russian). *Dokl. Akad. Nauk. SSSR*, 237(6):1486–1489.

Landes, K. K., 1973. Mother nature as oil polluter. *Am. Assoc. Petrol. Geol. Bull.*, 57:637–641.

Langnes, G. L., Robertson Jr., J. O., and Chilingar, G. V., 1972. *Secondary Recovery and Carbonate Reservoirs*, Am. Elsevier Publ. Co., Inc., New York, NY, 304 pp.

Leck, L. V., 1921. *Petroleum Resources of California*, California State Mining Bureau, California State Printing Office, Sacramento, 179 pp.

Lee, W. H., Wu, F. T., and Wang, S. C., 1978. A catolog of instrumentally determined earthquakes in China (magnitude ≥ 6) compiled from various sources. *Seismol. Soc. Am. Bull.*, 68(2):383–398.

Levy, M. and Salvadori, M., 1995. *Why Earth Quakes*. W. W. Norton and Co., New York–London, 215 pp.

Leythaeuser, D., Schaefer, R. G. and Yukler, A., 1980. Diffusion of light hydrocarbons through near-surface rocks. *Nature*, 284:522–525.

Link, W. K., 1952. Significance of oil and gas seeps in world oil exploration. *A.A.P.G. Bull.*, 36(8):1505–1541.

Luyendyk, B. P., Hornafius, J. S., Washburn, L., Trial, A. F., and Banerjee, S., 1996. Hydrocarbons in the marine environment: Relationships between natural hydrocarbon seepage and oil production. In *Use of New Technologies and Trade in Oil and Gas Industry,* Symposium, April 1996, Los Angeles, CA.

Mah, R. A. and Sussman, C., 1967. Microbiology of anaerobic sludge fermentation, I. Enumeration of the nonmethanogenic anaerobic bacteria. *Appl. Microbiol.*, 16:358–361.

Mah, R. A., Ward, D. M., Baresti, L., and Glass, T., 1977. Biogenesis of methane. *Annu. Rev. Microbiol.*, 31:109–131.

Marco, S. and Agnon, A., 1995. Prehistoric earthquake deformations near Massada, Dead Sea Graben. *Geology*, 23:695–698.

McNeilan, T. W., Rockwell, T. K., and Resnick, G. S., 1996. Style and rate of Holocene slip, Palos Verdes fault, southern California. *J. Geophysical Res.*, 101(B4):8317–8334.

Medvedev, S. V. (ed.), 1968. Sesmic zoning in the USSR (in Russian). *Acad. Sci. USSR Inst. Earth Physics*, Moscow. (Israel Program for Scientific Translations, Jerusalem, 1976, English Translation, 533 pp.)

Mogro-Campero, A. and Fleischer, R. L., 1980. Search for long-distance migration of subsurface radon. In T. F. Gessell and W. M. Lowder (eds.), *The Natural Radiation Environment III*. U.S. Dept. of Energy, Washington D.C., U.S.D.O.E. Rept. CONF-780422, pp. 72–83.

Muskat, M., 1937. *The Flow of Homogeneous Fluids Through Porous Media.* McGraw-Hill, New York, NY, 763 pp.

Natural Gas Processors Suppliers Men's Association, 1957. *Engineering Data Book*, N.G.P.S.A., Tulsa, OK, 7th ed. 174 pp.

Nechayeva, O. L., 1968. Hydrogen in gases dissolved in water of the western Siberian plain. *Akad. Nauk SSSR Doklady*, 179:197–199.

Nikonov, V. F., 1971. Distribution of methane homologs in gas and oil fields. *Akad. Nauk SSSR Doklady*, 206:234–236.

Oremland, R. S., 1981. Geochemcial prospecting for oil and gas: Microbiologists viewpoint (abs). *AAPG Bull.*, 65:967.

Ovchinnikov, L. N., Sokolov, V. A., Fridman, A. I., and Yanitskii, I. N., 1972. Gaseous geochemical methods in structural mapping and prospecting for ore deposits. *Proc. 4th International Geochem. Exploration Symp.*, London, pp. 177–183.

Poreda, R. J., Jeffrey, A. W. A., Kaplan, I. R., and Craig, H., 1988. Magmatic helium in subduction-zone natural gases. *Chemical Geology*, 71:199–210.

Powers, R. B., 1977. Assessment of oil and gas reserves in the Idaho-Wyoming thrust belt, in Rocky Mountain thrust belt; geology and resources: *Wyoming Geological Association Guidebook*, 29:629–637.

Price, L. C. and Schoell, M., 1995. Constraints on the origins of hydrocarbon gas from composition of gases at their site of origin. *Nature*, 378:368–371.

Pyre, A. 1977. Tectonics and major hydrocarbon accumulations. *Symp. Petrol. Evolution*, Gulf Research and Development Co., 11:665–679.

Reed, R. D., 1948. Types of oil. In *Geologic Formations and Economic Development of the Oil and Gas Fields of California, Bull. 118,* 2nd ed. Department of Natural Resources, Division of Mines, Ferry Building, San Francisco, CA, pp. 95–114.

Reimer, G. M. et al., 1979. Recent developments in uranium exploration using the U.S.G.S. mobile helium detector. *Geochemical Exploration*, 11: 1–12.

Rice, D. D. and Claypool, G. E., 1981. Generation, accumulation and resource potential of biogenic gas. *A.A.P.G. Bull.,* 65:5–25.

Roberts, H. A., 1981. Helium emanometry in exploring hydrocarbons—part II. *Unconventional Methods in Exploration of Petroleum and Natural Gas.* Dallas, Southwestern Methodist Univ., pp. 136–149.

Schoell, M., 1983. Genetic characterization of natural gases., *A.A.P.G. Bull.*, 67:2225–2238.

Schoell, M., Jenden, P. D., Beeunas, M. A., and Coleman, D. D., 1993. Isotope analysis of gases in gas field and gas storage operations. *SPE No. 26171, SPE Gas Technol. Symp.*, Calgary, Alberta, Canada, June 28–30, pp. 337–344.

Shannon, P. M. and Naylor, D., 1989. *Petrolum Basin Studies.* Graham and Trotman, London, 210 pp.

Shi Huixin, 1979. A note on carbon dioxide discharging zones and their seismicities in China. *Seismol. Geol.*, 1(2):86–93.

Sieh, K. and Levay, S., 1998. *The Earth in Turmoil.* W. H. Freeman and Company, New York, photo 17, p. 216.

Smith, J. E., Erdman, J. G., and Marris, D. A., 1971. Migration, accumulation and retention of petroleum in the earth. In *Proc. 8th World Petroleum Congress*, Moscow. Applied Science Publishers, London, pp. 13–26.

Sokolov, V. A., Alexeyev, F. A., Bars, E. A., Geodekyan, A. A., Mogilevskiy, G. A., Yorovskiy, Y. M., and Yasenev, B.P., 1959. Investigation into direct oil detection methods. *5th World Petrol. Cong. Proc.*, 1(36):667–687.

Sokolov, V. A. and Cheremisinov, O. A., 1971. Composition of gaseous hydrocarbons as an index of migration and distribution in oil and gas pools, *Geologiya Nefti i Gaza*, no. 9:40–43.

Steinitz, G., Vulkan, U., Lang, B., Gilat, A., and Zafrir, H., 1992. Radon emanation along border faults of the Rift in the Dead Sea area. *Isr. J. Earth Sci.*, 41:9–20.

Steinitz, G., Lang, B., Vulkan, U., and Avni, Y., 1995. Temporal variation of the radon flux in syenitic rocks along the Ramon fault, Makhtesh Ramon, Israel. *Isr. Geol. Surv. Rep.* GSI/41/95: 14 pp.

Steinitz, G., Vulkan, U., and Lang, B., 1996. Monitoring of the tectonically related radon flux in Israel. *Isr. Geol. Surv., Current Res.*, 10:148–153.

Tarr, A. C., 1974. *World Seismicity Map*, scale = 1:39,000,000. U.S. Geol. Surv., Washington, D.C.

Taylor, H. P., 1979. Oxygen and hydrogen isotope relationships in hydrothermal mineral deposits. In H. L. Barnes (ed.), *Geochemistry of Hydrothermal Ore Deposits,* pp. 236–277. J. Wiley and Sons, New York, NY.

Thompson, A. B., 1933. The economic value of surface petroleum manifestations. *London World Petrol. Congr. Proc.,* pp. 241–250.

Ulomov, V. I. and Mavashev, B. Z., 1967. On forerunners of strong earthquakes. *Dokl. Akad. Nauk SSSR*, 176:319.

Vernon, J. W. and Slater, R. A., 1963. Submarine tar mounds, Santa Barbara County, California. *Am Assoc. Pet. Geol. Bull.*, 47(8):1624–1627.

Wakita, H. and Sano, Y., 1983. $^3He/^4He$ ratios in CH_4–rich gases suggest magmatic origin. *Nature*, 305:792–794.

Ward, D. M., Mah, R. A., and Kaplan, I. R., 1978. Methogenesis from acetate: A nonmethanogenic bacterium from an anaerobic acetate enrichment. *App. Envirn. Microbiol.*, 35:1185–1192.

Weast, R. C., 1981. *Handbook of Chemistry and Physics.* CRC Press, Inc., Boca Raton, FL, 1000 pp.

Wilkinson, E. R., 1971. California offshore oil and gas seeps. *California Summary of Operations—Oil Fields*, California Division of Oil and Gas, Publ. 57(1):5–28.

Wilkinson, E. R., 1982. Oil production in urban environments. In *Proc. Land Use Planning in Urban Oil-Producing Areas.* Workshop, Feb. 25, Huntington Beach, California Division of Oil and Gas, Publ. TR31:5-13.

Wissler, S. G., 1948. Stratigraphic formations of the producing zones of the Los Angeles basin oil fields. In *Geologic Formations and Economic*

Development of the Oil and Gas Fields of California, Bull. 118, 2nd ed. Department of Natural Resources, Division of Mines, Ferry Building, San Francisco, CA, pp. 209–234.

Woese, C. R. and Wolfe, R. S. (eds.), 1985. *Archaeobacteria*. Vol. VIII of *Bacteria—A Treatise on Structure and Function*. Academic Press, New York, NY.

Wyss, M., 1991. Introduction. *Tectonophysics*, 193:253–254.

Yeats, R., Sieh, K., and Allen, C., 1997. *The Geology of Earth*. Oxford Univ. Press, New York-Oxford, 568 pp.

Yeats, R. F., McCulloh, T. H., Schoellhamer, J. E., and Vedder, J. G., 1965. *Geology of the Los Angeles Basin, California—An Introduction*, U.S.G.S. Prof. Paper 420-A:57 pp.

Zapata Valle, R., 1973. Aguas minerales del Peru. *Estud. Espec.* 2:345, Serv. de Geol. y Mineria del Peru, Lima.

PART II

Abduvaliev, A. K., Voitov, G. I., Rudakov, V. P., and Orlova, T. G., 1990. Changes in the subsurface radon atmosphere on the eve of shallow crustal earthquakes ($M \leq 5$). *Dokl. Akad. Nauk SSSR*, 312(6):1335–1338 (in Russian).

Abduvaliev, A. K., Voitov, G. I., and Rudakov, V. P., 1988. Some characteristics of radon concentration in the subsoil layer in seismically active areas (on the example of Western Fergan). *Dokl. Akad. Nauk SSSR*, 300(6): 1337–1340 (in Russian).

Abduvaliev, A. K., Voitov, G. I., and Rudakov, V. P., 1986. Radon precursor of some strong earthquakes in central Asia. *Dokl. Akad. Nauk SSSR,* 291(4):924–927 (in Russian).

Allen, D. R., Chilingar, G. V., Mayuga, M. N., and Sawabini, C. T., 1971. Studio e previsione della subsidenza. In A. Mondadori (ed.), *Enciclopedia della Scienza e della Technica*, Milano, Italy, pp. 281–292.

Asada T. (ed.), 1982. *Earthquake Prediction Techniques. Their Application in Japan*. University of Tokyo Press, Tokyo, 317 pp.

Barnes, I, Irwin, W. P., and Gibson, H. A., 1975. Geologic map showing springs rich in carbon dioxide or chloride in California, scale 1:1,500,000. *Water Resour. Invest. Open-File Map*. U.S. Geol. Survey, Washington, D.C.

Barnes, I., Irwin, W. P., and White, D. E., 1978. Global distribution of carbon dioxide discharges and major zones of seismicity, scale 1:40,000,000. *Water Resour. Invest. WRI 78-39,* U.S. Geol. Survey, Washington, D.C.

Behringer, R. P., 1997. The scientist in the sandbox: Time dependence, fractals and waves. *Internat. J. Bifurcation and Chaos*, 7(5):963–978.

Behringer, R. P., 1995. The scientist in the sandbox: Complexity and dynamics in granular flow. *Mat. Res. Soc. Symp. Proc.,* 367:461–469.

Berle, M. and Roeburt J., 1959. *Earthquake.* Random House, New York, NY, 245 pp.

Bolt, B. A., 1976. *Nuclear Explosions and Earthquakes: The Parted Veil.* W.H. Freeman and Company, San Francisco, CA.

Bolt, B.A., 1988. *Earthquakes.* W.H. Freeman and Co., New York, NY, 282 pp.

Bossu, R., Grasso, J. R., Plotnikova, L. M., Nurtaev, B., Frechet, J., and Moisy, M., 1996. Complexity of the intracontinental seismic faultings: The Gazli, Uzbekistan, Sequence. *Bull. Seism. Soc. Am.*, 86(4):959–971.

Chilingar, G. V., Adamson, L. G., Armstrong, R. A., and Beeson, C. M., 1965. Soils stabilized through electroosmosis. *Southwest Builder and Contractor,* 145:100–102.

Chilingar, G. V., Adamson, L. G., Armstrong, R. A., and Beeson, C. M., 1966. Electrokinetics of formations and interstitial waters. *Encyclopedia of Earth Sciences,* Reinhold Publ. Co., New York, NY.

Chilingar, G. V., El-Nassir, A., and Stevens, R. G., 1970. Effect of direct electric current on permeability of sandstone cores. *J. Petrol. Technol.*, 249(7):830–836.

Chilingar, G. V., Khilyuk, L. F., and Katz, S. A., 1996. Pronounced changes in upward natural gas migration as precursors of major seismic events. *J. Pet. Sci. Eng.*, 14:133–136.

Chilingar, G. V. and Terry, R. D., 1954. Relationship between porosity and chemical composition of carbonate rocks. *Pet. Eng.*, 26:341–342.

Chilingarian, G. V. and Wolf K. H. (eds.), 1975. *Compaction of Coarse-Grained Sediments, I.* Elsevier, Amsterdam, 552 pp.

Chilingarian, G. V., Mazzullo, S. J., and Rieke, H. H., 1992. *Carbonate Reservoir Characterizaion: A Geologic-Engineering Analysis, Part I.* Developments in Petroleum Science 30, Elsevier, Amsterdam, 639 pp.

Chilingarian, G. V., Donaldson, E. S., and Yen, T. F., 1995. *Subsidence Due to Fluid Withdrawal.* Elsevier, Amsterdam, 498 pp.

Chmura-Kraemer, H., Smith, B. E., and Levine, S., 1976. An animal behavior model for short term earthquake prediction. In *Abnormal Animal Behavior Prior to Earthquakes.* National Earthquake Hazard Reduction Program: 91–94.

Cornell, C. A. and Winterstein, S. R., 1988. Temporal and magnitude dependence in earthquake recurrence model. *Bull. Seism. Soc. Am.*, 78:1522–1537.

Decker, R. and Decker, B., 1982. *Volcanoes and the Earth's Interior.* W.H. Freeman and Co., San Francisco, CA, 141 pp.

Dickey, P. A. and Hunt, J. M., 1972. Geochemical and hydrgeologic methods of prospecting for stratigraphic traps. In *Stratigraphic Oil and Gas Fields—Classification, Exploration Methods and Case Histories.* AAPG Memoir 16—Soc. Expl. Geophys. Spec. Publ., 10:136–167.

Dott, R. H., Jr. and Batten, R. L., 1976. *Evolution of the Earth.* McGraw-Hill Co., New York, NY, 504 pp.

Endres, B., Chilingarian, G. V., and Yen, T. F., 1991. Environmental hazards of urban oil field operations. *J. Pet. Sci. and Engr.*, 6:95–106.

Galloway, J. M., 1992. *Natural Oil, Tar, and Gas Seeps of the Carpinteria Area.* Minerals Management Service, I.T.M., Field Trip, May 12th.

Gee, L. S., Neuhauser, D. S., Dreger, D. S., Pasyanos, M. E., Uhrhammer, R. A., and Romanovich B., 1996. Real-time seismology at UC Berkeley: The rapid earthquake data integration project. *Bull. Seism. Soc. Am.*, 86(4): 936–945.

Giusti, G. G., 1995. *How to Build Earthquake, Weather, and Solar Flare Monitors.* TAB Books, Division of McGraw-Hill, Inc., New York, NY, 290 pp.

Green II, H. W., 1994. Solving the paradox of deep earthquakes. *Scientific American*, Sept. 1994: 64–71.

Green II, H. W. and Burnley, P. C., 1989. A new self-organizing mechanism for deep-focus earthquakes. *Nature*, 341(6244):733–737.

Green II, H. W., Young, T. E., Walker, D., and Scholz, C. H., 1990. Anticrack-associated faulting at very high pressure in natural olivine. *Nature*, 348 (6303):720–722.

Gribbin, J. R. and Plageman, S. H., 1974. *The Jupiter Effect.* Walker and Company, New York, 136 pp.

Gubbins, D., 1990. *Seismology and Plate Tectonics.* Cambridge University Press, Cambridge, 339 pp.

Gurevich, A. E., Endres, B. L., Robertson, Jr., J. O., and Chilingarian, G. V., 1993. Gas migratrion from oil and gas fields and associated hazards. *J. Pet. Sci. Eng.*, 9:223–228.

Gurevich, A. E. and Chilingarian, G. V., 1993. Subsidence over producing oil and gas fields, and gas leakage to the surface. *J. Pet. Sci. Eng.*, 9: 239–250.

Guyod, H., 1944. Electric well logging; reprinted from the *Oil Weekly*, Aug. 7–Dec. 4 (16 art.), 103 pp.

Harden, D. R., 1998. *California Geology.* Prentice Hall, Inc., Upper Saddle River, NJ, 461 pp.

Helm, C. D., 1984. Field-based computational techniques for predicting subsidence. In T. L. Holzer (ed.), Man-Induced Land Subsidence. *Geol. Soc. Am., Rev. Eng. Geol.*, 1:1–23.

Holtzer, T. L., 1984. Ground failure induced by ground water withdrawal from unconsolidated sediments. In T. L. Holzer (ed.), Man-Induced Land Subsidence. *Geol. Soc. Am., Rev. Eng. Geol.*, 1:67–105.

Howell, D., Cole, F., Fanelli, M., and Wiese, K., 1993. Natural Gas. *Earth,* September, pp. 53–59.

Hubbert, M. K. and Rubey, W. W., 1959. Role of fluid pressure in mechanics of overthrust faulting. *Geol. Soc. Am. Bull.*, 70:115–166.

Iidaka, T. and Suetsugu, D.,1992. Seismological evidence for metastable olivine inside a subsiding slab. *Nature*, 356(6370):593–595.

Irwin, W. P. and Barnes I., 1980. Tectonic relations of carbon dioxide discharges and earthquakes. *J. Geophys. Res.*, 85(B6):3115–3122.

Ivanov, V. V., Ovchinnikov, A. M., and Yarotskiy, 1960. Map of underground mineral waters of the USSR, with explanatory notes, scale =1:7,500,000. *Ministry of Public Health, USSR Health Resort and Physiotherapy, State Res. Inst.*, Moscow.

Jaeger, H. M., Nagel, S. R., and Behringer, R. P., 1966. Granular solids, liquids, and gases. *Reviews of Modern Physics*, 68(4):1259–1273.

Jones, L. M., 1995. *Putting Down Roots in Earthquake Country*. Dept. of Interior, U.S.G.S. and the Southern California Earthquake Center (SCEC), University of Southern California, University Park, CA, 30 pp.

Karapetian, N. K., 1986. *Mechanisms of Origin of Earthquakes of Armenian Highlands (Seismic Setting).* lzd. A.N. Arm. SSR, Erevan, Armenia, 227 pp.

Karapetian N. K., 1991. *Earthquakes of the Armenian Highlands (Seismic Setting*). University of Southern California Publ., Los Angeles, CA, 234 pp. (Translated and edited by M. Agbabian, G. V. Chilingarian, and M. Garfunkel.)

Karapetian, B. K. and Karapetian N. K., 1981. *Conditions for Prognosticating Earthquakes as Related to Seismo-Resistant Constructions in the Armenian SSR.* Yerevan, Armenia, Izd. Ayastan, 169 pp. (in Russian).

Katz, S. A. and Aki, K., 1992. Experiments with a neural net based earthquake prediction. *Am. Geophys. Union Trans.*, 73:366.

Katz, S. A., Khilyuk, L. F., and Chilingarian, G. V., 1994. Interrelationships among subsidence owing to fluid withdrawal, gas migration to the surface and seismic activity. Environmental aspects of oil production in seismically active areas. *J. Pet. Sci. Eng.*, 11:103–112.

Khitarov, N. I., Voitov, G. I., and Lebedev, V. S., 1974. On geochemical precursors of earthquakes. In *Search for Earthquake Precursors on Forecasting Polygons*. Nauka, Moscow, 165–170 (in Russian).

King, P. B., 1969. *Tectonic Map of North America*, scale = 1:5,000,000, U.S. Geol. Surv., Washington, D.C.

Kious, W. J. and Tilling, R. I., 1995. *The Dynamic Earth: The Story of Plate Tectonics*. U.S. Geological Survey, ISBN 0-16-048 229-9, Denver, CO, 80225.

Koltov, B. A., Chmirev, V. A., Polyakov, V. A., Maldov, I. I., Azimi, N., and Alam, K. H., 1977. Minor elements in carbonated waters of central Afghanistan (in Russian). *Dokl. Akad. Nauk. SSSR*, 237(6):1486–1489.

Kouznetsov, O. L., Simkin, E. M., Chilingar, G. V., and Katz, S., 1998. Improved oil recovery by application of vibro-energy to waterflooded sandstones. *J. Pet. Sci. Eng.*, 19:191–200.

Kouznetsov, O. L., Sidorov, V. I., Katz, S., and Chilingarian, G. V., 1994. Interrelationships among seismic and short-term tectonic activity, oil and gas production, and gas migration to the surface. *J. Pet. Sci. Eng.*, 13:57–63.

Kouznetsov, O. L. and Simkin, E. M., 1990. *Transformation and Interaction of Geophysical Fields in the Lithosphere*. Nedra, Moscow, 226 pp.

Kravtsov, A. I. and Fridman, A. I., 1974. Geology and geochemistry of natural gases in the vicinity of deep faults and earthquake prediction. In *Search for Earthquake Precursors on Forecasting Polygons*. Nauka, Moscow, 158–164 (in Russian).

Levy, M. and Salvadori, M., 1995. *Why the Earth Quakes*. W.W. Norton and Co., New York-London, 215 pp.

Luyendyk, B. P., Hornafius, J. S., Washburn, L., Trial, A. F., and Banerjee, S., 1996. Hydrocarbons in the marine environment: Relationships between natural hydrocarbon seepage and oil production. In *Use of New Technologies and Trade in Oil and Gas Industry, Symposium,* April 1996, Los Angeles, CA.

Lee, W. H., Wu, F. T., and Wang, S. C., 1978. A catolog of instrumentally determined earthquakes in China (magnitude ≥ 6) compiled from various sources. *Seismol. Soc. Am. Bull.*, 68(2):383–398.

Ma, Z., Fu, Z., Zhang, Y., Wang, C., Zhang, G., and Liu, D., 1988. Earthquake prediction. *Nine Major Earthquakes in China (1966–1976)*. Seismological Press, Beijing, China. Springer-Verlag, Berlin-Heidelberg-New York-London-Paris-Tokyo, 333 pp.

Martin, J. C. and Serdengecti, S. C., 1984. Subsidence over oil and gas fields. In T. L. Holzer (ed.), *Man-Induced Land Subsidence. Geol. Soc. Am., Rev. Eng. Geol.*, 1:23–24.

McNeilan, T. W., Rockwell, T. K., and Resnick, G. S., 1996. Style and rate of Holocene slip, Palos Verdes fault, southern California. *J. Geolphysical Res.*, 101(B4):8317–8334.

Medvedev, S. V. (ed.), 1968. Sesmic zoning in the USSR (in Russian). *Acad. Sci. USSR Inst. Earth Physics*, Moscow. (Israel Program for Scientific Translations, Jerusalem, 1976, English Translation, 533 pp.)

Mei, S. (ed.), 1982. *Earthquake Prediction Techniques.* University of Tokyo Press, 317 pp.

Mogi, K., 1985. *Earthquake Prediction.* Academic Press, Inc., Tokyo, Japan, 355 pp.

Mogi, K., 1967. Earthquakes and fractures. *Tectonophysics*, 5:35–55.

Morrison, H. F., 1976. Electrical phenomena associated with rock strain: In *Abnormal Animal Behavior Prior to Earthquakes.* National Earthquake Hazard Reduction Progam, pp. 91–94.

O'Nions, R. R., Hamilton, P. J., and Evenson, N. M., 1980. The evolution of the earth's mantle. *Sci. Am.*, 242(5):132.

Pak, H. K. and Behringer, R. P., 1993. Surface waves in vertically vibrated granular material. *Physical Review Letters*, 71(12):1832–1835.

Panel of the public policy implications on earthquake prediction, 1975. *Earthquake Prediction and Public Policy.* National Academy of Sciences, Washington, D.C., 117 pp.

Rieke III, H. H. and Chiligarian, G. V., 1974. *Compaction of Argillaceous Sediments.* Elsevier, Amsterdam, 424 pp.

Rikitake, T., 1976. *Earthquake Prediction.* Elsevier, Amsterdam, 357 pp.

Rikitake, T., 1979. Classification of earthquake precursors. *Tectonophysics*, 54:293–309.

Riznichenko, Yu. V., 1976. Focus extent of a crustal earthquake as related to the seismic momentum. In *Studies in Physics of Earthquakes.* Nauka Publishers, Moscow (in Russian).

Riznichenko, Yu. V., 1963. Possibilities of calculating maximum earthquakes. In *Physics of Earthquakes and Explosion Seismology.* U.S. Department of Commerce. Office of Technical Services. Joint Publication Research Service: 3–18.

Russo, P., 1966. *Earthquakes.* Progress Publishers, Moscow (in Russian).

Sadovskiy, M. A., 1979. Natural fragmentation of rocks. *Dokl. Akad. Nauk SSSR*, 247(4):829–831 (in Russian).

Sadovskiy, M. A., Monakhov, F. I., and Semenov, A. N., 1977. Hydrogeodynamic precursors of South-Kuryl earthquakes. *Dokl. Akad. Nauk SSSR*, 236(1):50–53 (in Russian).

Sadovskiy, M. A., Monakhov, F. I., Kissin, I. G., and Shyrokov, B. D., 1984. Short-term hydrogeodynamic precursors of earthquakes. In *Earthquake Prediction*, TERRAPUB, Tokyo; UNESCO, Paris: 233–242.

Serruya, C., Picard, L., and Chilingarian, G. V., 1967. Possible role of electric currents and potentials during diagenesis ("Electrodiagenesis"). *J. Sediment. Petrol.*, 37(2):695–698.

Stein, R. S., Barka, A. A., and Dietrich, J. H., 1997. Progressive failure on the North Anatolian fault since 1939 by earthquake stress triggering. *Geophys. J. Int.*, 128:594–604.

Surguchev, M. L., Kouznetsov, O. L., and Simkin, E. M., 1975. *Hydrodynamic, Accoustic, and Thermal Cyclic Processing of Oil-Bearing Formations*. Nedra Publ., Moscow, 320 pp. (in Russian).

United States Geol. Survey, 1993. *Earthquake Research at Parkfield, California, for 1993 and Beyond.* National Earthquake Prediction Evaluation Council Working Group, 1994, U.S. Geological Survey Circular 1116.

United States Geol. Survey, 1990. *Probabilities of Large Earthquakes in the San Francisco Bay Region, California*, Working Group on California Earthquake Probabilities, US Geol. Surv. Circular 1053.

Uyeda, S., 1996. Introduction to the VAN method of earthquake prediction. In *A Critical Review of VAN.* World Scientific, Singapore-New Jersey-London-Hong Kong: 3–26.

Varotsos, P. and Lazaridou, M., 1991. Latest aspects of earthquake prediction in Greece based on seismic electric signals I. *Tectonophysics*, 188, 321–343.

Varotsos, P., Alexopoulos, K., and Lazaridou, M., 1993. Latest aspects of earthquake prediction in Greece based on seismic electric signals II. *Tectonophysics*, 224, 1–37.

Varotsos, P., Lazaridou, M., Eftaxias, K., Antonopoulos, G., Makris, J., and Kopanas, J., 1996. Short-term earthquake prediction in Greece by seismic electric signals. In Sir James Lighthill (ed.), *A Critical Review of VAN. Earthquake Prediction from Seismic Electrical Signals.* Univ. College, London, World Scientific Publ., Singapore, 376 pp.

Vernadskiy, V. I., 1934. *Outline of Geochemistry*, GosGeoNefteIzdat. (in Russian).

Vernadskiy, V. I., 1940. *Living Matter and the Biosphere*. Nauka Publishers, Moscow (in Russian).

Wakita H., 1977. Ecochemistry as a tool for earthquake prediction. In C. Kisslinger and Z. Suzuki (eds.). *Earthquake Precursors*. Supplemental Issue of *J. of Physics of the Earth*. Center for Academik Publications, Tokyo: 175–183.

Whittaker, B. N. and Reddish, D. J., 1989. *Subsidence: Occurrence, Prediction and Control.* Elsevier, Amsterdam, 528 pp.

Wiens, D. A., McGuire, J. J., and Shore, P. J., 1993. Evidence for transformational faulting from a deep double seismic zone in Tonga. *Nature,* 364(6440):790–793.

Willis B., 1921. *The Next Earthquake.* Trans. Commonwealth Club Calif. San Francisco, CA, 16(10):375–379.

Yeats, R., Sieh, K., and Allen, C., 1997. *The Geology of Earth.* Oxford Univ. Press, New York-Oxford, 568 pp.

Zapata Valle, R., 1973. Aguas minerales del Peru. *Estud. Espec.* 2: 345 pp. Serv. de Geol. y Mineria del Peru, Lima.

PART III

Allen, A. A. and Schlueter, R. S., 1969. Estimates of Surface Pollution Resulting from Submarine Oil Seeps at Platform "A" and Coal Oil Point: General Research Corp., Memo 1230; in Wilkinson, E. R. (1971).

Aschenbrenner, B. C. and Achauer, C. W., 1960. Minimum conditions for migration of oil in water-wet carbonate rocks. *A.A.P.G. Bull.*, 44(2):235-243.

Barends, F. B. J., Brouwer, F. J. J., and Schroder, F. H. (eds.), 1995. *Land Subsidence*, Proc. Fifth Internat. Symp. on Land Subsidence, 16–20 Oct. The Hague, Netherlands, A.A. Balkema, Rotterdam.

Blocker, P. C., 1964. Spreading and evaporation of petroleum products on water. *Proceedings of the Fourth International Harbor Conference*, June 1954, Antwerp, Belgium.

Brown, K. E., 1967. *Gas Lift Theory and Practice*, The Petroleum Publishing Co., Tulsa, OK, 924 pp.

Carlisle, C. T., Bayliss, G. S., and Van Delinder, D. G., 1975. Distribution of light hydrocarbons in sea floor sediments: Correlations between geochemistry, seismic structure and possible reservoir oil and gas. In *Proc. 1975 Conf. Offshore Technology,* Dallas, TX, 3:65–70.

Chapman, R. E., 1983. *Petroleum Geology.* Elsevier, Amsterdam, pp. 245–251.

Chilingar, G. V. and Beeson, C. M., 1969. *Surface Operations in Petroleum Production*, Am. Elsevier Publ. Co., Inc., New York, NY, 397 pp.

Chilingarian, G. V. and Vorabutr, P., 1983. *Drilling and Drilling Fluids.* Elsevier, Amsterdam, pp. 71–79.

Chilingarian, G. V., Robertson Jr., J. O., and Kumar, S., 1987. *Surface Operations in Petroleum Production*, 1, Developments in Petroleum Science, 19A, Elsevier, Amsterdam, pp. 146–147; 416–420.

Chilingarian, G. V., Mazzullo, S. J., and Rieke, H. H., 1992. *Carbonate Reservoir Characterization: A Geologic Engineering Analysis*, Part 1, 30. Elsevier, Amsterdam, 639 pp.

Chilingarian, G. V., Katz, S. A., and Khilyuk, L. F., 1995. Neural network based upon prediction of subsidence, surface gas migration, and seismic activity. In F. B. J. Barends, F. J. J. Brouwer, and F. H. Schroder (eds.), Land Subsidence, Proc. Fifth Internat. Symp. on Land Subsidence, The Hague, Netherlands, 16–20 Oct., pp. 41–46.

Chilingarian, G. V., Donaldson, E. C., and Yen, T. F., 1995. *Subsidence Due to Fluid Withdrawal*, Developments in Petroleum Science, 41, Elsevier, Amsterdam, 498 pp.

Claypool, G. E. and Kaplan, I. R., 1974. The origin and distribution of methane in marine sediments. In I. R. Kaplan (ed.), *Natural Gases in Marine Sediments*, Plenum Press, New York, NY, pp. 99–140.

Craig, H. 1953. The geochemistry of stable carbon isotopes. *Geochim. Cosmochim. Acta*, 3:53–92.

Culberson, O. L. and McKetta, J. J., Jr., 1951. Phase equilibrium in hydrocarbon-water systems. III. The solubility of methane in water at pressures to 10,000 psi. *Trans. AIME*, 192:223–226.

Dodson, C. R. and Standing, M. B., 1944. Pressure-volume-temperature and solubility relations for natural gas-water mixtures. *Drilling and Production Practices*, API, pp. 173–179.

Endres, B., Chilingarian, G. V., and Yen, T. F., 1992. Environmental hazards of urban oil field operations, *J. Pet. Sci. Engr.*, 6:95-106.

Frick, T. C., 1962. *Petroleum Production Handbook, Volume 2*, Society of Petroleum Engineers of AIME, Dallas, TX, 1000 pp.

Geertsma, J., 1973. Land subsidence above compacting oil and gas reservoir. *J. Petrol. Technol.*, 25(6):734–744.

Gold, T. and Soter, S., 1980. The deep-earth gas hypothesis. *Sci. Amer.*, 242: 154–161.

Gold, T. and Soter, S., 1982. Abiogenic methane and the origin of petroleum. *Energy Explor. Exploit.*, 1:89–104.

Gurevich, A. E., 1970. Processes of groundwater, oil and gas migration in the petroliferous basins of Mangyshlak. Ph.D. Diss., VNIGRI, St. Petersburgh, Russia.

Gurevich, A. E. and Chilingarian, G. V., 1995. Possible impact of subsidence on gas leakage to the surface from subsurface oil and gas reservoirs. In G. V. Chilingarian, E. C. Donaldson and T. F. Yen (eds.), *Subsidence Due to Fluid Withdrawal*. Developments in Petroleum Science, 41, Elsevier Amsterdam, 498 pp.

Hunt, J. M., 1979. *Petroleum Geochemistry and Geology*. Freeman and Co., San Francisco, CA, 617 pp.

Jenden, P. D., Kaplan, I. R., Poreda, R. J., and Craig, H., 1988. Origin of nitrogen-rich gases in the California Great Valley: Evidence from helium, carbon and nitrogen isotope ratios. *Geochimica et Cosmochimica Acta*, 52(4):851–861.

Jenden, P. D. and Kaplan, I. R., 1989. Origin of natural gas in Sacramento basin, California. *A.A.P.G. Bull.*, 73(4):431–453.

Jenden, P. D., Drazan, D. J., and Kaplan, I. R., 1994. Mixing of thermogenic natural gases in Northern Applachian basin. *A.A.P.G. Bull.*, (6):980–998.

Jenden, P. D., Hilton, D. R., Kaplan, I. R., and Craig, H., 1993. Abiogenic hydrocarbons and mantle helium in oil and gas fields. In D. G. Howell (ed.), *The Future of Energy Gases*. U.S. Geol. Surv. Prof. Paper 1570:31-36.

Jones V. T. and Drozd, R. J., 1983. Predictions of oil or gas potential by near-surface geochemistry. *A.A.P.G. Bull.*, 67(6):932–952.

Kaplan, I. R., 1994. Identification of formation process and source of biogenic gas seeps. *Isr. J. Earth Sci.*, 43:297–308.

Langnes, G. L., Robertson Jr., J. O., and Chilingar, G. V., 1972. *Secondary Recovery and Carbonate Reservoirs*, Am. Elsevier Publ. Co., Inc., New York, NY, 304 pp.

Link, W. K., 1952. Significance of oil and gas seeps in world oil exploration. *A.A.P.G. Bull.*, 36(8):1505–1540.

Mah, R. A. and Sussman, C., 1967. Microbiology of anaerobic sludge fermentation, I. Enumeration of the nonmethanogenic anaerobic bacteria. *Appl. Microbiol.*, 16:358–361.

Mah, R. A., Ward, D. M., Baresti, L., and Glass, T., 1977. Biogenesis of methane. *Annu. Rev. Microbiol.*, 31:109–131.

Natural Gas Processors Suppliers Men's Association, 1957. *Engineering Data Book*, N.G.P.S.A., Tulsa, OK, 7th ed. 174 pp.

Poreda, R. J., Jeffrey, A. W. A., Kaplan, I. R., and Craig, H., 1988. Magmatic helium in subduction-zone natural gases. *Chemical Geology*, 71:199–210.

Rice, D. D. and Claypool, G. E., 1981. Generation, accumulation and resource potential of biogenic gas. *A.A.P.G. Bull.*, 65:5–25.

Riegle, J. R., 1953. Gas storage in the Playa del Rey oil field. In *Summary of Operations, California Oil Fields*. State of California Division of Oil and Gas, San Francisco, CA, 39(2):17–33.

Schoell, M., 1983. Genetic characterization of natural gases., *A.A.P.G. Bull.*, 67:2225–2238.

Schoell, M., Jenden, P. D., Beeunas, M. A., and Coleman, D. D., 1993. Isotope analysis of gases in gas field and gas storage operations. *SPE No. 26171, SPE Gas Technol. Symp.*, Calgary, Alberta, Canada June 28–30, pp. 337–344.

Schumacher, D. and Abrams, M. A. (eds.), 1994. *Hydrocarbon Migration and Its Near-Surface Expression.* AAPG Memoir, 66, 350 pp.

Smith, J. E., Erdman, J. G., and Marris, D. A., 1971. Migration, accumulation and retention of petroleum in the earth. In *Proc. 8th World Petroleum Congress*, Moscow. Applied Science Publishers, London, pp. 13–26.

Tek, M. R., 1987. *Underground Storage of Natural Gas*, Contributions in Petroleum Geology and Engineering, no. 3, G. V. Chilingar (ed.), Gulf Publishing Co., Houston, TX, 389 pp.

Ward, D. M., Mah, R. A., and Kaplan, I. R., 1978. Methanogenesis from acetate: A nonmethanogenic bacterium from an anaerobic acetate enrichment. *App. Envirn. Microbiol.*, 35:1185–1192.

Wilkinson, E. R., 1982. Oil production in urban environments. In *Proc. Land Use Planning in Urban Oil-Producing Areas*. Workshop, Feb. 25, Huntington Beach, California Division of Oil and Gas, Publ. TR31: 5-13.

Woese, C. R. and Wolfe, R. S. (eds.), 1985. *Archaeobacteria*. Vol. VIII of: *Bacteria—A Treatise on Structure and Function*. Academic Press, New York, NY.

PART IV

Allen, D. R., Chilingar, G. V., Mayuga, M. N., and Sawabini, C. T., 1971. Studio e previsione della subsidenza. In A. Mondadori (ed.), *Enciclopedia della Sicenza e della Tecnica*, Milano, Italy, pp. 281–292.

Allen, D. R., 1973. Subsidence, rebound and surface strain associated with oil-producing operations, Long Beach, California. *Assoc. Eng. Geol., Spec. Publ.*

Ammosov, S. M., Voitov, G. I., Korobeinic, G. S., Kouznetsov, O. L., Nikolaev, A. V., and Fedorova, G. S., 1987. Two types of gas-geochemical phenomena in the field of seismic vibration source. *Dokl. Akad. Nauk SSSR*, 301 (1):62–66.

Andronopoulos, B., Rozos, D., and Hadzinakos, I., 1991. Subsidence phenomena in the industrial area of Thessaloniki, Greece. In A. I. Johnson (ed.), *Land Subsidence*. IAHS Publ., 200:59–70.

Balestri, M. and Villani, B., 1991. Study of the subsidence in the Bolognese area. In A. I. Johnson (ed.), *Land Subsidence*. IAHS Publ., 200:71–80.

Beckwith, G. H., Slemmons, D. B., and Weeks, R. E., 1991. Use of low-sun angle photography for identification of subsidence-induced fissures. In A. I. Johnson (ed.), *Land Subsidence*. IAHS Publ., 200:261–269.

Bell, J. S. and Dusseault, M. B., 1991. Scale effects and the use of borehole breakouts as stress indicators. In A. Pinta da Cunha (ed.), *Scale Effects in Rock Masses*. Balkema, Rotterdam, pp. 327–337.

Bergado, D. T., Balasubramaniam, A. S., and Apaipong, W., 1984. Field monitoring of subsidence effects in AIT campus, Bangkok, Thailand. *Proc. 3rd Inter. Symp. on Land Subsidence, Venice, Italy, 19–25 Mar.* Inter. Asso. Hydrological Sci., IAHS Pub. No. 151:391–404.

Biot, M. A., 1941. General theory of three-dimensional consolidation. *J. Appl. Phys*., 12:426–430.

Brand, E. W. and Balasubramaniam, A. S., 1976. Soil compressibility and land subsidence in Bangkok. *Proc. 2nd Inter. Symp. on Land Subsidence, Anaheim, CA, USA, 13–17 Dec.* Inter. Asso. Hydrological Sci., IAHS-AISH Pub. No. 121:365–374.

Bravo, R., Rogers, J. R., and Cleveland, T. G., 1991. A new three-dimensional finite difference model of ground water flow and land subsidence in the Houston area. In A. I. Johnson (ed.), *Land Subsidence*. IAHS Pubi., 200: 15–26.

Brighenti, G., Borgia, G. C., and Mesini, E., 1995. Subsidence studies in Italy. In G. V. Chilingarian, E. C. Donaldson, and T. F. Yen, *Subsidence due to Fluid Withdrawal.* Elsevier Sci. Pub., Amsterdam: 248–253.

Carbognin, L., Gatto, P., and Mozzi, G., 1976. New trend in the subsidence of Venice. *Proc. 2nd Inter. Symp. on Land Subsidence, Anaheim, CA, USA, 13–17 Dec.* Inter. Asso. Hydrological Sci., IAHS-AISH Pub. No. 121: 65–81.

Carbognin, L. and Gatto, P., 1984. An overview of the subsidence of Venice. *Proc. 3rd Inter. Symp. on Land Subsidence, Venice, Italy, 19–25 Mar.* Inter. Asso. Hydrological Sci., IAHS Pub. No. 151:321–328.

Chilingarian, G. V. and Endres, B., 1997. Land subsidence due to fluid withdrawal. *Second Annual Symp. on Groundwater Resources and Groundwater Quality Protection,* Jan 9–10, Natl. Cheng Kung Univ., Tainan, Taiwan, pp. 1–13.

Christensen, S. O., Janbu, N., and Jones, M. E., 1989. Subsidence due to oil-gas production. *Erdoel Kohle, Erdgas, Petrochem*., 42 (5):185–189.

Contaldo, G. J. and Mueller, J. E., 1991. Earth fissures and land subsidence of the Mimbres basin, Southwestern New Mexico, U.S.A. In A. I. Johnson (ed.), *Land Subsidence*. IAHS Publ., 200:301–310.

Deflache, A. P., 1979. Land subsidence versus head decline in Texas. In S. K. Saxena (ed.), *Evaluation and Prediction of Subsidence*. ASCE Publ., New York, NY, pp. 320–331.

Donaldson, E. C. and van Domselaar, H., 1983. *Forum on Subsidence Due to Fluid Withdrawals*. CONF-821199, National Technical Information Service, Springfield, VA, 22161, 141 pp.

Donaldson, E. C. and Chilingarian, G. V., 1997. Review of subsidence due to groundwater withdrawal. *Second Annu. Symp. Groundwater Resources and Groundwater Quality Protection,* Jan 9–10, Natl. Cheng Kung Univ., Tainan, Taiwan, pp. 247–264.

Dusseault, M. B., 1983. Identifying reservoirs susceptible to subsidence due to fluid withdrawal. In E. C. Donaldson and H. van Domselaar (eds.). *Proceedings 1982 Forum on Subsidence due to Fluid Withdrawals*. CONF-821199, National Technical Information Service, Springfield, VA, 22161, pp. 6–14.

Enever, J. R., Walton, R. J., and Wold, M. B., 1990. Scale effects influencing hydraulic fracture and overcoming stress measurements. In A. Pinta da Cunha (ed.), *Scale Effects in Rock Masses*. Balkema, Rotterdam, pp. 317–326.

Erickson, R. C., 1977. Subsidence control and urban oil production—A case history, Beverly Hills (East) oilfield, California. *Proc. Internat. Symp. Land Subsidence,* Anaheim, California, Internal. Assoc. Hydrological Sciences Publ., 121:285–297.

Esaki, T., Shikata, K., Aoki, K., and Kimura, T., 1991. Surface subsidence in natural gas fields. In A. I. Johnson (ed.), *Land Subsidence*. IAHS Publ., 200:109–118.

Fairchild, J. B. and Wiebe, K. H., 1976. Subsidence of organic solids and salinity barrier design in coastal Orange County, California. *Proc. 2nd Inter. Symp. on Land Subsidence, Anaheim, CA, USA, 13–17 Dec.* Inter. Asso. Hydrological Sci., IAHS-AISH Pub. No. 121:334–346.

Fjaer, E., Holt, R. M., Horsrud, P., Raaen, A. M., and Risnes, R., 1992. *Petroleum Related Rock Mechanics*. Elsevier, Amsterdam, 338 pp.

Gabrish, R. K. and Holzer, T. L., 1978. Fault control of subsidence, Houston, Texas. Discussion. *Ground Water*, 16(1):51–55.

Gabrysch, R. K., 1976. Land-surface subsidence in the Houston-Galveston region, Texas. *Proc. 2nd Inter. Symp. on Land Subsidence, Anaheim, CA, USA, 13–17 Dec.* Inter. Asso. Hydrological Sci., IAHS-AISH Pub. No. 121: 16–24.

Gambolati, G., Ricceri, G., Bertom, W., Brighenti, G., and Villermin, E., 1991. Numerical analysis of land subsidence at Ravenna due to water and gas removal. In A. I. Johnson (ed.), *Land Subsidence*. IAHS Publ., 200: 119–128.

Garbognin, L., Gatto, P., Mozzi, G., and Gambolati, G., 1979. Land subsidence of Ravenna and its similarities with the Venice case. In S. K. Saxena (ed.), *Evaluation and Prediction of Subsidence*. ASCE, Publ., New York, NY, pp. 254–266.

Geertsma, J., 1957. The effect of fluid-pressure decline on volumetric changes of rocks. *Trans AIME*, 210:331–339.

Geertsma, J., 1973. A basic theory of subsidence due to reservoir compaction: The homogeneous case. *Verh. Kon. Ned. Geol. Mijnbouwk. Gen.*, 28:43–62.

Geertsma, J., 1973. Land subsidence above compacting oil and gas reservoir. *J. Petrol. Technol.,* 25(6):734–744.

Guacci, G., 1979. The Pixley fissure, San Joaquin Valley, California. In S. K. Saxena (ed.), *Evaluation and Prediction of Subsidence*. ASCE Publ., New York, NY, pp. 303–319.

Guangxiao, D. and Yiaoqi, Z., 1984. Land subsidence in China. *Proc. 3rd Inter. Symp. on Land Subsidence, Venice, Italy, 19–25 Mar.* Inter. Asso. Hydrological Sci., IAHS Pub. No. 151: 405–414.

Gurevich, A. E., 1969. *Processes of Ground Water, Oil, and Gas Migration*. Nedra Publishers, Leningrad, 112 pp. (in Russian).

Gurevich, A. E., Endres, B. L., Robertson, Jr., J. O., and Chilingarian, G. V., 1993. Gas migration from oil and gas fields and associated hazards. *J. Pet. Sci. Eng.* 9:233–238.

Gurevich, A. E. and Chilingarian, G. V., 1993. Subsidence over producing oil and gas fields, and gas leakage to the surface. *J. Petrol. Sci. and Eng.*, 9(3):239–250.

Gurevich, A. E., Chilingarian, G. V., and Bou-Rabee, F., 1998. Subsidence caused by fluid withdrawal: The need for a systems approach. In J. W. Borchers (ed.), *Land Subsidence Case Studies and Current Reseach, Proc. J. F. Poland Symp.,* Assoc. Engr. Geol. and Star Publ. Co., pp. 479–486.

Gurevich, A. E., 1969. *Processes of Ground-Water, Oil, and Gas Migration*. Nedra Publishers, Leningrad, 112 pp. (in Russian).

Gurevich, A. E., 1980. *Handbook of Ground-Water Motion Exploration*, Nedra Publishers, Leningrad, 216 pp. (in Russian).

Gurevich, A. E., Kapchenko, L. N., and Kruglikov, N. M., 1972. *Theoretical Principles of Petroleum Hydrogeology*. Nedra Publishers, Leningrad, 272 pp. (in Russian).

Gurevich, A. E. et al., 1987. *Formation Fluid Pressure*. Nedra Publishers, Leningrad, 223 pp. (in Russian).

Haneberg, W. C., Reynolds, C. B., and Reynolds, I. B., 1991. Geophysical characterization of soil deformation associated with earth fissures near San Marcial and Deming, New Mexico, In A. I. Johnson (ed.), *Land Subsidence*. IAHS Publ., 200:271–280.

Holdahl, S. R., Zilkoski, D. B., and Holzschuh, J. C., 1991. Subsidence at Houston, Texas, 1973–1987. In A. I. Johnson (ed.), *Land Subsidence*. IAHS Publ., 200:3–14.

Holzer, T. L., 1984. Ground failure induced by groundwater withdrawal from unconsolidated sediments. In T. L. Holzer (ed.), *Man-induced Land Subsidence*. Rev. Eng. Geol. VI, Geol. Soc. Am., Boulder, Col., pp. 67–105.

Holzer, T. L., 1990. Land subsidence caused by withdrawal of oil and gas in the Gulf Coastal Plain—the Houston, Texas, case history. *A.A.P.G. Bull.*, 74(9):1497–1498.

Holzer, T. L. and Thatcher, W., 1979. Modeling deformation due to subsidence faulting. In: S.K. Saxena (ed.), *Evaluation and Prediction of Subsidence*. ASCE Publ., New York, NY, pp. 349–357.

Ishii, M., Kuramochi, F., and Endo, T., 1976. Recent tendencies of the land subsidence in Tokyo. *Proc. 2nd Inter. Symp. on Land Subsidence, Anaheim., CA, USA, 13–17* Dec. Inter. Asso. Hydrological Sci., IAHS-AISH Pub. No. 121: 25–34.

Ito, T., Hayashi, K., and Abe, H., 1990. Scale effect in breakdown pressure of hydraulic fracturing stress measurements. In A. Pinta da Cunha (ed.), *Scale Effects in Rock Masses*. Balkema, Rotterdam, pp. 289–295.

Jachens, R. C. and Holzer, T. L., 1979. Geophysical investigation of ground failure related to ground-water withdrawal—Picacho Basin, AZ. *Ground Water*, 17(6):574–585.

Jaeger, C., 1979. *Rock Mechanics and Engineering*. Cambridge University Press, Cambridge, 523 pp.

Jaeger, J. C. and Cook, N. G. W., 1969. *Fundamentals of Rock Mechanics*. Mathuen, London, 513 pp.

Johnson, A. I., Carbognin, L., and Ubertini, L., 1986. *Proc. 3rd Inter. Symp. on Land Subsidence*. IAHS, Institute of Hydrology, Wallingford, OX10 8BB, UK, IAHS Pub. No. 151: 939 pp.

Katz, S., Khilyuk, L., and Chilingarian, G. V., 1994. Interrelationships among subsidence owing to fluid withdrawal, gas migration to the surface and seismic activity. Environmental aspects of oil production in seismically active areas. *J. Pet. Sci. Engr.*, 11(2):103–112.

Keaton, J. R. and Shlemon, R. J., 1991. The Fort Hancock earth fissure system, Hudspeth County, Texas: Uncertainties and implications. In A. I. Johnson (ed.), *Land Subsidence*. IAHS Publ., 200:281–290.

Kosloff, D., Scott, R. F., and Scranton, J., 1980. Finite element simulation of Wilmington oil field subsidence: 1. Linear modeling. *Tectonophysics*, 65:339–368.

Kouznetsov, O. L., Sidorov, V. I., Katz, S., and Chilingarian G. V., 1994. Interrelationships among seismic and short-term tectonic activity, oil and gas production, and gas migration to the surface. *J. Pet. Sci. Eng.*, 13: 57–63.

Kouznetsov, O. L. and Simkin, E. M., 1990. *Transformation and Interaction of Geophysical Fields in the Lithosphere*. Nedra, Moscow, 226 pp.

Kovach, R. L., 1974. Source mechanism for Wilmington oil field, California, subsidence earthquakes. *Bull. Seismol. Soc. Am.,* 64:699–711.

Kreitler, C. W., 1977. Faulting and land subsidence from ground-water and hydrocarbon production, Houston–Galveston, Texas. *Int. Assoc. Hydrol. Sci. Symp. Land Subsidence*, Paris, pp. 435–446.

Krestnikov, V. N., Belousov, T. P., and Shtange, D. V., 1988. Seismotectonic conditions for triggering Gazli earthquakes. *Physics of the Earth*, 9:12–28.

Kumar, M., 1979. Geodetic leveling in land subsidence studies. In S. K. Saxena (ed.), *Evaluation and Prediction of Subsidence*. ASCE Publ., New York, NY, pp. 538–548.

Lee, K. L., 1979. Subsidence earthquake at a California oil field. In S. K. Saxena (ed.), *Evaluation and Prediction of Subsidence*. ASCE Publ., New York, NY, pp. 549–564.

Li, F., 1990. The comparison of the results obtained from different techniques such as overcoring, hydrofracturing, breakouts, and focal mechanism solution. In A. Pinta da Cunha (ed.), *Scale Effects in Rock Masses*. Balkema, Rotterdam, pp. 279–287.

Liao, J. S., Pan, K. L., and Haimson, B. C., 1991. The monitoring and investigation of ground subsidence in Southern Taiwan. In A. I. Johnson (ed.), *Land Subsidence*. IAHS Publ., 200:81–96.

Lister, L. A. and Secrest, C. D., 1985. Giant desiccation cracks and differential surface subsidence, Red Lake Playa, Mojave County, Arizona. *Am. Assoc. Eng. Geol. Bull.*, 22:299–314.

Lofgren, B. E., 1979. Changes in aquifer-system properties with groundwater depletion. In S. K. Saxena (ed.), *Evaluation and Prediction of Subsidence*. ASCE Publ., New York, NY, pp. 26–46.

Love, D. W., Reimers, R. F., Hawley, J. W., Johnpeer, G. D., and Bobrow, D. J., 1987. Summary of geotechnical investigations near Espanola,

New Mexico. In C. Menges (ed.), *Quaternary Tectonics, Landform Evolution, Soil Chronologies, and Glacial Deposits*. New Mexico Dep. Geol., Albuquerque, NM, pp. 133–157.

Matsuoka, I., 1984. Land subsidence of an inland basin, northeastern Japan—the Yamagat Basin. *Proc. 3rd Inter. Symp. on Land Subsidence, Venice, Italy, 19–25 Mar.* Inter. Asso. Hydrological Sci., IAHS Pub. No. 151: 373–379.

Mess, K. W., 1979. On the interpretation of core compaction behavior. In S. K. Saxena (ed.), *Evaluation and Prediction of Subsidence*. ASCE Publ., New York, NY, pp. 76–91.

Meyer, R. F. and Powly, D. E., 1988. Subsidence and the petroleum industry: An overvew. *A.A.P.G. Bull.*, 72(2):223.

Morales Y. M., Muillo-Fernandez, R., and Hernandez-Rubio, A., 1991. Subsidence of the former Texcoco Lake. In A. I. Johnson (ed.), *Land Subsidence*. IAHS Publ., 200:35–44.

Murria, J., 1991. Subsidence due to oil production in western Venezuela: Engineering problems and solutions. In A. I. Johnson (ed.), *Land Subsidence*. IAHS Publ., 200:129-139.

Nicolas, A., 1987. *Principles of Rock Deformation*. Reidel, Dordrecht, 208 pp.

Pakiser, L. C., Eaton, J. P., Healy, J. M., and Raleigh, C. B., 1969. Earthquake prediction and control. *Sci.*, 166:1464–1467.

Pampeyan, E. H., Holzer, T. L., and Clarke, M. M., 1988. Modern ground failure in the Garlock fault zone, Fremont Valley, California. *Geol. Soc. Am. Bull.*, 100:677–691.

Pewe, T. L., Raymond, R. H., and Schumann, H. H., 1987. Land subsidence and earth-fissure formation in eastern Phoenix metropolitan area, Arizona. In G. H. Davis and E. M. Van den Dolder, *Geologic Diversity of Arizona and Its Margins—Excursions to Choice Areas*. Ariz. Bur. Geol. Min. Tech., Geol. Surv. Branch, Spec. Pap., 5:199–211.

Piancharoen, C., 1976. Ground water and land subsidence in Bangkok, Thailand. *Proc. 2nd Inter. Symp. on Land Subsidence, Anaheim, CA, USA, 13–17 Dec.* Inter. Asso. Hydrological Sci., IAHS-AISH Pub. No. 121: 355–364.

Poland, J. F. and Davis, G. H., 1969. Land subsidence due to withdrawal of fluids. Rev. Eng. Geol., II *Geol. Soc. Am.*, pp. 187–269.

Poland, J. F. and Working Group 8.4, 1984. *Guidebook to Studies of Land Subsidence Due to Ground-water Withdrawal*. UNESCO, Place de Fontenoy, 75700 Paris, 305 pp.

Poland, J. F., 1984. Case History No. 9.14, Santa Clara Valley, California. In J. F. Poland (ed.) *Guidebook to Studies of Land Subsidence due to Ground-water Withdrawal.* United Nations Educational, Scientific and Cultural Org., Fontenoy, Paris: 279–290.

Poland, J. F., Lofgren, B. E., and Woodward-Clyde Consultants, 1984. Case History No. 9.13. San Joaquin Valley, California. In J. F. Poland (ed.) *Guidebook to Studies of Land Subsidence due to Ground-water Withdrawal.* United Nations Educational, Scientific and Cultural Org., Fontenoy, Paris: 263–277.

Pottgens, J. J. E. and Brouwer, F. J. J., 1991. Land subsidence due to gas extraction in the northern part of The Netherlands. In A. I. Johnson (ed.), *Land Subsidence.* IAHS Publ., 200:99–108.

Pratt, W. E. and Johnson, D. W., 1926. Local subsidence of the Goose Creek Oil Field. *Geology*, XXXIV (7, Part 1): 577–590.

Prokopovich, N. P., 1976. Origin and treatment of hydrocompaction in the San Joaquin Valley, CA. *Proc. 2nd Inter. Symp. on Land Subsidence, Anaheim, CA, USA, 13–17 Dec.* Inter. Asso. Hydrological Sci., IAHS-AISH Publ. No. 121: 537–546.

Prokopovich, N. P., 1991. Detection of aquifer susceptibility to land subsidence. In A. I. Johnson (ed.), *Land Subsidence.* IAHS Publ., 200:27–34.

Ratigan, J. L., 1990. Scale effects in the hydraulic fracture test associated with the estimation of tensile strengh. In A. Pinta da Cunha (ed.), *Scale Effects in Rock Masses.* Balkema, Rotterdam, pp. 297–306.

Richter, C. F., 1958. *Elementary Seismology.* Freeman, San Francisco, CA, 768 pp.

Rivera, A., Ledoux, E., and Marsily, G. de, 1991. Nonlinear modeling of ground-water flow and total subsidence of the Mexico City aquifer-aquitard system. In A. I. Johnson (ed.), *Land Subsidence.* IAHS Publ., 200: 45–58.

Scott, R. F., 1979. Subsidence—A review. In A. I. Johnson (ed.), *Land Subsidence.* IAHS Publ., 200:125.

Strehle, R. W., 1989. Subsidence hazards—A history. In S. M. Testa (ed.), *Environmental Concerns in the Petroleum Industry.* A.A.P.G. Publ., pp. 107–116.

Terzaghi, K., 1943. *Theoretical Soil Mechanics*, Wiley, New York, NY, 510 pp.

Terzaghi, K. and Peck, R. B., 1967. *Soil Mechanics in Engineering Practice,* John Wiley, New York, NY, 729 pp.

Whittaker, B. N. and Reddish, D. J., 1989. *Subsidence: Occurrence, Prediction and Control*, Elsevier, Amsterdam, 528 pp.

Van Sickle, V. R. and Groat, C. G., 1981. Subsidence and induced faulting: Key environmental issues in geopressured-geothermal resource development. *5th Louisiana Geol. Surv., Louisiana State Univ., U.S. Dep. Energy U.S. Gulf Coast Geopressured—Geothermal Eng. Conf. Proc.*, Paris, pp. 325–330.

Vega, G. E. F., 1976. Subsidence of the City of Mexico: A historical review. *Proc. 2nd Inter. Symp. on Land Subsidence, Anaheim, CA, USA, 13–17 Dec.* Inter. Asso. Hydrological Sci., IAHS-AISH Pub. No. 121: 35–38.

Vega, G. E. F., 1984. Case History No. 9.8. Mexico, D. F., Mexico. In J. F. Poland (ed.), *Guidebook to Studies of Land Subsidence due to Groundwater Withdrawal.* United Nations Educational, Scientific and Cultural Org., Fontenoy, Paris: 217–230.

Wetmiller, R. J., 1986. Earthquakes near Rocky Mountain House, Alberta, and their relationship to gas production facilities. *Can. J. Earth Sci.*, 2(23): 172–181.

Wu, Chian-Min, 1976. Groundwater depletion and land subsidence in Taipei Basin. *Proc. 2nd Inter. Symp. on Land Subsidence, Anaheim, CA, USA, 13–17 Dec.* Inter. Asso. Hydrological Sci., IAHS-AISH Pub. No. 121: 389–398.

Yamamoto, S., 1976. Recent trend of land subsidence in Japan. *Proc. 2nd Inter. Symp. on Land Subsidence, Anaheim, CA, USA, 13–17 Dec.* Inter. Asso. Hydrological Sci., IAHSAISH Pub. No. 121: 9–15.

Author Index

Note: Italic page numbers indicate complete citation.

Subject Index

D

G

H

I

J

K

L

M

N

Q

R

S

T

U

V

W

Y

Z

About the Authors

Leonid F. Khilyuk, Ph.D., is a consultant in mathematical modeling of environmental processes and former chairman of the Department of Computer Sciences and Applied Mathematics in the Kiev Technological University. He has more than 100 publications worldwide in mathematical modeling of environmental processes, control theory, probability, and statistics.

George V. Chilingar, Ph.D., is a professor of civil and petroleum engineering. He is an author of 45 books and 500 articles in the field of petroleum engineering and geology. Dr. Chilingar has received more than 100 medals and awards from various countries and organizations, including foreign member-ship in the Russian Academy of Sciences. During his career, he was chief of petroleum and chemicals laboratory at Wright Patterson Air Force Base, Dayton, Ohio, and senior advisor to the United Nations. He is the Dir-ector of Dr. Rudolf W. Gunnerman Energy and Environment Laboratory in the University of Southern California.

Bernard Endres, Ph.D., is an environmental consultant specializing in the fields of oil and gas migration, geological site characterization, and evaluation of chemicals and gas constituents relating to source identification within the earth's geological structure. His scientific publications have included the environmental hazards associated with oil and gas migration, and evaluation of regulatory compliance.

John O. Robertson, Jr, Ph.D., has more than 30 years' experience in the area of petroleum and environmental engineering. Dr. Robertson has co-authored 9 books and 20 articles in leading scientific journals. He has been awarded three gold medals and many international honors for his work. He has served as president of Earth Engineerings, Inc. for the past 20 years.